Electromagnetics of Superconductor/Paramagnet Heterostructures

INTERNATIONAL SERIES OF MONOGRAPHS ON PHYSICS

---

175. Y. Genenko, H. Rauh: *Electromagnetics of Superconductor/Paramagnet Heterostructures*
174. H. Li, F. R. W. McCourt: *Transport Properties and Potential Energy Models for Monatomic Gases*
173. G. H. Fredrickson, K. T. Delaney: *Field-Theoretic Simulations in Soft Matter and Quantum Fluids*
172. J. Kübler: *Theory of itinerant electron magnetism, Second edition*
171. J. Zinn-Justin: *Quantum field theory and critical phenomena, Fifth edition*
170. V. Z. Kresin, S. G. Ovchinnikov, S. A. Wolf: *Superconducting state - mechanisms and materials*
169. P. T. Chruściel: *Geometry of black holes*
168. R. Wigmans: *Calorimetry – Energy measurement in particle physics, Second edition*
167. B. Mashhoon: *Nonlocal gravity*
166. N. Horing: *Quantum statistical field theory*
165. T. C. Choy: *Effective medium theory, Second edition*
164. L. Pitaevskii, S. Stringari: *Bose-Einstein condensation and superfluidity*
163. B. J. Dalton, J. Jeffers, S. M. Barnett: *Phase space methods for degenerate quantum gases*
162. W. D. McComb: *Homogeneous, isotropic turbulence - phenomenology, renormalization, and statistical closures*
160. C. Barrabés, P. A. Hogan: *Advanced general relativity - gravity waves, spinning particles, and black holes*
159. W. Barford: *Electronic and optical properties of conjugated polymers, Second edition*
158. F. Strocchi: *An introduction to non-perturbative foundations of quantum field theory*
157. K. H. Bennemann, J. B. Ketterson: *Novel superfluids, Volume 2*
156. K. H. Bennemann, J. B. Ketterson: *Novel superfluids, Volume 1*
155. C. Kiefer: *Quantum gravity, Third edition*
154. L. Mestel: *Stellar magnetism, Second edition*
153. R. A. Klemm: *Layered superconductors, Volume 1*
152. E. L. Wolf: *Principles of electron tunneling spectroscopy, Second edition*
151. R. Blinc: *Advanced ferroelectricity*
150. L. Berthier, G. Biroli, J.-P. Bouchaud, W. van Saarloos, L. Cipelletti: *Dynamical heterogeneities in glasses, colloids, and granular media*
149. J. Wesson: *Tokamaks, Fourth edition*
148. H. Asada, T. Futamase, P. Hogan: *Equations of motion in general relativity*
147. A. Yaouanc, P. Dalmas de Réotier: *Muon spin rotation, relaxation, and resonance*
146. B. McCoy: *Advanced statistical mechanics*
145. M. Bordag, G. L. Klimchitskaya, U. Mohideen, V. M. Mostepanenko: *Advances in the Casimir effect*
144. T. R. Field: *Electromagnetic scattering from random media*
143. W. Götze: *Complex dynamics of glass-forming liquids - a mode-coupling theory*
142. V. M. Agranovich: *Excitations in organic solids*
141. W. T. Grandy: *Entropy and the time evolution of macroscopic systems*
140. M. Alcubierre: *Introduction to 3+1 numerical relativity*
139. A. L. Ivanov, S. G. Tikhodeev: *Problems of condensed matter physics - quantum coherence phenomena in electron-hole and coupled matter-light systems*

138. I. M. Vardavas, F. W. Taylor: *Radiation and climate*
137. A. F. Borghesani: *Ions and electrons in liquid helium*
135. V. Fortov, I. Iakubov, A. Khrapak: *Physics of strongly coupled plasma*
134. G. Fredrickson: *The equilibrium theory of inhomogeneous polymers*
133. H. Suhl: *Relaxation processes in micromagnetics*
132. J. Terning: *Modern supersymmetry*
131. M. Mariño: *Chern–Simons theory, matrix models, and topological strings*
130. V. Gantmakher: *Electrons and disorder in solids*
129. W. Barford: *Electronic and optical properties of conjugated polymers*
128. R. E. Raab, O. L. de Lange: *Multipole theory in electromagnetism*
127. A. Larkin, A. Varlamov: *Theory of fluctuations in superconductors*
126. P. Goldbart, N. Goldenfeld, D. Sherrington: *Stealing the gold*
125. S. Atzeni, J. Meyer-ter-Vehn: *The physics of inertial fusion*
123. T. Fujimoto: *Plasma spectroscopy*
122. K. Fujikawa, H. Suzuki: *Path integrals and quantum anomalies*
121. T. Giamarchi: *Quantum physics in one dimension*
120. M. Warner, E. Terentjev: *Liquid crystal elastomers*
119. L. Jacak, P. Sitko, K. Wieczorek, A. Wojs: *Quantum Hall systems*
117. G. Volovik: *The Universe in a helium droplet*
116. L. Pitaevskii, S. Stringari: *Bose–Einstein condensation*
115. G. Dissertori, I. G. Knowles, M. Schmelling: *Quantum chromodynamics*
114. B. DeWitt: *The global approach to quantum field theory*
112. R. M. Mazo: *Brownian motion - fluctuations, dynamics, and applications*
111. H. Nishimori: *Statistical physics of spin glasses and information processing - an introduction*
110. N. B. Kopnin: *Theory of nonequilibrium superconductivity*
109. A. Aharoni: *Introduction to the theory of ferromagnetism, Second edition*
108. R. Dobbs: *Helium three*
105. Y. Kuramoto, Y. Kitaoka: *Dynamics of heavy electrons*
104. D. Bardin, G. Passarino: *The Standard Model in the making*
103. G. C. Branco, L. Lavoura, J. P. Silva: *CP Violation*
101. H. Araki: *Mathematical theory of quantum fields*
100. L. M. Pismen: *Vortices in nonlinear fields*
99. L. Mestel: *Stellar magnetism*
98. K.H. Bennemann: *Nonlinear optics in metals*
94. S. Chikazumi: *Physics of ferromagnetism*
91. R. A. Bertlmann: *Anomalies in quantum field theory*
90. P. K. Gosh: *Ion traps*
87. P. S. Joshi: *Global aspects in gravitation and cosmology*
86. E. R. Pike, S. Sarkar: *The quantum theory of radiation*
83. P. G. de Gennes, J. Prost: *The physics of liquid crystals*
73. M. Doi, S. F. Edwards: *The theory of polymer dynamics*
69. S. Chandrasekhar: *The mathematical theory of black holes*
51. C. Møller: *The theory of relativity*
46. H. E. Stanley: *Introduction to phase transitions and critical phenomena*
32. A. Abragam: *Principles of nuclear magnetism*
27. P. A. M. Dirac: *Principles of quantum mechanics*
23. R. E. Peierls: *Quantum theory of solids*

Electromagnetics of Superconductor/Paramagnet Heterostructures

Yuri Genenko and Hermann Rauh

Institute of Materials Science Technical University of Darmstadt Germany

OXFORD
UNIVERSITY PRESS

OXFORD
UNIVERSITY PRESS

Great Clarendon Street, Oxford, OX2 6DP,
United Kingdom

Oxford University Press is a department of the University of Oxford.
It furthers the University's objective of excellence in research, scholarship,
and education by publishing worldwide. Oxford is a registered trade mark of
Oxford University Press in the UK and in certain other countries

Published in the United States of America by Oxford University Press
198 Madison Avenue, New York, NY 10016, United States of America

British Library Cataloguing in Publication Data
Data available

Library of Congress Control Number: 2024947280

ISBN 9780198714576

DOI: 10.1093/9780191782855.001.0001

Printed and bound by
CPI Group (UK) Ltd, Croydon, CR0 4YY

FSC
www.fsc.org
MIX
Paper | Supporting responsible forestry
FSC® C013604

The manufacturer's authorised representative in the EU for product safety is Oxford University Press España S.A. of el Parque Empresarial San Fernando de Henares, Avenida de Castilla, 2 – 28830 Madrid (www.oup.es/en).

In memory of Ernst Helmut Brandt

With reverence for his exceptional sincerity
and praise for his landmark contributions
to superconductor research

Preface

The electromagnetics of superconductor/paramagnet heterostructures represents a subject of high topicality in physics and electrical engineering research. Broadly, it is about the conditioning of magnetic fields, a novel route to improve type-II superconductor performance in high-current and high-field applications, directed, for example, at increasing the current-carrying capability and the critical fields of such heterostructures as well as reducing their hysteretic AC loss.

Recalling established superconductor terminology may serve to elucidate what the conditioning of magnetic fields precisely means here. When speaking of a superconductor's critical current in the traditional sense, its critical current density is usually addressed. Keeping these two notions apart seems irrelevant, because in the full critical state, with the current flowing almost uniformly over the superconductor bulk, the critical current and the critical current density enjoy an essentially trivial relationship, mediated by the cross section which the current pervades. The situation, however, looks different for superconductors shaped as thin strips or tapes and exposed to magnetically susceptible environments, making up the sort of superconductor/paramagnet heterostructures largely dealt with throughout this book. The correspondence between the critical current and the critical current density thereby becomes nontrivial and significant, since the distribution of the current in the superconductor constituent is nonuniform and dependent on the particulars of the magnetic shields. As a result of the coupling of the current density to the magnetic field, the latter fact can be utilized for favourably shaping the distribution of the current, e.g. so as to raise the total loss-free current that can be carried by the strip. The conditioning of magnetic fields thus intends to enhance, or even optimize, electromagnetic properties of superconductor/paramagnet heterostructures by a judicious choice of their materials' characteristics and geometries.

The idea for the said route, contrived by the first author while visiting the University of Göttingen about two decades ago, has been developed and substantiated through ensuing research in close collaboration with the second author upon joining his group at the Technical University of Darmstadt. This work was followed by numerous investigations aimed at translating the method for conditioning magnetic fields into applications for practical use.

The main purpose of this book is to offer a systematic phenomenological account of the remarkable electromagnetic properties of superconductor/paramagnet heterostructures as captured by Maxwell's electrodynamics together with the generalized London theory and Bean's model of the critical state, a goal which suggests rigorous analytical methods and numerical procedures, too. Exact closed-form solutions of the resultant governing equations, obtained for some idealized heterostructural characteristics, admit straightforward (and often intuitive) physical interpretations. These, in turn, promote a fundamental understanding of the magnetization behaviour of basic superconductor/paramagnet heterostructures and guide further realizations of more complex heterostructures to boot. Finite-element simulations aid the theoretical analysis by enabling investigations of heterostructural features due to functional requirements and thus help pave the way for brand-new superconductor-led technologies.

An introductory survey on research activities devoted to superconductor/paramagnet heterostructures, past and present, sets the scene. This is followed by an outline of the theoretical key ingredients underlying the present text and by a summary of the principles for the conditioning of magnetic fields. The treatise itself begins with a rationale and discussion of the elements of idealized superconductor/paramagnet heterostructures and leads up to modellings of broader, realistic scope. Investigations of the superconductor constituents primarily focus on strips or tapes, filaments and tubes, with an imposed transport current or an applied magnetic field. Geometrical and material aspects of both the magnetic shielding effect and the hysteretic AC loss undergo detailed analysis which, interestingly enough, makes it possible to identify the conditions for nondissipative critical, or even overcritical, states to exist. Crucial issues such as the barrier against the penetration of magnetic flux at superconductor/paramagnet interfaces or the nucleation of magnetic vortex loops equally find their place. Although the overall subject involves heterostructures with superconducting and paramagnetic constituents, some attention is also paid to magnetically shielded structures which include diamagnetic constituents instead.

The book employs a methodical style of presentation, whereby each of the model calculations is given a self-contained form and supplemented with discussion of the achieved results, sometimes indicating routes which potential applications might take. Assuming a natural grasp of the phenomenon of superconductivity and relevant knowledge of the classical theory of fields, the statements provided here thus could be of interest to a wide readership from academia (graduates and scientists) as well as industry (practitioners and engineers), i.e. to those pursuing basic or applied research in physics, electrical engineering, materials science, and possibly other disciplines too.

On writing this monograph, we made extensive use of our own research, whose presentation has been adapted to the book's concept and frequently enriched with new, previously unpublished results. Works of others received due quotes where appropriate to the text. We thus are obliged to our coauthors L. Alff, A. Badía-Majós, H. C. Freyhardt, D. Gölden, P. Krüger, S. Kurdi, N. Narayanan, J. Ogawa, T. Oka, A. V. Pan, S. T. Ranecky, H. S. Ruiz, A. Snezhko, A. Usoskin, H. Watanabe, and S. V. Yampolskii for their contributions within the scope of various tasks. We are also indebted to many colleagues—E. H. Brandt, A. M. Campbell, J. R. Clem, E. W. Collings, S. X. Dou, F. Grilli, A. Gurevich, H. Heese, J. Horvat, R. P. Huebener, H. Jarzina, C. Jooss, R. Kress, V. V. Moshchalkov, E. B. Sonin, M. D. Sumption, M. I. Tsindlekht, A. V. Ustinov, and V. M. Vinokur among them—for stimulating discussions and helpful remarks. Special thanks, last but not least, go to our wives Olena and Sigrid, whose constant love and devotion has always meant the greatest support.

Darmstadt — Yuri Genenko
March 2024 — Hermann Rauh

Acknowledgements

We are grateful to the following organizations and publishers for permission to reuse, in adapted forms, disclosed material of which they hold the copyright.

American Institute of Physics, in respect of the articles:
Genenko, Y. A. and Rauh, H. (2003). *Appl. Phys. Lett.* **82**, 2115.
Genenko, Y. A., Rauh, H. and Krüger, P. (2011). *Appl. Phys. Lett.* **98**, 152508.
Genenko, Y. A., Rauh, H. and Kurdi, S. (2015). *J. Appl. Phys.* **117**, 243909.
Genenko, Y. A. and Snezhko, A. V. (2002). *J. Appl. Phys.* **92**, 357.
Genenko, Y. A., Yampolskii, S. V. and Pan, A. V. (2004). *Appl. Phys. Lett.* **84**, 3921.
American Physical Society, in respect of the articles:
Genenko, Y. A. (2002). *Phys. Rev.* B **66**, 184520.
Genenko, Y. A., Snezhko, A. V. and Freyhardt, H. C. (2000). *Phys. Rev.* B **62**, 3453.
Genenko, Y. A., Usoskin, A. and Freyhardt, H. C. (1999). *Phys. Rev. Lett.* **83**, 3045.
Yampolskii, S. V. and Genenko, Y. A. (2005). *Phys. Rev.* B **71**, 134519.
Elsevier, in respect of the articles:
Genenko, Y. A. (2002). *Physica* C **372–6**, 1847.
Genenko, Y. A. and Rauh, H. (2007). *Physica* C **460–2**, 1264.
Genenko, Y. A., Rauh, H. and Snezhko, A. V. (2002). *Physica* C **372–6**, 1389.
Genenko, Y. A., Snezhko, A. V. and Usoskin, A. (2004). *Physica* C **401**, 236.
Genenko, Y. A., Usoskin, A., Snezhko, A. V. and Freyhardt, H. C. (2000). *Physica* C **341–8**, 1063.
Yampolskii, S. V., Genenko, Y. A. and Rauh, H. (2004). *Physica* C **415**, 151.
IOP Publishing, in respect of the articles:
Genenko, Y. A. and Rauh, H. (2006). *J. Phys.: Conf. Ser.* **43**, 568.
Genenko, Y. A. and Rauh, H. (2010). *Supercond. Sci. Technol.* **23**, 075007.
Genenko, Y. A., Rauh, H., Krüger, P. and Narayanan, N. (2009). *Supercond. Sci. Technol.* **22**, 055001.
Genenko, Y. A., Rauh, H. and Snezhko, A. V. (2001). *Supercond. Sci. Technol.* **14**, 699.
Genenko, Y. A., Rauh, H. and Yampolskii, S. V. (2005). *J. Phys.: Condens. Matter* **17**, L93.
Genenko, Y. A., Snezhko, A. V. and Usoskin, A. (2004). *IOP Conf. Ser.* **181**, 2599.

Genenko, Y. A., Usoskin, A., Snezhko, A. V. and Freyhardt, H. C. (2000). *IOP Conf. Ser.* **167**, 483.

Yampolskii, S. V., Genenko, Y. A. and Rauh, H. (2004). *IOP Conf. Ser.* **181**, 2335.

Yampolskii, S. V., Genenko, Y. A., Rauh, H. and Snezhko, A. V. (2006). *J. Phys.: Conf. Ser.* **43**, 576.

Contents

List of symbols

The following common symbols are used throughout this book to the greatest possible extent. Additional symbols or different meanings are, if required, put forward locally in the text.

$\mathbf{A}$	magnetic vector potential
$\hat{\mathbf{A}}$	magnetic vector potential in transformed coordinates
$\tilde{\mathbf{A}}$	Fourier transformed magnetic vector potential
A_n	normal component of the magnetic vector potential
a	distance between the edge/plane of a superconductor strip and a paramagnet or radius of a magnetic vortex loop
$\mathbf{B}$	magnetic induction
$< \mathbf{B} >$	thermodynamic average magnetic induction
$\mathbf{B}_{\mathrm{a}}$	magnetic induction due to an applied magnetic field
B_{a}	strength of the magnetic induction due to an applied magnetic field
B_{n}	normal component of the magnetic induction
B_{t}	tangential component of the magnetic induction
b	half-width of a flux-free region
C	core of a magnetic vortex or integral contour
c	distance between the centre of a superconductor strip and a paramagnet
$\mathbf{D}$	electric displacement
D	cross-sectional domain
$\mathrm{d}\mathbf{l}$	vector line element
$\mathrm{d}\mathbf{s}$	vector area element

d	thickness of a superconductor strip or a superconducting tubular wire
d_{e}	upper bound for the thickness of a superconductor coating
d_{m}	thickness of a paramagnetic sheath or support
dl	line element
dv	volume element
$\mathbf{E}$	electric field
$\mathbf{E}_{\mathrm{a}}$	applied electric field
E_{a}	amplitude or strength of an applied electric field
E	complete elliptic integral of the second kind
E_{n}	normal component of the electric field
E_{t}	tangential component of the electric field
$\overline{E}_{\mathrm{t}}$	average tangential component of the electric field
E_V	self-energy per unit length of a straight magnetic vortex
$\mathbf{e}_r, \mathbf{e}_\phi, \mathbf{e}_z$	cylindrical unit vectors
$\mathbf{e}_x, \mathbf{e}_y, \mathbf{e}_z$	Cartesian unit vectors
e	elementary charge
e	Euler's number
$\mathbf{F}_{\mathrm{L}}$	Lorentz-like force per unit length
$\mathbf{F}_{\mathrm{b}}$	barrier force per unit length
$\mathbf{F}_{\mathrm{d}}$	drag force per unit length
$\mathbf{F}_{\mathrm{p}}$	maximum pinning force per unit length
$\mathbf{F}_{\mathrm{s}}$	force per unit length between magnetic vortices
F_{V}	self-energy of an arbitrary magnetic vortex
G	Gibbs free energy or Green's function
$\mathbf{H}$	magnetic field
$\hat{\mathbf{H}}$	magnetic field in transformed coordinates
$\tilde{\mathbf{H}}$	Fourier transformed magnetic field
$\mathbf{H}_{\mathrm{a}}$	applied magnetic field

H_a	amplitude or strength of an applied magnetic field
H_c	characteristic magnetic field
H_{c1}	lower critical field
H_{c2}	upper critical field
H_{c3}	surface nucleation field
H_{cth}	thermodynamic critical field
H_f	full penetration field
H_i	induced magnetic field
H_m	effective reduced magnetic field
H_{max}	maximum value of the magnetic field
H_n	normal component of the magnetic field
H_p	surface penetration field
H_t	tangential component of the magnetic field
h	cut-off length
$\hbar$	Planck's constant
I	total current
I_b	barrier controlled total current
I_c	critical current
Im	imaginary part of a complex function
I_{max}	maximum total current
I_n	modified Bessel function of the first kind of order n
I_p	surface penetration current
I_{tr}	total transport current
i	imaginary unit
J	sheet current or total current per unit length
$\bar{J}$	average sheet current
$\hat{J}$	sheet current in transformed coordinates
J_b	barrier value of the sheet current
J_c	critical current per unit length or critical sheet current
J_e	sheet current for entry of magnetic flux

J_{edge}	edge sheet current
J_{max}	maximum sheet current
$\bar{J}_{\mathrm{max}}$	maximum average sheet current
J_{min}	minimum sheet current
$\mathbf{j}$	current density
$\mathbf{j}_{\mathrm{s}}$	supercurrent density
j_{GL}	Ginzburg–Landau depairing critical current density
j_{c}	critical current density
j_{max}	maximum current density
j_{p}	surface penetration current density
j_{tr}	transport current density
K	complete elliptic integral of the first kind
K_n	modified Bessel function of the second kind of order n
$\mathbf{k}$	wavevector
$\mathbf{k}_{\perp}$	planar projection of the wavevector
k	wavenumber
L	Lobachevsky's function
L_n	modified Struve function of order n
$\mathbf{l}$	delineation of a magnetic vortex core
l	depth of penetration of magnetic flux
lim	limit
$\mathbf{M}$	magnetization
$\langle \mathbf{M} \rangle$	thermodynamic average magnetization
$\mathbf{m}$	magnetic dipole moment or magnetic dipole moment per unit length
m_{s}	effective mass of a superconducting particle
n_0	superconducting particle density in the London limit
n_{s}	superconducting particle density
n_{V}	number of flux lines per unit area

P	electromagnetic power per unit volume
p	parameter of a conformal map
p_{m}	magnetic pressure
Q	integral kernel or total electric charge per unit length
$Q_{\max}$	maximum total electric charge per unit length
q_{s}	electric charge of a superconducting particle
Re	real part of a complex function
$\mathbf{r}$	position vector
$\mathbf{r}_{\perp}$	planar projection of the position vector
r, ϑ, ϕ	spherical coordinates
r, ϕ, z	cylindrical coordinates
r_{s}	radius of a superconductor filament or a superconducting tubular wire
res	residual of a least-squares approximation
s	depth of penetration of magnetic flux
$s = x + iy$	complex variable
sgn	sign function
sn	Jacobian elliptic function
T	absolute temperature
T_{c}	critical temperature
$\mathbf{t}$	tangential unit vector
t	time
t_{s}	kinetic energy density of a supercurrent
U_{ac}	hysteretic AC loss per unit area or unit length
$\mathbf{u}$	velocity of linear viscous fluid flow
$\mathbf{V}$	vorticity
$\tilde{\mathbf{V}}$	Fourier transformed vorticity

V	volume
$\mathbf{v}$	velocity of rotational incompressible fluid flow
W_{L}	London energy per unit length
W_{V}	electromagnetic energy of a magnetic vortex
w	half-width of a superconductor slab or strip
w_{m}	half-width of a paramagnetic strip
x, y, z	Cartesian coordinates
α	proportionality constant in magnetostatic-electrostatic analogues
β	opening half-angle of cusp- or wedge-shaped magnetic shields
Γ	Euler's gamma function
γ	Euler's constant or half-angle of flux penetration
γ_{a}	anisotropy parameter
Δ	scalar or vector Laplacian
δ	Dirac's delta function
ε	electric permittivity
ε_0	electric permittivity of a vacuum
η	magnetic dipole-layer strength
η_{V}	friction coefficient
θ	Heaviside's step function or phase of the superfluid wavefunction
κ	image current strength
λ_{J}	Josephson penetration depth

λ_{L}	London penetration depth
$\lambda_{\perp}$	transverse penetration depth
μ	magnetic permeability
μ_0	magnetic permeability of a vacuum
ξ, η, ζ	Cartesian coordinates
ξ_{GL}	Ginzburg–Landau coherence length
Π	complete elliptic integral of the third kind
ρ	volume charge density
ρ_{ff}	flux-flow resistivity
ρ_{n}	normal-state resistivity
σ	surface charge density
$\sigma = \xi + i\eta$	complex variable
σ_{b}	barrier value of the surface charge density
σ_{c}	critical surface charge density
σ_{d}	defect size
σ_{r}	surface roughness
Φ	Lerch's transcendent or magnetic flux
Φ_0	quantum of magnetic flux
Φ_{a}	magnetic flux due to an applied magnetic field
Φ_{s}	magnetic flux due to an induced supercurrent
φ	phase difference of superfluid wavefunctions or scalar electric potential
φ_{a}	scalar potential of an applied electric field
χ	linear response function

$\hat{\chi}$	linear response function in transformed coordinates
χ_{m}	magnetic loss reduction factor
Ψ	superfluid wavefunction
ψ	Euler's psi function or scalar magnetic potential

1
Introduction

First brought to light by Kamerlingh Onnes (1911), superconductivity has fascinated and puzzled researchers for more than a hundred years. This macroscopic quantum phenomenon with its distinct 'science-fiction potential', (to quote Dew-Hughes' metaphor) looks back on a long history: technical exploitations always happened before a profound understanding of the physical mechanisms was gained. Following a number of fundamental discoveries such as the Meissner effect (Meissner and Ochsenfeld 1933) and the mixed state, or Shubnikov phase (Shubnikov, Hotkevich, Shepelev, and Ryabinin 1937), at least two kinds of theoretical approach to comprehending superconductivity at a macro- or mesoscopic scale—the London theory (London and London 1935) and the Ginzburg–Landau theory (Ginzburg and Landau 1950)—were developed consecutively and employed fruitfully before low-temperature superconductivity was eventually resolved, at a microscopic scale, by the BCS theory (Bardeen, Cooper, and Schrieffer 1957). The rapidly growing terrain of high-temperature superconductivity, sparked off by the breakthrough discovery of superconducting cuprates by Bednorz and Müller (1986), is no exception to this trend. Recent observations of superconducting transition temperatures gradually approaching room temperature, even though at very high pressures (Drozdov et al. 2015, Snider et al. 2021), indicated a further prospective group of dense hydrogen-rich materials. Whereas dozens of technical devices (ranging from microstrip resonators or logical switches for electronics applications to miles-long cables and tapes for high-current or high-field applications) involve high-temperature superconducting constituents, the physics underlying the phenomenon of high-temperature superconductivity itself still waits for a generally accepted explanation. Fortunately, phenomenological theories based on the firm ground of physical observables (such as the London theory and the Ginzburg–Landau theory referred to above, perhaps with generalizations or due complements) suffice for many investigations aimed at superconductor-based technology, an asset which immensely adds to their appeal. They not only allow for analysis of the macro- or mesoscopic structure of superconducting states, but also offer scope for new and far-reaching conceptual developments. The Ginzburg–Landau theory is a spectacular case in point, as it makes it possible to predict type-II superconductivity (Abrikosov 1957) and weak superconductivity (Josephson 1962) to some extent.

Electromagnetics of Superconductor/Paramagnet Heterostructures. Yuri Genenko and Hermann Rauh, Oxford University Press.
 DOI: 10.1093/9780191782855.003.0001

Type-II superconductors of great technological importance include the intermetallic compounds Nb_3Sn and Nb_3Ge, having respective critical temperatures of $T_c = 18$ K and $T_c = 23$ K, as well as high-temperature materials like the cuprate oxides $YBa_2Cu_3O_7$ and $Bi_2Sr_2Ca_2Cu_3O_{10}$, scoring $T_c = 92$ K and $T_c = 110$ K, respectively. Such materials are widely used for their exceptional properties which set them apart from conductors in the normal state, e.g. the ability to carry large electrical currents I with only minimum hysteretic AC loss U_{ac} and the capacity to withstand high magnetic fields. Although these properties do not go unlimited, improvements of, for instance, the critical current I_c, i.e. the maximum total current I that can be carried without loss, are commonly achieved through the introduction of suitable pinning centres which suppress the flow of magnetic flux and the consequential dissipation of electromagnetic energy (Campbell and Evetts 1972). There are, however, further ways of increasing current-carrying capability, namely those which take advantage of the shielding effect of magnetically susceptible environments. It is for this reason that the present monograph concentrates on a phenomenological analysis of electromagnetic properties of type-II superconductor/paramagnet heterostructures, with superconducting constituents predominantly in the shape of thin strips or tapes, elements for both large-scale power and microelectronic device applications deemed particularly promising.

The idea of guiding the magnetic flux with the help of magnetic shields so as to improve superconductor performance has been advanced and substantiated in a number of works (Genenko, Usoskin, and Freyhardt 1999; Genenko, Snezhko, and Freyhardt 2000; Genenko, Usoskin, Snezhko, and Freyhardt 2000a,b; Glowacki and Majoros 2000; Majoros, Glowacki, and Campbell 2000a,b; Genenko 2002c). Thus, magnetically sheathed single or composite superconductor cables stand out for weakening the (applied and self-induced) magnetic fields exerted upon their superconductor filaments by decoupling these constituents (Glowacki and Majoros 2000; Majoros, Glowacki, and Campbell 2000a,b). A resort to para- or soft-magnetic environments of special shapes in conjunction with flat superconductors like thin superconductor strips admits intricate control of the magnetic flux and the current density as well (Genenko, Usoskin, and Freyhardt 1999; Genenko, Snezhko, and Freyhardt 2000; Genenko, Usoskin, Snezhko, and Freyhardt 2000a,b; Genenko 2002c). Lowering the high density of the magnetic flux near the edges of such strips by means of convex magnetic shields entails pronounced mitigations of the effect self-induced magnetic fields have on the strips. Applying the method of conditioning of magnetic fields to partially flux-filled superconductor strips may even create nondissipative overcritical states delineated by transport currents much higher than those in isolated strips (Genenko, Snezhko, and Freyhardt 2000). Hence, from a theoretical point of view, transport currents may be strongly redistributed in superconductor strips when exposed to magnetic (Genenko, Usoskin, and Freyhardt 1999; Genenko, Snezhko, and Freyhardt 2000; Genenko, Usoskin, Snezhko, and Freyhardt 2000a,b; Genenko 2002c) or superconducting (Genenko 2002b; Genenko and Rauh 2003) shields, bringing about substantial enhancements in the total loss-free current I as compared to the critical current I_c attainable by conventional means. Although originally predicted within the framework of analytical investigations for idealized paramagnetic environments, nondisspative overcritical states have been found in numerical computations for more realistic and practically relevant shielding geometries, too (Genenko, Rauh, Krüger, and Narayanan 2009).

Whereas nondissipative overcritical states in superconductor strips, for an imposed transport current, do not seem to have been directly observed so far, they have manifested in a magnetically shielded superconducting tape subject to a transverse magnetic field, with values of the local current density exceeding the critical limit j_c roughly thrice (Jarzina, Jooss, and Freyhardt 2002). Notwithstanding the fact that further empirical evidence for overcritical states was sparse, and still is to date, the method for conditioning magnetic fields in superconductor/para- or ferromagnet heterostructures has since been implemented successfully worldwide. Experimental studies and finite-element simulations of single-filament MgB_2/Fe composite cables by Collings' group (Ohio State University, USA) showed that, for an applied magnetic field $\mathbf{H}_a$ with a strength of the corresponding magnetic induction below 0.2 T, both the penetration of magnetic flux and the consequential hysteretic AC losses U_{ac} in the superconductor core of these cables were suppressed, findings characteristically termed 'pseudo Meissner effect', (Sumption et al. 2002). The interpretation of the magnetization data obtained, and hence the theoretical estimate of the hysteretic AC losses deduced, relied, however, on the assumption that the superconductor core was exposed to a magnetic field equal to the applied magnetic field shielded by the magnets, when the superconductor core were absent, a simplification which renders the assessment of the hysteretic AC losses questionable to a certain extent (Genenko, Yampolskii, and Pan 2004). Similar results for hysteretic AC losses in multifilament MgB_2 cables covered with iron sheaths have been reported by Campbell's and Glowacki's group (University of Cambridge, UK) in several works (Glowacki and Majoros 2000; Majoros, Glowacki, and Campbell 2000a,b; Glowacki and Majoros 2009). Their experiments revealed that magnetic coatings of even a small thickness d_m (20–40 % of the radius r_s of the superconductor filaments) and a medium relative permeability ($\mu/\mu_0 \simeq 50$ in the case of pure iron at the temperature $T = 40$ K) can give rise to considerable shielding effects, in agreement with theoretical assertions (Genenko and Snezhko 2002). A possibility of the maximum reduction of the energy losses due to a nearly full decoupling of the superconductor filaments in multifilament MgB_2 wires was demonstrated by Ruiz' group (University of Leicester, UK) in recent works (Fareed and Ruiz 2021; Kapolka and Ruiz 2022).

Enhancements of the critical current in a single-filament MgB_2/Fe composite cable have been claimed by Dou's group (University of Wollongong, Australia) in various works (Horvat, Wang, Soltanian, and Dou 2002; Horvat, Soltanian, and Yeoh 2005; Horvat, Yeoh, and Miller 2005). It was shown that at the temperature $T = 30$ K, the critical current remains virtually independent of the magnetic induction $\mathbf{B}_a$ for strengths up to 0.7 T, a fact which could not be attributed to magnetic shielding by the iron sheath alone (Horvat, Wang, Soltanian, and Dou 2002; Horvat, Yeoh, and Miller 2005). Comparable behaviour of the critical current has been monitored experimentally and simulated numerically by Gömöry's group (Slovak Academy of Sciences, Bratislava, Slovakia) for Bi-2223/Ag/Fe tapes exposed to magnetic inductions $\mathbf{B}_a$ of a strength up to 0.4 T, conceding the possible occurrence of significant hysteretic AC losses due to the magnetic hysteresis of the iron sheath (Kováč et al. 2003; Gömöry 2006; Gömöry et al. 2006). Enhancements of the critical current, utterly seen as a magnetic shielding effect, with the hysteretic AC losses left unchanged, have been established by Han's group (Tsinghua University of Beijing, China) and reported in several works (Alamgir, Gu, and Han 2005; Gu, Alamgir, Qu, and Han 2007).

The penetration of magnetic flux in superconductor/para- or ferromagnet heterostructures of various set geometries has been studied magneto-optically by Freyhardt's and Jooss's group at the University of Göttingen, Germany (Jarzina, Jooss, and Freyhardt 2002; Jooss, Brinkmeier, and Heese 2005), by Dou's group at the University of Wollongong, Australia (Pan, Zhou, Liu, and Dou 2003; Pan and Dou 2004; Pan, Zhou, and Dou 2004), and by Johansen's group at the University of Oslo, Norway (Yurchenko et al. 2007), with the aim of exploring the interaction between superconductor and magnetic constituents. Their research clearly uncovered strong redistributions of the magnetic flux due to the presence of the magnetic environments. Adopting this result backed up by the information on supplementary magnetic measurements, some authors concluded that nondissipative overcritical currents, as theory allows (Genenko, Snezhko, and Freyhardt 2000; Genenko, Usoskin, Snezhko, and Freyhardt 2000a,b; Genenko 2002c), had really been discerned (Pan and Dou 2004). Such an inference should, however, be treated with caution since the materials used to furnish the magnetic constituents exhibit nonlinear, hysteretic behaviour for the magnetic field strengths under testing (Pan and Dou 2004; Horvat, Soltanian, and Yeoh 2005; Horvat, Yeoh, and Miller 2005), an aspect not fully taken into account in the analysis. The effect of magnetic shielding of thin superconductor films with grain boundary defects (Jooss, Brinkmeier, and Heese 2005) may, on the other hand, well be interpreted in the spirit of results for long Josephson junctions exposed to magnetic shields (Genenko 2002a).

A different way to theoretically pursue the goal of improving the current-carrying capability of superconductor strips by means of magnetic shields has been suggested by Kress's group (University of Göttingen, Germany). Rather than proceeding from a shielding geometry a priori assigned, the inverse problem of determining the shape of the magnetic shields so as to procure the maximum loss-free current $I_{\max}$ was addressed. To this end, the feasibility of the integral equation approach, developed for solving the proper inverse boundary value problem, was established first (Heese 2004); the method was then used to calculate the distribution of the magnetic field and to analyse empirical data relating to elaborate magnetic environments (Jooss, Brinkmeier, and Heese 2005).

Experimental results for hysteretic AC losses in bilayer superconductor/soft-magnet heterostructures (such as magnetically coated superconductor tapes) obtained by Duckworth et al. (2003) at the Oak Ridge National Laboratory, Oak Ridge, USA and by Suenaga and Li (2006), as well as Suenaga et al. (2008) at the Brookhaven National Laboratory, Upton, USA, have been appraised by Mawatari (2008) at the National Institute of Advanced Industrial Science and Technology, Tsukuba, Japan with the model assumptions of the magnetic coatings being infinitesimally thin and their permeability being infinitely large. The intricate effects of substrate magnetism on hysteretic AC losses of high-temperature superconductor tapes for power transmission conductors and high-field magnetic coils, with a field-dependent substrate permeability μ implied, have been assessed by a number of research groups: by Grilli, Ashworth, and Civale (2007); Nguyen, Sastry, and Schwartz (2007); Nguyen, Ashworth, and Willis (2009); and Nguyen et al. (2010) at the Los Alamos National Laboratory, Los Alamos, USA and the Karlsruhe Institute of Technology, Karlsruhe, Germany; by Claassen (2007); and Claassen and Thieme (2008) at the Naval Research Laboratory, Washington, USA; furthermore by Amemiya and Nakahata (2007); and Amemiya, Nakahata, Fujiwara, and Shiohara (2010) at the University of Kyoto, Japan; and by Miyagi, Amadutsumi, and Takahashi, (2007); Miyagi, Umabuchi, Takahashi, and Tsukamoto (2007); Miyagi, Umabuchi, and Takahashi (2008a,b); and Umabuchi, Miyagi, Takahashi, and Tsukamoto (2008) at the University of Okayama, Japan. Interpretations of the role of the magnetic substrate on hysteretic AC losses for both imposed transport currents and applied magnetic fields, however, remained contradictory.

The effect of enhancements in the critical current of thin superconductor strips in open configurations with soft-magnetic shields has been confirmed through transport measurements and finite-element simulations of full critical states for critical current densities j_c depending on the local magnetic induction $\mathbf{B}$ (Alamgir, Gu, and Han 2005; Gömöry 2006; Gömöry et al. 2007; Gu, Alamgir, Qu, and Han 2007). The finite-element approach has, furthermore, been used for clarifying the problem of hysteretic AC losses in planar superconductor/soft-magnet heterostructures with simple, albeit realistic shielding geometries and arbitrary permeabilities of their magnetic constituents, rather than for identifying nondissipative overcritical states brought about by sophisticated magnetic environments (Glowacki and Majoros 2000; Majoros, Glowacki, and Campbell 2000b; Sumption et al. 2002; Gömöry et al. 2007; Krüger, Grilli, and Farinon 2011; Krüger et al. 2013, Grilli et al. 2014).

Using mathematical tools which Maxwell's and London's phenomenological theories provide, we address the electromagnetics of superconductor/paramagnet heterostructures with a view to potential realizations of nondissipative overcritical states. This ansatz implies omitting fine electronic effects as would arise, e.g. from the interaction between superconductivity and ferromagnetism at a microscopic scale—an issue thoroughly described elsewhere (Lyuksyutov and Pokrovsky 2005; Aladyshkin, Silhanek, Gillijns, and Moshchalkov 2009)—while bearing in mind that both hard- and soft-magnetic environments can also lead to geometry-dependent effects at a macroscopic scale (Lange, Van Bael, Bruynseraede, and Moshchalkov 2003; Laiho, Lähderanta, Sonin, and Traito 2003; Touitou et al. 2004; Vodolazov et al. 2005; Morelle and Moshchalkov 2006; Silhanek et al. 2006; de Souza Silva et al. 2007; Gillijns et al. 2007; Silhanek et al. 2008; Aladyshkin, Volodin, and Moshchalkov 2010; Del-Valle, Navau, Sanchez, and Chen 2011; Del-Valle, Navau, Sanchez, and Dinner 2012; Vlasko-Vlasov et al. 2015; Vlasko-Vlasov et al. 2017a,b; Vlasko-Vlasov et al. 2022a,b; Zhitlukhina et al. 2023). Chapter 2 reviews the fundamentals of both the generalized London theory and Bean's model of the critical state, thus paving the way for later theoretical analysis in conjunction with type-II superconductor constituents; the chapter also explains the idea of conditioning magnetic fields by paramagnetic shields. Singling out geometrical contingencies, Chapter 3 attends to the flux-free Meissner state of a thin superconductor strip placed in various infinite, or semi-infinite, paramagnetic environments with infinite permeability, adopting the most idealized premises. The partially flux-filled critical state of a thin superconductor strip exposed to some of these environments is considered in Chapter 4. Offering investigations of the Meissner state and the critical state of a thin superconductor strip in perfectly diamagnetic environments, Chapter 5 digresses from the actual theme while treating a topic intimately related to that of paramagnetic shields. Chapter 6 goes a step further towards a more realistic account of paramagnetic environments of thin superconductors of planar, tubular, and annular shapes by allowing for restricted shielding geometries and finite magnetic permeabilities. Chapter 7 examines the magnetic shielding effect on bulk superconductors of cylindrical shapes, again in order to establish the role of geometrical contingencies. Finite-element simulations of the Meissner state and the critical state of thin superconductors in paramagnetic environments of arbitrary shape and permeability are at the heart of Chapter 8. A summary of the main results together with an outlook for possible further research is given in Chapter 9. Some mathematical results underlying select issues of the book are derived in Appendices A and B.

2 Phenomenological concepts in superconductivity

When modelling electromagnetic properties of superconductor/soft-magnet heterostructures, the natural question that arises straight away is how to adequately describe the two types of constituents. While the ansatz of a linear magnetization law, or a constant (possibly even infinite) magnetic permeability μ, typifying the magnetic constituents as para- or diamagnetic elements often proves a fair choice, representations of the superconducting constituents are necessarily more involved: the genuinely quantum mechanical phenomenon of superconductivity manifests itself up to macroscopic scales and calls for profound, yet essentially simple, concepts nevertheless. We shall adopt the generalized London theory (London 1948, 1950) supplemented with Maxwell's electrodynamics throughout major parts of the present text. Based on a macroscopic quantum concept, this theory encompasses the original London theory (London and London 1935) that was able to reveal the fundamental superconductor properties of zero DC resistivity and perfect diamagnetism, i.e. the flux-free Meissner state. Moreover, it extends the latter theory consistently into the quantum regime by taking the phase coherence of superconducting states into account, and hence yields predictions like the quantization of magnetic flux or the existence of magnetic vortices in a type-II superconductor's mixed state. It also permits a phenomenological analysis of weak superconductivity, i.e. the Josephson effects. The generalized London theory relates to the Ginzburg–Landau theory (Ginzburg and Landau 1950) by virtue of common quantum mechanical roots and the allowance for thermodynamic ground, equally uniting electromagnetic with thermal traits. The Ginzburg–Landau theory, at last, turns out to be a limiting result of the microscopic BCS theory (Bardeen, Cooper, and Schrieffer 1957), which is widely perceived a reference theory of conventional, low-temperature superconductivity.

Observables of foremost significance for technical realizations of superconductor/paramagnet heterostructures include the distributions of electrical currents and magnetic fields as well as the critical current and the hysteretic AC loss in their type-II superconductor constituents, when a transport current is imposed or a magnetic field applied. Predictions of such quantities with sufficient accuracy usually do not require a comprehensive knowledge of, for instance, variations on the length scale

Electromagnetics of Superconductor/Paramagnet Heterostructures. Yuri Genenko and Hermann Rauh, Oxford University Press.
 DOI: 10.1093/9780191782855.003.0002

of individual magnetic vortices which together make up the mixed state. Hence, distributions of electrical currents and magnetic fields in this state are conveniently described by averages over regions much larger than the spacing between individual vortices, an assertion which defines the level of sophistication implicitly assumed to a large extent in this book, unless properties of the vortices themselves are explicitly addressed. The problem of penetration of magnetic flux into the superconducting constituents is treated by taking a step of coarsening further; Bean's model of the critical state (Bean 1962, 1964) here furnishes the right tool for appropriately describing the mixed state if strong pinning of magnetic vortices prevails. Its use in studies on the effect of paramagnetic constituents on guiding the magnetic flux aimed at improving superconductor performance by the conditioning of magnetic fields in the partially flux-filled critical state, or the extreme of the completely flux-free Meissner state, recommends itself (Genenko, Usoskin, and Freyhardt 1999; Genenko 2002c; Genenko, Rauh, Krüger, and Narayanan 2009; Genenko and Rauh 2010).

These phenomenological concepts will be outlined in some detail next.

2.1 Generalized London theory

Let us begin with the notion that the superconducting state at a given temperature T below the critical temperature T_c must be a highly ordered, collective ground state, separated from excited states by a finite energy gap and resistant against weak external perturbations such as a uniform applied magnetic field $\mathbf{H}_a$ below a certain critical strength (London 1948, 1950). The latter constraint holds in the case of type-I bulk superconductivity which is characterized by a *positive* energy of the interface between the superconducting and normal phases: a single upper bound on the magnitude of the applied magnetic field H_a, the so-called thermodynamic critical field H_{cth}, exists above which a transition from the flux-free Meissner state to the normal state takes place. A distinction, however, must be drawn in the case of type-II bulk superconductivity which is characterized by a *negative* energy of the interface between the superconducting and normal phases. Whereas here the flux-free Meissner state persists up to the lower critical field H_{c1}, a coexistence of the superconducting and normal phases sets in when penetration of magnetic flux creates the mixed state for H_a between H_{c1} and the upper critical field H_{c2}, followed by a transition to the normal state—surface superconductivity with its still higher surface nucleation field, H_{c3}, exempt. These hallmarks of the two types of superconductivity are obvious when considering the thermodynamic average magnetic induction $\langle B \rangle$ of a superconductor in response to the applied magnetic field $\mathbf{H}_a$. If the latter relationship is unique, and hence reversible, as Fig. 2.1's depiction of a thermal equilibrium state suggests, the superconductor would be called 'ideal'. Yet, we add parenthetically that, since for type-II superconductors of *restricted* size the Bean–Livingston energy barrier and possibly another, geometrical surface barrier inevitably exist, the first entry of magnetic flux into the flux-free superconductor bulk is generally delayed up to some surface penetration field H_p above

H_{c1}, a fact which already shows nonequilibrium traits, and hence introduces irreversibility alone, even if the superconductor were perfect otherwise. Naturally, all the critical fields vanish as T approaches T_c. Their definitions hinge upon the superconductor's shape and orientation with respect to the applied magnetic field, apart from temperature itself. We note that bulk superconductors well describable by just one coordinate of space (such as an infinitely extended slab or cylinder subject to a longitudinal magnetic field) are conventionally assumed, unless stated otherwise.

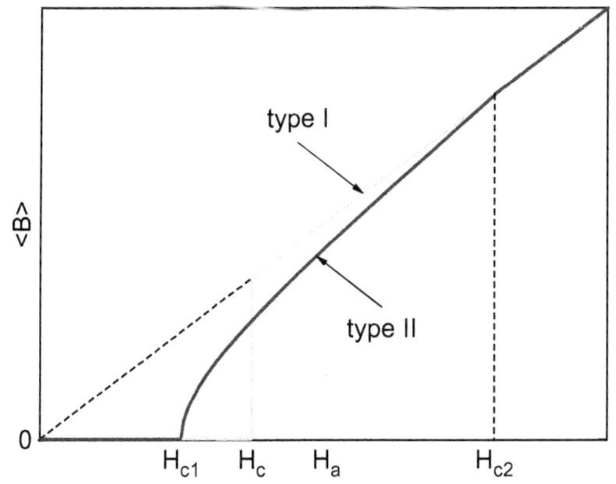

Fig. 2.1 Thermodynamic average magnetic induction $\langle B \rangle$ of ideal type-I and type-II superconductors as a function of the magnitude of an applied magnetic field H_a (schematic view). The dashed lines extrapolate or delimit the normal-state regimes and merely serve as guides to the eye.

2.1.1 Basic conceptual elements

Actual realizations of the phenomenological theory proceed from the assumption that the carriers of superconductivity in a homogeneous, isotropic superconductor can be conceived as superconducting particles ('Cooper pairs') having effective mass m_s, electric charge q_s, and zero total spin; they are present in macroscopic numbers and act coherently, forming a superfluid, or condensate. Macroscopic occupancy ensures the collective state can be assigned a (complex) macroscopic quantum wavefunction Ψ depending on the location $\mathbf{r}$ and time t to describe the behaviour of the condensate when exposed to an electromagnetic field. Since this wavefunction is to match the wavefunctions of all the superconducting particles of the condensate, it must obey the Schrödinger-like equation

$$i\hbar \frac{\partial}{\partial t} \Psi(\mathbf{r}, t) = \frac{1}{2m_s} \left(\frac{\hbar}{i} \nabla - q_s \mathbf{A}(\mathbf{r}, t) \right)^2 \Psi(\mathbf{r}, t) + q_s \varphi(\mathbf{r}, t) \Psi(\mathbf{r}, t) \qquad (2.1)$$

with the scalar electric potential φ, the magnetic vector potential $\mathbf{A}$, and Planck's constant $\hbar \simeq 1.054 \times 10^{-34}\, Js$; the electric field $\mathbf{E}$ and the magnetic induction $\mathbf{B}$

herein derive from the relations

$$\mathbf{E} = -\nabla\varphi - \frac{\partial \mathbf{A}}{\partial t} \tag{2.2}$$

and

$$\mathbf{B} = \nabla \times \mathbf{A}. \tag{2.3}$$

This choice of the potentials is not unique, of course; however, if they were subjected to the simultaneous gauge transformations

$$\varphi \to \varphi' \equiv \varphi - \frac{\partial \chi}{\partial t} \quad \text{and} \quad \mathbf{A} \to \mathbf{A}' \equiv \mathbf{A} + \nabla\chi \tag{2.4}$$

with an arbitrary single-valued and differentiable function of space and time, χ, the electromagnetic field itself would be conserved. If, in addition, the superfluid wavefunction is transformed according to

$$\Psi(\mathbf{r}, t) \to \Psi'(\mathbf{r}, t) \equiv \Psi(\mathbf{r}, t) \exp\left[i\left(q_\mathrm{s}/\hbar\right)\chi(\mathbf{r}, t)\right], \tag{2.5}$$

gauge invariance of the governing equation (2.1), and hence of measurable physical quantities like the supercurrent density

$$\mathbf{j}_\mathrm{s} = \frac{q_\mathrm{s}}{2m_\mathrm{s}}\left[\Psi^*\left(\frac{\hbar}{i}\nabla - q_\mathrm{s}\mathbf{A}\right)\Psi - \Psi\left(\frac{\hbar}{i}\nabla + q_\mathrm{s}\mathbf{A}\right)\Psi^*\right] \tag{2.6}$$

deriving from it, will be secured (Orlando and Delin 1991).

In consistency with the premise of macroscopic occupancy of the collective superconducting state, the square of the modulus of the superfluid wavefunction Ψ can be interpreted as the local superconducting particle density n_s. The representation

$$\Psi(\mathbf{r}, t) = \sqrt{n_\mathrm{s}(\mathbf{r}, t)} \exp\left(i\theta(\mathbf{r}, t)\right), \tag{2.7}$$

which expounds n_s and the local phase θ as two separate entities, therefore suggests itself. The latter quantity obviously undergoes the renormalization

$$\theta(\mathbf{r}, t) \to \theta'(\mathbf{r}, t) \equiv \theta(\mathbf{r}, t) + \left(q_\mathrm{s}/\hbar\right)\chi(\mathbf{r}, t) \tag{2.8}$$

upon applying the gauge transformation, eqn. (2.5). By substituting eqn. (2.7) into eqn. (2.6), the supercurrent density takes the succinct form

$$\mathbf{j}_\mathrm{s} = \frac{n_\mathrm{s} q_\mathrm{s}}{m_\mathrm{s}}\left(\hbar\nabla\theta - q_\mathrm{s}\mathbf{A}\right), \tag{2.9}$$

where the difference in brackets reflects the kinetic momentum of the superconducting particles, made up of their canonical momentum $\hbar\nabla\theta$ and the field momentum $q_\mathrm{s}\mathbf{A}$. Naturally, since the superfluid wavefunction Ψ is confined to sufficiently slow variations in space for this hydrodynamic approach to hold, it follows that both the superconducting particle density n_s and the phase θ could only slightly vary in space too. A lower-bound criterion here certainly involves the Ginzburg–Landau coherence length of the condensate ξ_GL.

If we entirely disregard spatial changes of the superconducting particle density and, in line with the endorsed concept, also assume that temporal fluctuations are too fast to be meaningful for the issues looked at here, the superconducting particle density n_s may be approximated by a constant, n_0 say, only depending on temperature T. The supercurrent density, eqn. (2.9), then can be cast into the form

$$\mathbf{j}_s = \frac{1}{\mu_0 \lambda_L^2} \left(\frac{\hbar}{q_s} \nabla\theta - \mathbf{A} \right), \tag{2.10}$$

upon introducing the London penetration depth λ_L, another characteristic associated with the superconducting particle properties of the condensate defined by

$$\lambda_L = \sqrt{\frac{m_s}{\mu_0 n_0 q_s^2}}, \tag{2.11}$$

where μ_0 denotes the magnetic permeability of a vacuum. We shall base our further applications of the macroscopic quantum concept on this limiting case. Type-II superconductivity can be roughly said to occur if $\xi_{GL} < \lambda_L$, an inequality which also reflects the ordering of the critical fields $H_{c1} < H_{c2}$.

Equation (2.10) extends the corresponding result of the original London theory by the phase term on the right-hand side and therefore enables, e.g., understanding the phenomenon of quantization of magnetic flux or modelling electromagnetic properties of magnetic vortices. If the phase θ reveals a single-valued dependence of space, as in the case of a compact superconductor domain, eqn. (2.10) with eqn. (2.3) yields the classical London result

$$\mathbf{B} + \mu_0 \lambda_L^2 \, \nabla \times \mathbf{j}_s = 0 \tag{2.12}$$

(London and London 1935). Equation (2.12) may serve to describe the shielding of a superconductor constituent against penetration of magnetic flux, i.e. the experimentally observed Meissner effect (Meissner and Ochsenfeld 1933). This reveals that the London penetration depth λ_L typifies the length scale of spatial changes of the magnetic induction $\mathbf{B}$ in the superconductor bulk. It is the only characteristic which the London theory as presented here contains. Owing to the fact that it introduces temperature T through the superconducting particle density n_0, the paramcter λ_L renders the theoretical frame applicable to the full temperature range below T_c. Generalizations accounting for possible effective-mass anisotropy (Orlando and Delin 1991) or nonlocal effects (Ketterson and Song 1999; Tinkham 2004) could, in addition, be addressed.

2.1.2 Quantization of magnetic flux

We resume the macroscopic quantum concept by considering a homogeneous, isotropic superconductor of a given shape at a fixed temperature T, subject to an applied magnetic field $\mathbf{H}_a$. Thermal equilibrium shall exist, so that the total current flowing in the superconductor is entirely the supercurrent induced therein. If we choose a cross-sectional superconductor domain D, the magnetic flux due to the applied magnetic field, Φ_a, threading this domain is defined by the surface integral

$$\Phi_a = \int_D \mathrm{d}\mathbf{s} \cdot \mathbf{B} \tag{2.13}$$

with the magnetic induction in the superconductor $\mathbf{B}$. Referring to the vector potential representation, eqn. (2.3), and exploiting Stokes' theorem as well as eqn. (2.10) yields

$$\int_D \mathrm{d}\mathbf{s} \cdot \mathbf{B} + \Phi_s = \frac{\hbar}{q_s} \oint_{\partial D} \mathrm{d}\mathbf{l} \cdot \nabla\theta, \tag{2.14}$$

where Φ_s denotes the magnetic flux due to the induced supercurrent,

$$\Phi_s = \mu_0 \lambda_L^2 \oint_{\partial D} \mathrm{d}\mathbf{l} \cdot \mathbf{j}_s, \tag{2.15}$$

the line integrals extending (in the positive sense) about the contour ∂D of the identified domain. In order to evaluate the right-hand side of eqn. (2.14), we note that the superfluid wavefunction Ψ, as a solution of eqn. (2.1), must possess a unique dependence on space. This, by inspection of eqn. (2.7), however, leaves open a multitude of values of the phase θ differing within modulo 2π of its principal value still. We therefore conclude that

$$\oint_{\partial D} \mathrm{d}\mathbf{l} \cdot \nabla\theta = 2\pi k; \quad k = 0, 1, 2, \ldots \tag{2.16}$$

From eqns. (2.13) to (2.16), the total magnetic flux, or fluxoid, Φ associated with the domain D is thus seen to be given by

$$\Phi = \Phi_a + \Phi_s = \pm k\, \Phi_0 \quad \text{with } \Phi_0 = 2\pi\hbar/q_s; \quad k = 0, 1, 2 \ldots, \tag{2.17}$$

i.e. by integer multiples of the quantum of magnetic flux Φ_0, where the plus/minus sign relates to positive or negative values of q_s, respectively. This represents an immediate generalization of the predicted London result (London 1948, 1950).

Quantization of the total magnetic flux Φ, eqn. (2.17), directly testifies to the phase coherence of the superconducting state; the number k of flux quanta Φ_0 contained reflects the change of the phase when tracing the contour ∂D once. Although manifesting itself in superconductors with multiply connected domains D, this phenomenon extends beyond such circumstances alone: it occurs whenever a superconductor is permeated by magnetic flux, while the supercurrent encompasses one, or several, nonsuperconducting parts, even pits. The penetration of magnetic flux into type-II superconductors by Abrikosov vortices (Abrikosov 1957) and the spread of magnetic flux across superconducting tunnel junctions by Josephson vortices (Ferrell and Prange 1963) are prominent examples of that. If, by contrast, the supercurrent only circulates around superconducting parts, as with superconductors with simply connected domains D, zero total magnetic flux ensues, in which case eqn. (2.17) reverts to the integral form of eqn. (2.12), exhibiting the diamagnetic property of the Meissner state.

Some more remarkable aspects of eqn. (2.17) exist. The magnetic flux due to an applied magnetic field, Φ_a, will not normally obey a quantization rule. The supercurrent therefore adjusts in a way that the fluxoid Φ equals integer multiples of the quantum of magnetic flux Φ_0. If most of the supercurrent flows inside a thin sheath below the contour of the domain D, as could happen with a simply connected superconductor bulk in the Meissner state for linear dimensions of D much larger than the London penetration depth λ_L, the supercurrent contributes little to the total magnetic flux Φ, the dominant share being furnished by the applied magnetic field. On the other hand, quantization of magnetic flux becomes distinct if D is sufficiently small for few flux quanta to be involved. Suppose now the domain D includes a tubular cavity, i.e. a nonsuperconducting part, and the originally applied magnetic field is switched off; a persistent current flowing around the cavity then may ensure that the established magnetic flux is invariably maintained (or 'trapped').

The phenomenon of quantization of the fluxoid Φ has been confirmed experimentally both for conventional, low-temperature superconductors (Deaver and Fairbank 1961; Doll and Näbauer 1961) and for cuprate, high-temperature superconductors (Gammel et al. 1987; Witt 1988).

Empirical findings attribute the numerical value $\Phi_0 \simeq 2.067 \times 10^{-15}\ Wb$ to the quantum of magnetic flux which, from eqn. (2.17), is consistent with the superconducting particles carrying twice the elementary charge, $|q_s| = 2e$. The latter fact clearly demonstrates the nature of supercurrents due to Cooper pairs. Whereas these pairs generally involve two correlated electrons rendering $q_s = -2e$, conventional superconductivity implied, many high-temperature superconductors prove that there are two correlated holes, resulting in $q_s = +2e$ (Buckel and Kleiner 2012).

Since type-II superconductors are capable of accepting a certain amount of magnetic flux without losing their superconducting property, it follows that quantization of magnetic flux plays a vital role. Magnetic flux enters these materials through magnetic vortices—each carrying one quantum of magnetic flux—which, depending on the strength of the applied magnetic field, form a more-or-less densely packed vortex lattice with often triangular symmetry. If, for a regular array of magnetic vortices, n_V denotes the number of flux lines per unit area intersecting a chosen domain D of the superconductor bulk, the thermodynamic average magnetic induction brought about thus is $\langle B \rangle = n_V \Phi_0$, a result which can be directly obtained from eqn. (2.17). Vortex lattices have been observed experimentally in conventional, low-temperature superconductors, where the flux tubes of the Abrikosov vortices usually extend over many lattice spacings (Essmann and Träuble 1967; Yoshida et al. 1999); they have also been imaged in cuprate, high-temperature superconductors (Maggio-Aprile et al. 1995; Loudon and Midgley 2009), with the flux tubes of the 'pancake' vortices predicted for layered superconductors to occupy one lattice spacing alone (Efetov 1979; Feigel'man, Geshkenbein, and Larkin 1990; Belevtsov, Genenko, Ivanchenko, and Medvedev 1991; Clem 1991; Ivanchenko, Belevtsov, Genenko, and Medvedev 1992; Clem, Pe, and Benkraouda 1997).

Readers interested to know further details of the quantization of magnetic flux may consult some specialized literature (Gallop 1991; Portis 1993); those who want to learn more about magnetic vortex lattices and magnetic flux structures in type-II superconductors might find a number of textbooks (Orlando and Delin 1991; de Gennes 1994; Tilley and Tilley 2003; Huebener 2010) or a review (Brandt 1995b) of great use.

2.1.3 Straight magnetic vortices

An Abrikosov vortex in a type-II superconductor subject to a uniform applied magnetic field $\mathbf{H}_a$ may be envisaged as a tubular singularity, with a normal core of a radial extent given by the coherence length ξ_{GL}, where the superfluid wavefunction Ψ recedes, and a region of supercurrent circulation around the core, of a radial extent delimited by the London penetration depth λ_L, where the superfluid wavefunction Ψ prevails. This also marks the range of the magnetic induction associated with the supercurrent, which outwardly decays from the central vortex core. Comparatively simple descriptions based on the London concept exist in the practically interesting limit of extreme type-II superconductivity, i.e. when the critical fields are far apart, $H_{c1} \ll H_{c2}$, and hence the characteristic lengths obey $\xi_{GL} \ll \lambda_L$, a limit often referred to in this book. The idea of letting the vortex core shrink to a linear singularity then suggests itself, admitting though that the macroscopic quantum approach at hand loses its validity for distances smaller than about ξ_{GL}. The reduction to a line therefore is just an artifice for obtaining quantities like the distributions of currents and fields of magnetic vortices conveniently. This, however, requires due measures like cut-offs in order to avoid divergences, as we shall see.

2.1.3.1 Isolated magnetic vortex

We first consider an infinitely extended, straight magnetic vortex in a homogeneous, isotropic superconductor that fills all space, the flux line intersecting the x, y-plane at the origin $\mathbf{r}_\perp = 0$, with an orientation along the z-axis of a Cartesian coordinate system x, y, z. The vortex is supposed to carry one quantum of magnetic flux, as noted in section 2.1.2. An equation governing the magnetic induction $\mathbf{B}$ of the vortex can be derived straight away when going back to the fundamental current density relation, eqn. (2.10). By taking the curl of eqn. (2.10) and using Stokes' theorem as well as eqn. (2.16), interpreted for a contour which embraces the vortex core, we conclude that the formal substitution regarding the phase θ of the superfluid wavefunction Ψ,

$$\nabla \times \nabla\theta \to 2\pi\delta\left(\mathbf{r}_\perp\right)\mathbf{e}_z, \tag{2.18}$$

is permissible, with Dirac's delta function δ and the unit vector in the z-direction $\mathbf{e}_z$, the geometrical and physical conditions referred to above understood. Employing Ampère's law for supercurrent flow in a nonmagnetic medium

$$\mu_0\mathbf{j}_s = \nabla \times \mathbf{B} \tag{2.19}$$

and observing the derivation of the magnetic induction $\mathbf{B}$, eqn. (2.3), as well as the definition of the quantum of magnetic flux Φ_0, eqn. (2.17), finally yields the generalized London equation

$$\mathbf{B} + \lambda_{\mathrm{L}}^2 \nabla \times (\nabla \times \mathbf{B}) = \mathbf{V} \tag{2.20}$$

with the vorticity

$$\mathbf{V} = \pm \Phi_0 \, \delta\left(\mathbf{r}_{\perp}\right) \mathbf{e}_z \tag{2.21}$$

(Schmidt 1997). The plus/minus sign in eqn. (2.21) relates to a positive/negative vortex, i.e. to the magnetic induction $\mathbf{B}$ pointing along the positive/negative z-direction and the supercurrent, of density $\mathbf{j}_{\mathrm{s}}$, hence circulating in the positive or negative sense, respectively. Equation (2.20) with eqn. (2.21) generalizes eqn. (2.12) for a multiply connected superconductor domain, as a resort to eqn. (2.19) shows, proving that eqn. (2.17) represents the integral form of eqn. (2.12) confined to a single quantum of magnetic flux.

Because of its linearity, eqn. (2.20) furnishes a direct extension to an array of infinite, straight magnetic vortices i, if the centres of their cores intersect the x, y-plane at the respective points $\mathbf{r}_{\perp} = \mathbf{r}_{\perp,i}$, with an orientation parallel to the z-axis of the Cartesian coordinate system x, y, z. The total vorticity, by due application of eqn. (2.21), then reads

$$\mathbf{V} = \Phi_0 \sum_i \alpha_i \, \delta\left(\mathbf{r}_{\perp} - \mathbf{r}_{\perp,i}\right) \mathbf{e}_z, \tag{2.22}$$

the coefficients α_i herein taking values $+1$ or -1. Remarkably, if a system of magnetic vortices i of *finite* length and arbitrary shape delineated by $\mathbf{r} = \mathbf{l}_i$ were analyzed instead, eqn. (2.20) would still hold, but with the right-hand side now given by

$$\mathbf{V} = \Phi_0 \sum_i \alpha_i \int_{C_i} \mathrm{d}\mathbf{l}_i \, \delta\left(\mathbf{r} - \mathbf{l}_i\right), \tag{2.23}$$

a representation that involves the coefficients α_i too, apart from line integrals defined along the cores C_i (Brandt 1995b).

The magnetic induction $\mathbf{B}$ from eqns. (2.20) and (2.21) fully determines the electromagnetic, or London, energy per unit length, W_{L}, associated with the straight magnetic vortex in the otherwise vortex-free superconductor bulk. There are two contributions which together make up W_{L}: the field energy density, or magnetic pressure, p_{m} arising from the magnetic induction of the vortex and the kinetic energy density of the supercurrent t_{s} linked to the field,

$$p_{\mathrm{m}} = \frac{1}{2\mu_0} \mathbf{B}^2 \quad \text{and} \quad t_{\mathrm{s}} = \frac{\mu_0}{2} \lambda_{\mathrm{L}}^2 \mathbf{j}_{\mathrm{s}}^2, \tag{2.24}$$

ignoring the contribution of the thin normal vortex core. Calling upon eqn. (2.19) thus leads to the form

$$W_{\mathrm{L}} = \frac{1}{2\mu_0} \int ds \left(\mathbf{B}^2 + \lambda_{\mathrm{L}}^2 \left(\nabla \times \mathbf{B}\right)^2\right), \tag{2.25}$$

where the surface integral extends over the whole x, y-plane. The decay to 0 of the magnetic induction $\mathbf{B}$ far from the central vortex core makes it possible to simplify eqn. (2.25) considerably. If we return to eqn. (2.20) and utilize some vector analysis we arrive at

$$W_{\mathrm{L}} = \frac{1}{2\mu_0} \int ds \, (\mathbf{V} \cdot \mathbf{B}) \,, \tag{2.26}$$

an expression which can be evaluated most conveniently (Orlando and Delin 1991).

Equation (2.20) with eqn. (2.21) has the appropriate particular solution

$$\mathbf{B} = \pm \frac{\Phi_0}{2\pi\lambda_{\mathrm{L}}^2} K_0 \left(\frac{|\mathbf{r}_\perp|}{\lambda_{\mathrm{L}}} \right) \mathbf{e}_z \tag{2.27}$$

(Abrikosov 1957) established by the modified Bessel function of the second kind of order zero, K_0 (Abramowitz and Stegun 1972). This represents the magnetic induction $\mathbf{B}$ of a vortex of either sign in the region outside its core $|\mathbf{r}_\perp| \geq \xi_{\mathrm{GL}}$, revealing the maximum strength

$$B_{\mathrm{max}} \simeq \frac{\Phi_0}{2\pi\lambda_{\mathrm{L}}^2} \ln \left(\frac{\lambda_{\mathrm{L}}}{\xi_{\mathrm{GL}}} \right), \quad \xi_{\mathrm{GL}} \ll \lambda_{\mathrm{L}} \tag{2.28}$$

at the boundary of the core $|\mathbf{r}_\perp| = \xi_{\mathrm{GL}}$. Clearly, $\mathbf{B}$ must be continuous across this boundary and, since the core is in the normal state, can be assumed constant for $|\mathbf{r}_\perp| \leq \xi_{\mathrm{GL}}$, of a magnitude estimated by eqn. (2.28). The supercurrent density $\mathbf{j}_{\mathrm{s}}$ associated with the magnetic induction $\mathbf{B}$ follows from eqns. (2.19) and (2.27),

$$\mathbf{j}_{\mathrm{s}} = \pm \frac{\Phi_0}{2\pi\mu_0\lambda_{\mathrm{L}}^3} K_1 \left(\frac{|\mathbf{r}_\perp|}{\lambda_{\mathrm{L}}} \right) \mathbf{e}_\phi, \tag{2.29}$$

in terms of the modified Bessel function of the second kind of order one, K_1 (Abramowitz and Stegun 1972), and the azimuthal unit vector $\mathbf{e}_\phi$. Again, eqn. (2.29) holds in the region outside the core $|\mathbf{r}_\perp| \geq \xi_{\mathrm{GL}}$, unfolding the maximum strength

$$j_{\mathrm{s,max}} \simeq \frac{\Phi_0}{2\pi\mu_0\lambda_{\mathrm{L}}^2\xi_{\mathrm{GL}}}; \quad \xi_{\mathrm{GL}} \ll \lambda_{\mathrm{L}} \tag{2.30}$$

at the boundary $|\mathbf{r}_\perp| = \xi_{\mathrm{GL}}$, but vanishing inside the core, i.e. for $|\mathbf{r}_\perp| \leq \xi_{\mathrm{GL}}$. Equation (2.30) reproduces the parameter dependence of the depairing critical current density j_{GL}—the intrinsic maximum current density which a superconductor is still able to sustain without loss—as consistently derived from the Ginzburg–Landau theory,

$$j_{\mathrm{GL}} = \frac{\Phi_0}{3\sqrt{3}\pi\mu_0\lambda_{\mathrm{L}}^2\xi_{\mathrm{GL}}} \tag{2.31}$$

(Bardeen 1962), albeit numerically overestimating the latter result. Substitution of eqns. (2.21) and (2.27) into eqn. (2.26) yields the self-energy per unit length of the vortex,

$$E_{\rm V} \simeq \frac{\Phi_0^2}{4\pi\mu_0\lambda_{\rm L}^2} \ln\left(\frac{\lambda_{\rm L}}{\xi_{\rm GL}}\right); \quad \xi_{\rm GL} \ll \lambda_{\rm L}, \tag{2.32}$$

cutting off the magnetic induction at the boundary $|\mathbf{r}_\perp| = \xi_{\rm GL}$ and neglecting its contribution within the core $|\mathbf{r}_\perp| \leq \xi_{\rm GL}$. We comment that from $H_{\rm c1} = E_{\rm V}/\Phi_0$, this energy also determines the lower critical field for the appearance of straight magnetic vortices in the superconductor bulk.

2.1.3.2 Interacting magnetic vortices

Let us now further develop the analysis and consider two infinitely extended, straight magnetic vortices, the centres of their cores intersecting the x, y-plane at the respective points $\mathbf{r} = \mathbf{r}_{\perp,1}$ and $\mathbf{r} = \mathbf{r}_{\perp,2}$, with an orientation parallel to the z-axis of a Cartesian coordinate system x, y, z, as before. How do these vortices mutually interact? The total vorticity, from eqn. (2.22), obviously is

$$\mathbf{V} = \Phi_0 \left[\delta\left(\mathbf{r}_\perp - \mathbf{r}_{\perp,1}\right) \pm \delta\left(\mathbf{r}_\perp - \mathbf{r}_{\perp,2}\right)\right] \mathbf{e}_z, \tag{2.33}$$

allowing the two vortices to have equal or opposite signs. The appropriate particular solution of eqn. (2.20) with this vorticity can be readily inferred from eqn. (2.27):

$$\mathbf{B} = \frac{\Phi_0}{2\pi\lambda_{\rm L}^2} \left[K_0\left(\frac{|\mathbf{r}_\perp - \mathbf{r}_{\perp,1}|}{\lambda_{\rm L}}\right) \pm K_0\left(\frac{|\mathbf{r}_\perp - \mathbf{r}_{\perp,2}|}{\lambda_{\rm L}}\right)\right] \mathbf{e}_z. \tag{2.34}$$

Equation (2.34) describes the magnetic induction $\mathbf{B}$ around the two vortices outside their respective cores, i.e. for both $|\mathbf{r}_\perp - \mathbf{r}_{\perp,1}| \geq \xi_{\rm GL}$ and $|\mathbf{r}_\perp - \mathbf{r}_{\perp,2}| \geq \xi_{\rm GL}$. Like before, continuity of $\mathbf{B}$ across the boundary of either core must be ensured. Substitution of eqns. (2.33) and (2.34) into eqn. (2.26) yields the total London energy per unit length of the vortices,

$$W_{\rm L,12} = 2E_{\rm V} \pm \frac{\Phi_0^2}{2\pi\mu_0\lambda_{\rm L}^2} K_0\left(\frac{|\mathbf{r}_{\perp,12}|}{\lambda_{\rm L}}\right), \tag{2.35}$$

upon denoting the vortex separation $|\mathbf{r}_{\perp,12}| = |\mathbf{r}_{\perp,1} - \mathbf{r}_{\perp,2}|$ and invoking constancy of the magnetic induction within the regions of both cores. The first term in eqn. (2.35) represents twice the self-energy per unit length of a single vortex, eqn. (2.32), whereas the second term reflects the interaction between the two vortices adding to, or subtracting from, the former when the vortices have equal or opposite signs, respectively, and hence repel each other, or attract.

The electromagnetic energy associated with the mutual interaction of the two vortices resolves the force per unit length $\mathbf{F}_{\rm s,12}$ that the vortex located at $\mathbf{r} = \mathbf{r}_{\perp,1}$ exerts upon the vortex located at $\mathbf{r} = \mathbf{r}_{\perp,2}$. Thus, taking the directional derivative of $W_{\rm L,12}$ from eqn. (2.35) along the separation of the two vortices gives

$$\mathbf{F}_{\rm s,12} = \pm\frac{\Phi_0^2}{2\pi\mu_0\lambda_{\rm L}^3} K_1\left(\frac{|\mathbf{r}_{\perp,12}|}{\lambda_{\rm L}}\right) \mathbf{e}_{\rm r} \tag{2.36}$$

with the radial unit vector $\mathbf{e}_{\rm r}$. This finding elucidates repulsion for vortices of equal signs and attraction for vortices of opposite signs. Remarkably enough, eqn. (2.36) can be cast into a still more instructive form by reference to the supercurrent density $\mathbf{j}_{\mathrm{s},12}$ of the vortex located at $\mathbf{r} = \mathbf{r}_{\perp,1}$ surviving at the site of the other, $\mathbf{r} = \mathbf{r}_{\perp,2}$:

$$\mathbf{j}_{\mathrm{s},12} = \pm \frac{\Phi_0}{2\pi\mu_0\lambda_{\mathrm{L}}^3} K_1\left(\frac{|\mathbf{r}_{\perp,12}|}{\lambda_{\mathrm{L}}}\right) \mathbf{e}_\phi. \tag{2.37}$$

Using eqn. (2.37) in eqn. (2.36) constitutes the Lorentz-like relationship

$$\mathbf{F}_{\mathrm{s},12} = \mathbf{j}_{\mathrm{s},12} \times (\Phi_0 \mathbf{e}_z), \tag{2.38}$$

which brings together the force per unit length exerted by the vortex at $\mathbf{r} = \mathbf{r}_{\perp,1}$ with vortex strength Φ_0 and its supercurrent density $\mathbf{j}_\mathrm{s}$ evaluated at $\mathbf{r} = \mathbf{r}_{\perp,2}$. Clearly, if the force per unit length $\mathbf{F}_{\mathrm{s},21}$ that the vortex located at $\mathbf{r} = \mathbf{r}_{\perp,2}$ exerts upon the vortex located at $\mathbf{r} = \mathbf{r}_{\perp,1}$ were addressed instead, the structure of eqn. (2.38) would be retained, albeit with $\mathbf{F}_{\mathrm{s},21} = -\mathbf{F}_{\mathrm{s},12}$. Hence, when dwelling on this result and identifying a central magnetic vortex in a regular array of magnetic vortices of equal signs, the total force per unit length exerted upon that particular vortex by all the other vortices would obviously sum up to give null. It should be commented that, although obtained for straight, parallel magnetic vortices, eqn. (2.38) *mutatis mutandis* applies to curved magnetic vortices too, with a local tangential unit vector $\mathbf{t}$ in place of the unit vector $\mathbf{e}_z$ (Brandt 1995b).

The insights gained by analytical means will now be explained with the help of graphical plots. Figure 2.2 presents an example of the distribution of the magnetic induction $\mathbf{B}$ around two straight, parallel magnetic vortices of equal and opposite signs, with the separation of the vortices kept fixed. It appears from Fig. 2.2(a) for vortices of equal signs that the density of the lines of constant $\mathbf{B}$, which here coincide

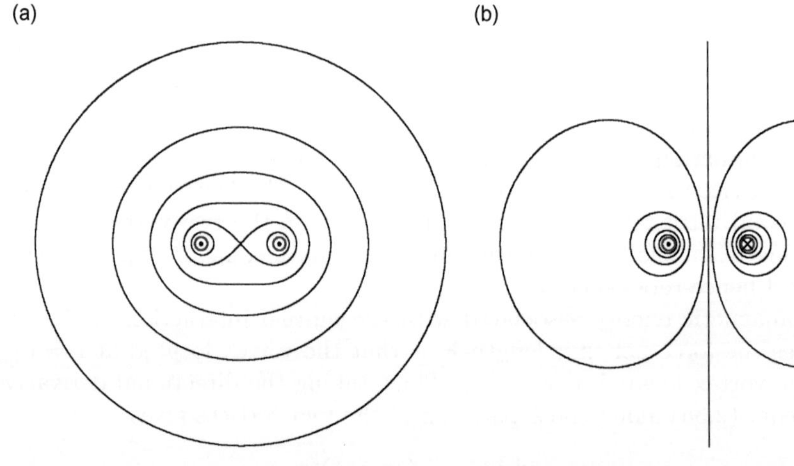

Fig. 2.2 Lines of constant magnetic induction $\mathbf{B}$ coinciding with the streamlines of the supercurrent density $\mathbf{j}_\mathrm{s}$ around two straight, parallel magnetic vortices of (a) equal and (b) opposite signs for the normalized separation of the vortices $|\mathbf{r}_{\perp,12}|/\lambda_\mathrm{L} = 1/2$, when $\lambda_\mathrm{L}/\xi_{\mathrm{GL}} = 25$ (cross-sectional view).

with the streamlines of the supercurrent density $\mathbf{j}_s$, decreases to 0 halfway between the vortices, whilst it increases outside as compared with the situation of isolated vortices. Conservation of the London energy density $p_m + t_s$ given by eqn. (2.24) addressed to the respective vortex field, eqn. (2.34), thus demands a simultaneous reduction of the magnetic pressure exerted upon the cores of the vortices from outside and an amplification inside, with the net result of an outwardly directed, *repulsive* force on each of the two vortices. If, in contrast to that, the vortices have opposite signs, as in Fig. 2.2(b), the density of the lines of constant $\mathbf{B}$, and hence the supercurrent density $\mathbf{j}_s$ associated with it, increases between the vortices, whilst it decreases outside as compared with the situation of isolated vortices. Conservation of the London energy density $p_m + t_s$ therefore entails a simultaneous amplification of the magnetic pressure exerted upon the cores of the vortices from outside and a reduction inside, with the net result of an inwardly directed, *attractive* force on each of the two vortices. Precisely the same conclusion would be reached when considering a single, straight magnetic vortex parallel to the flat surface of a semi-infinite superconductor instead: the vortex and its image (of opposite sign) in a vacuum adjacent to the superconductor attract, which means the vortex itself experiences attraction towards the boundary.

These characteristics are also evident from the variation of the normalized London energy $W_{L,12}/E_V$ of the two straight, parallel magnetic vortices with their normalized separation $|\mathbf{r}_{\perp,12}|/\lambda_L$ shown in Fig. 2.3. If the vortices are far apart, they virtually take on the self-energy that two noninteracting vortices would afford. As the separation of the vortices abates, their electromagnetic energy drops monotonically, reaching zero with the annihilation of the vortices in the case of opposite

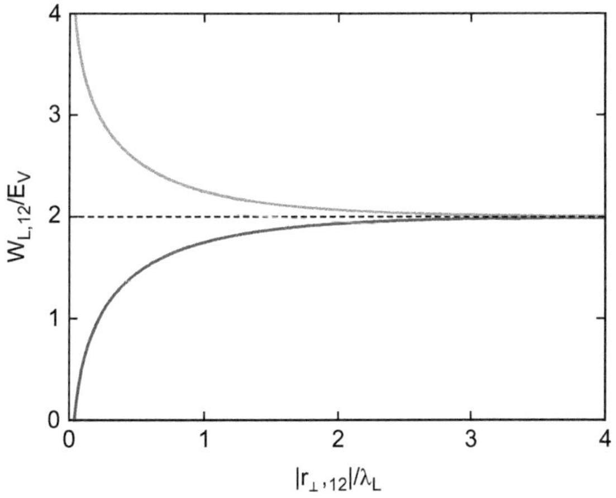

Fig. 2.3 Normalized London energy $W_{L,12}/E_V$ of two straight, parallel magnetic vortices of equal (upper curve) and opposite (lower curve) signs as a function of the normalized separation of the vortices $|\mathbf{r}_{\perp,12}|/\lambda_L$, when $\lambda_L/\xi_{GL} = 25$. The dashed line separates the two energy regimes and merely serves as a guide to the eye.

vortex signs; by contrast, their electromagnetic energy builds up monotonically, attaining the fourfold value of the self-energy of a single vortex with the fusion of the two vortices to yield a single vortex of doubled vorticity in the case of equal vortex signs. Magnetic vortices therefore tend to be confined to the minimal vortex strength of one quantum of magnetic flux.

A wealth of further information about the structure and electrodynamics of magnetic vortices as well as their mutual interactions is offered in the salient literature (de Gennes 1994; Ketterson and Song 1999; Tilley and Tilley 2003; Tinkham 2004; Poole, Farach, Creswick, and Prozorov 2007; Huebener 2010).

2.2 Bean's model of the critical state

The observations of section 2.1.3 show that straight, parallel magnetic vortices in a type-II superconductor exert a Lorentz-like force upon each other, whence vortices of equal signs repel and vortices of opposite signs attract. An array of like magnetic vortices, with the external magnetic field switched off, thus would disperse, seeing the vortices attracted towards the superconductor's boundary. A uniform applied magnetic field $\mathbf{H}_a$, on the other hand, counteracts this effect by exerting a magnetic force that overcomes the mutual repulsion of the vortices. It is the interplay between attraction and repulsion which, in an ideal, homogeneous type-II superconductor filling all space, stabilizes a regular lattice of vortices to form a mixed thermal equilibrium, and hence reversible, state. The imposition of a transport current would drive the vortices and eventually give rise to electromagnetic energy being converted into heat. Since no loss-free transport current—let alone the high electrical currents as well as the strong magnetic fields which many practical uses demand—could therefore persist in such a state, we examine metastable states caused by structural inhomogeneities acting as pinning centres for magnetic vortices. These nonequilibrium states invariably entail irreversibility through their suppression of the flow of magnetic flux, a prerequisite for current transport without loss. The latter condition paves the way for the critical-current concept; current flow when an electrical resistance is absent directly involves hysteresis of the magnetization curve. Both traits are united in Bean's model of the critical state.

2.2.1 Flow and pinning of magnetic flux

In studying the flow and pinning of magnetic flux associated with a type-II superconductor's *inhomogeneous* mixed state, we can distinguish three types of forces that act on individual magnetic vortices: (i) a Lorentz-like force brought about by imposed transport currents; (ii) a magnetic force arising from the repulsion of the residual vortices; and (iii) a pinning force due to the presence of localized defects such as small precipitates of a normal metallic phase or extended defects like irradiation-induced columnar tracks, apart from intergranular and outer boundaries (Portis 1993). The applied magnetic field does not explicitly come into play here, since its integrative effect on the system of vortices only enters through a thermodynamic analysis.

To discuss matters quantitatively, let us identify an infinitely extended, straight magnetic vortex within an array of like vortices constituting the mixed state. The orientation of its flux line as well as that of the applied magnetic field $\mathbf{H}_\mathrm{a}$ shall be parallel to the z-axis of a Cartesian coordinate system x, y, z. When a transport current, of density $\mathbf{j} = j_y \mathbf{e}_y$, flowing in the (positive) y-direction is imposed, the average influence of the magnetic force due to other vortices will be to render this current density virtually uniform on a length scale on the order of the spacing between the vortices, but allow for spatial variations otherwise. It is therefore appropriate to comprehend the effect of the external current in terms of the Lorentz-like force per unit length of the identified vortex,

$$\mathbf{F}_\mathrm{L} = \mathbf{j} \times (\Phi_0 \mathbf{e}_z), \tag{2.39}$$

analogous to what eqn. (2.38) states for the supercurrent density created by a vortex at the site of another (Orlando and Delin 1991). Dissipation of electromagnetic energy during the motion of the vortices which this force provokes has two sources at least: temporal changes of the magnetic flux induce an electric field whose action on the normal vortex cores generates Joule heat (Bardeen and Stephen 1965); and repeated phase transitions away from equilibrium between the superconducting and the normal state entail additional loss owing to the irreversibility of the processes invoked (Tinkham 1964). These types of dissipation can be usefully conceived by postulating a phenomenological drag that tends to impede the motion of the vortices. If assuming linear viscous flow characterized by the velocity $\mathbf{u} = u_x \mathbf{e}_x$, the drag force per unit length of the identified vortex reads

$$\mathbf{F}_\mathrm{d} = -\eta_\mathrm{V} \mathbf{u} \tag{2.40}$$

with the friction coefficient η_V. The motion of the vortices is furthermore suppressed by their attractive interaction with structural defects. Immobilization or 'pinning' of magnetic vortices, entailing $\mathbf{u} = 0$, determines the critical current density j_c, i.e. the maximum strength of the current density $\mathbf{j}$ obtainable without loss (Kunzler, Buehler, Hsu, and Wernick 1961), a quantity which, in practice, stays well below the theoretically limiting depairing critical current density j_GL. The maximum pinning force per unit length of the identified vortex thus results from its balance with the Lorentz-like force, eqn. (2.39), when $\mathbf{j} = j_\mathrm{c} \mathbf{e}_y$:

$$\mathbf{F}_\mathrm{p} = -(j_\mathrm{c} \mathbf{e}_y) \times (\Phi_0 \mathbf{e}_z). \tag{2.41}$$

Gathering eqns. (2.39) to (2.41) yields the total force per unit length acting on the identified vortex,

$$\mathbf{F}_\mathrm{tot} = \mathbf{F}_\mathrm{L} + \mathbf{F}_\mathrm{d} + \mathbf{F}_\mathrm{p}. \tag{2.42}$$

When requiring $\mathbf{F}_\mathrm{tot} = 0$ in eqn. (2.42), as with an overall balance of force, and also considering the vortex to rest if $|\mathbf{F}_\mathrm{L}| \leq |\mathbf{F}_\mathrm{p}|$, the velocity of steady-state flow of magnetic flux,

$$\mathbf{u} = \begin{cases} 0 & \text{if } 0 \leq j_y \leq j_c \\ \dfrac{1}{\eta_V}(j_y - j_c)\,\Phi_0\,\mathbf{e}_x & \text{if } j_y > j_c, \end{cases} \tag{2.43}$$

ensues. Proceeding from this result, a description of electromagnetic properties of the whole system of vortices can be given in the sense of averages over regions with linear dimensions much larger than the spacing between the vortices. Thus, eqns. (2.2) and (2.3) show that the local electric field $\mathbf{E} = E_y\mathbf{e}_y$ induced by the moving vortices and the local magnetic induction $\mathbf{B} = B_z\mathbf{e}_z$ generated by them are connected via

$$\mathbf{E} = \mathbf{B} \times \mathbf{u}. \tag{2.44}$$

Substituting eqn. (2.43) into eqn. (2.44) therefore yields the quasi-Ohmic relationship

$$\mathbf{E} = \begin{cases} 0 & \text{if } 0 \leq j_y \leq j_c \\ \rho_{ff}\,(j_y - j_c)\,\mathbf{e}_y & \text{if } j_y > j_c \end{cases} \tag{2.45}$$

with the flux-flow resistivity

$$\rho_{ff} = \left(\frac{\Phi_0}{\eta_V}\right) B_z, \tag{2.46}$$

indicating a transition from the loss-free state to a lossy state when $j_y = j_c$. The friction coefficient η_V in eqn. (2.46) may be expressed through the normal-state resistivity ρ_n appearing for strengths of the applied magnetic field H_a as high as the upper critical field H_{c2}, and thus permits the distinct form

$$\frac{\rho_{ff}}{\rho_n} = \frac{B_z}{B_{c2}}, \tag{2.47}$$

with $B_{c2} = \mu_0 H_{c2}$, a correspondence well observed by conventional superconductors in the low-temperature limit $T \to 0$ (Kim, Hempstead, and Strnad 1964). The local electromagnetic power per unit volume lost as Joule heat finally ensues

$$P = \mathbf{j} \cdot \mathbf{E}, \tag{2.48}$$

with the electric field $\mathbf{E}$ given by eqn. (2.45).

The pinning of magnetic vortices at crystal defects implies that the spatial distribution of the vortices will deviate from their distribution which thermal equilibrium would prescribe; the local magnetic induction $\mathbf{B}$ brought about by the vortices cannot therefore be uniform, and hence be associated with a current-free state. It rather generates a local current density $\mathbf{j}$, as Ampère's law for current flow in a nonmagnetic medium claims:

$$\nabla \times \mathbf{B} = \mu_0 \mathbf{j}. \tag{2.49}$$

Clearly, if $|\mathbf{j}| > j_c$, the superconductor will pass into the resistive state. On the other hand, if $|\mathbf{j}| < j_c$, the impact of pinning is less distinct than it could at most be. This suggests hypothesizing, for superconductors with strong pinning of magnetic vortices, that the latter are distributed in such a way so as to produce the maximum pinning force everywhere,

$$|\nabla \times \mathbf{B}| = \mu_0 j_c, \tag{2.50}$$

i.e. constraining the magnetic induction **B** to always generate the critical current density j_c, regardless of the direction in which the current flows. Equation (2.50) is the constitutive equation of Bean's model of the critical state (Bean 1962, 1964); a simple, but powerful macroscopic approach capturing the principal features of the magnetic behaviour of superconductors which reveal the said property, for current-field configurations describable by reference to merely one coordinate of space. Its only parameter, j_c, hinges upon temperature T and, strictly, the magnetic induction **B**. We shall disregard the second dependence in keeping with the assumptions underlying eqn. (2.11), nevertheless, for field strengths much smaller than the upper critical field H_{c2}, a situation addressed throughout this text.

Type-II superconductors that demonstrate strong pinning of magnetic vortices are often termed 'hard' (or 'high-field') superconductors; a thorough account of their irreversible properties is given in several books (Campbell and Evetts 1972; Ullmaier 1975; Rose-Innes and Rhoderick 1994; Waldram 1996; Gurevich, Mints, and Rakhmanov 1997; Poole, Farach, Creswick, and Prozorov 2007; Huebener 2010).

2.2.2 Critical state of a superconductor slab

We give a simple example by investigating the critical state of an infinitely extended type-II superconductor slab with thickness $2w$ limited by the range $-w \leq x \leq w$ of a Cartesian coordinate system x, y, z and subject to either an imposed transport current or an applied magnetic field. It will turn out in the latter case that measurable quantities like the critical current density and the hysteretic AC loss of the slab are directly linked to characteristics of the thermodynamic average magnetization loop.

2.2.2.1 Imposed transport current

Let us first assume the slab carries an oscillating transport current with maximum strength J_c, per unit length of the slab, along the z-direction of the Cartesian coordinate system x, y, z in the absence of an applied magnetic field. For this geometry, the magnetic induction and the current density display respective nonzero y- and z-components only, both depending on the coordinate x. When adopting the quasistatic approximation and making recourse to eqn. (2.50), the instantaneous distribution of the magnetic induction B_y within the slab is governed by the differential equation

$$\left| \frac{dB_y}{dx} \right| = \mu_0 j_c \tag{2.51}$$

together with the boundary condition

$$B_y = \pm \mu_0 H_i \quad \text{at } x = \pm w, \tag{2.52}$$

where the upper/lower sign relates to positive or negative values of x, respectively, and $H_\mathrm{i} = J/2$ denotes the induced magnetic field corresponding to the actual value of the imposed transport current per unit length J. A convenient simplification, for amplitudes $H_\mathrm{c} = J_\mathrm{c}/2$ of this field much larger than the lower critical field $H_{\mathrm{c}1}$, is to model the magnetic flux as entering the slab at imposed transport current $J \to 0$. Then, given a solution of eqns. (2.51) and (2.52), the transport current density j_z within the slab derives from

$$\mu_0 j_z(x) = \frac{dB_y}{dx}. \tag{2.53}$$

We follow up the evolution of these observables during cycling of the imposed transport current for $H_\mathrm{c} = H_\mathrm{f}$, with the full penetration field $H_\mathrm{f} = w j_\mathrm{c}$, in which case a saturated steady state is reached.

Initial transient. Range $0 \leq H_\mathrm{i} \leq H_\mathrm{f}$: beginning with the flux-free virgin state of the superconductor slab, magnetic flux of opposite sign penetrates from the two surfaces up to the depth $s = H_\mathrm{i}/j_\mathrm{c}$, leaving a central region of half-width $b = w - s$ devoid of magnetic flux, before H_i reaches the value H_f. The distribution of the magnetic induction thus is

$$B_y(x) = \begin{cases} 0 & \text{for } 0 \leq |x| \leq b \\ \mu_0 j_\mathrm{c} x \pm \mu_0 (H_\mathrm{i} - H_\mathrm{f}) & \text{for } b \leq |x| \leq w, \end{cases} \tag{2.54}$$

consistent with the current density $j_z(x)$ assuming the value j_c in both flux-filled marginal parts of the slab, but vanishing otherwise unless H_i adopts the maximum H_f, where the current density, still at j_c, occupies the whole slab.

Steady-state cycle. Range $H_\mathrm{f} > H_\mathrm{i} \geq -H_\mathrm{f}$: proceeding from the state of maximum flux-filling of the superconductor slab, magnetic flux of either sign leaves through the surfaces up to the depth $s_- = (H_\mathrm{f} - H_\mathrm{i})/2j_\mathrm{c}$, complying with the decreased (and possibly even reversed) induced magnetic field but preserving trapped magnetic flux in a central region of half-width $b_- = w - s_-$, before H_i reaches the value $-H_\mathrm{f}$. The distribution of the magnetic induction thus is

$$B_y(x) = \begin{cases} \mu_0 j_\mathrm{c} x & \text{for } 0 \leq |x| \leq b_- \\ -\mu_0 j_\mathrm{c} x \pm \mu_0 (H_\mathrm{i} + H_\mathrm{f}) & \text{for } b_- \leq |x| \leq w, \end{cases} \tag{2.55}$$

consistent with the current density $j_z(x)$ adopting the value $-j_\mathrm{c}$ in the flux-reduced marginal parts of the slab and the value j_c otherwise. Range $-H_\mathrm{f} < H_\mathrm{i} \leq H_\mathrm{f}$: proceeding from the state of maximum flux-filling of the superconductor slab, magnetic flux of opposite sign leaves through the surfaces up to the depth $s_+ = (H_\mathrm{f} + H_\mathrm{i})/2j_\mathrm{c}$, complying with the increased (and possibly again reversed) induced magnetic field but preserving trapped magnetic flux of opposite sign in a central region of half-width $b_+ = w - s_+$, before H_i reaches the value H_f. The distribution of the magnetic induction thus is

$$B_y(x) = \begin{cases} -\mu_0 j_\mathrm{c} x & \text{for } 0 \leq |x| \leq b_+ \\ \mu_0 j_\mathrm{c} x \pm \mu_0 (H_\mathrm{i} - H_\mathrm{f}) & \text{for } b_+ \leq |x| \leq w, \end{cases} \tag{2.56}$$

consistent with the current density $j_z(x)$ adopting the value j_c in the flux-reduced marginal parts of the slab and the value $-j_\mathrm{c}$ otherwise.

The evolution of the profile of the normalized magnetic induction B_y/B_f with $B_\mathrm{f} = \mu_0 H_\mathrm{f}$ during cycling of the imposed transport current is shown in Fig. 2.4. This bears evidence of the *antisymmetry* of the magnetic induction with respect to the midplane $x = 0$ of the slab and, hence, of the *symmetry* of the distribution of the transport current density across the slab.

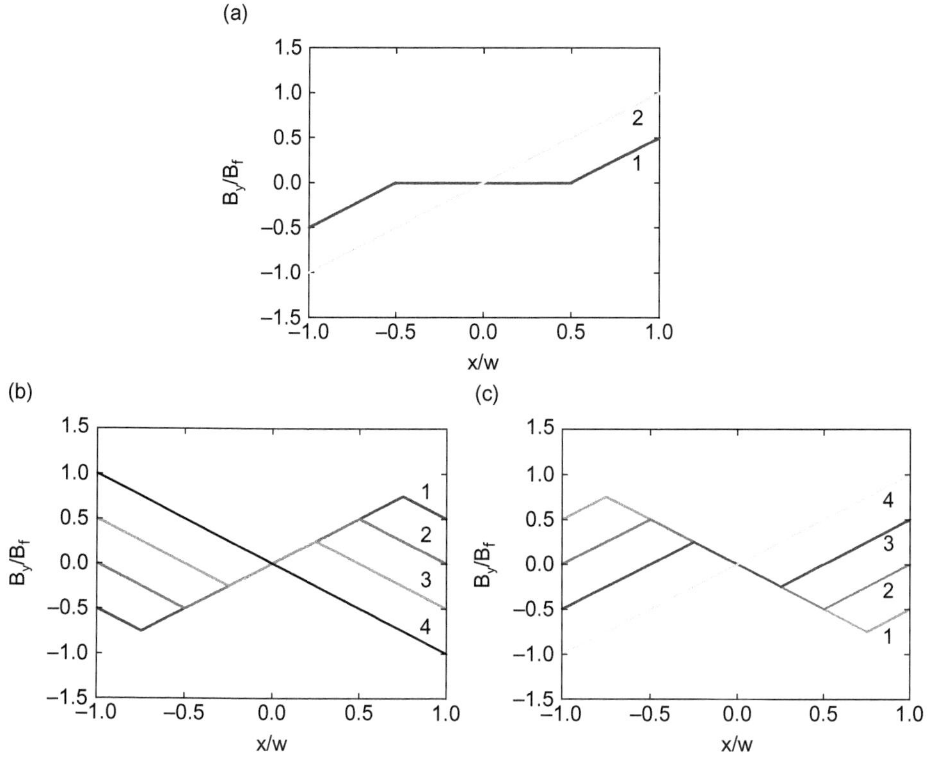

Fig. 2.4 Distribution of the normalized magnetic induction $B_\mathrm{y}/B_\mathrm{f}$ as a function of the normalized coordinate x/w over the thickness of the superconductor slab for different values of the normalized induced magnetic field $H_\mathrm{i}/H_\mathrm{f}$ during (a) the initial transient with increasing field, when $H_\mathrm{i}/H_\mathrm{f} = 0.5$, 1, (b) a steady-state half-cycle with decreasing field, when $H_\mathrm{i}/H_\mathrm{f} = 0.5$, 0, -0.5, -1, and (c) a steady-state half-cycle with increasing field, when $H_\mathrm{i}/H_\mathrm{f} = -0.5$, 0, 0.5, 1, understanding $H_\mathrm{c}/H_\mathrm{f} = 1$.

Making recourse to the local electromagnetic power per unit volume lost as Joule heat, eqn. (2.48), makes it possible to establish the hysteretic AC loss, per unit area of the slab, U_ac, incurred during a steady-state cycle of the imposed transport current,

$$U_\mathrm{ac} = \oint_{\mathrm{f.c.}} dt \int_{-w}^{w} dx\, j_z(x) E_z(x)\,, \tag{2.57}$$

by integrating over the time t of a full cycle of the evolving magnetic induction profile (Carr 2001). The induced electric field E_z herein follows from Faraday's law,

$$\frac{dE_z}{dx} = \dot{B}_y(x) \tag{2.58}$$

given that E_z vanishes in the midplane $x = 0$ of the slab (Norris 1970). Hence, using eqns. (2.53) and (2.58) together with the results, eqns. (2.55) and (2.56), in eqn. (2.57) yields the characteristic form for $H_c = H_f$,

$$U_{ac} = 2\mu_0 w J_c^2/3, \tag{2.59}$$

which reveals a quadratic dependence on J_c.

2.2.2.2 Applied magnetic field

Let us now assume the slab is exposed to an oscillating applied magnetic field with amplitude H_a oriented parallel to the z-axis of a Cartesian coordinate system x, y, z. For this geometry, the current density and the magnetic induction display respective nonzero y- and z-components only, both depending on the coordinate x. When adopting the quasistatic approximation and making recourse to eqn. (2.50), the instantaneous distribution of the magnetic induction B_z within the slab is governed by the differential equation

$$\left|\frac{dB_z}{dx}\right| = \mu_0 j_c \tag{2.60}$$

together with the boundary condition—absence of surface currents understood—

$$B_z = \mu_0 H_z \quad \text{at } x = \pm w, \tag{2.61}$$

where H_z denotes the actual value of the applied magnetic field. A convenient simplification, for amplitudes of this field much larger than the lower critical field H_{c1}, is to model the magnetic flux as entering the slab at applied magnetic field $H_z \to 0$. Then, given a solution of eqns. (2.60) and (2.61), the induced current density j_y within the slab derives from

$$\mu_0 j_y(x) = -\frac{dB_z}{dx}, \tag{2.62}$$

and the thermodynamic average magnetization $\langle M_z \rangle$ of the slab obtains from

$$\mu_0 \langle M_z \rangle = \frac{1}{2w} \int_{-w}^{w} dx\, B_z(x) - \mu_0 H_z \tag{2.63}$$

(Tamm 1979). We follow up the evolution of these observables during cycling of the applied magnetic field for $H_a \geq H_f$, with the full penetration field $H_f = w j_c$, in which case a saturated steady state is reached.

Initial transient. Range $0 \leq H_z \leq H_f$: beginning with the flux-free virgin state of the superconductor slab, magnetic flux of equal sign penetrates from the two surfaces up to the depth $s = H_z/j_c$, leaving a central region of half-width $b = w - s$ devoid of magnetic flux, before H_z reaches the value H_f. The distribution of the magnetic induction thus is

$$B_z(x) = \begin{cases} 0 & \text{for } 0 \leq |x| \leq b \\ \mu_0 j_c |x| + \mu_0 (H_z - H_f) & \text{for } b \leq |x| \leq w, \end{cases} \tag{2.64}$$

consistent with the current density $j_y(x)$ assuming respective values j_c and $-j_c$ in the flux-filled marginal parts of the slab, but vanishing otherwise. The thermodynamic average magnetization, given by

$$\langle M_z \rangle = -H_z \left(1 - \frac{H_z}{2H_f}\right), \tag{2.65}$$

reveals a *diamagnetic* property of the slab, tending towards saturation as H_z approaches H_f. Range $H_f < H_z \leq H_a$: magnetic flux continues to enter the fully flux-filled superconductor slab, increasing the magnetic induction

$$B_z(x) = \mu_0 j_c |x| + \mu_0 (H_z - H_f) \quad \text{for } 0 \leq |x| \leq w \tag{2.66}$$

furthermore, yet sustaining a central minimum, until H_z reaches the maximum H_a, while the current density $j_y(x)$, still at j_c and $-j_c$, occupies both halves of the slab. The thermodynamic average magnetization persists in *diamagnetic* saturation,

$$\langle M_z \rangle = -\frac{H_f}{2}. \tag{2.67}$$

Steady-state cycle. Range $H_a > H_z > H_a - 2H_f$: proceeding from the state of maximum flux-filling of the superconductor slab, magnetic flux of like sign leaves through the surfaces up to the depth $s_- = (H_a - H_z)/2j_c$, complying with the decreased (and possibly even reversed) external magnetic field but preserving trapped magnetic flux in a central region of half-width $b_- = w - s_-$, before H_z reaches the value $H_a - 2H_f$. The distribution of the magnetic induction thus is

$$B_z(x) = \begin{cases} \mu_0 j_c |x| + \mu_0 (H_a - H_f) & \text{for } 0 \leq |x| \leq b_- \\ -\mu_0 j_c |x| + \mu_0 (H_z + H_f) & \text{for } b_- \leq |x| \leq w, \end{cases} \tag{2.68}$$

consistent with the current density $j_y(x)$ adopting respective values $-j_c$ and j_c in the flux-reduced marginal parts of the slab and also values of reversed signs otherwise. The thermodynamic average magnetization, given by

$$\langle M_z \rangle = (H_a - H_z)\left(1 - \frac{H_a - H_z}{4H_f}\right) - \frac{H_f}{2}, \tag{2.69}$$

shows a transition from a *diamagnetic* to a *paramagnetic* property of the slab, tending towards saturation as H_z approaches $H_a - 2H_f$. Range $H_a - 2H_f \geq H_z \geq -H_a$: magnetic flux of opposite sign forms in the interior and spreads across the superconductor slab, decreasing the magnetic induction

$$B_z(x) = -\mu_0 j_c |x| + \mu_0 (H_z + H_f) \quad \text{for } 0 \le |x| \le w \tag{2.70}$$

furthermore, yet sustaining a central maximum, until H_z reaches the minimum $-H_a$, while the current density $j_y(x)$, at $-j_c$ and j_c, occupies both halves of the slab. The thermodynamic average magnetization remains in *paramagnetic* saturation,

$$\langle M_z \rangle = \frac{H_f}{2}. \tag{2.71}$$

Range $-H_a < H_z < -H_a + 2H_f$: proceeding from the state of maximum flux-filling of the superconductor slab, magnetic flux of opposite sign leaves through the surfaces up to the depth $s_+ = (H_a + H_z)/2j_c$, complying with the increased (and possibly again reversed) external magnetic field but preserving trapped magnetic flux of opposite sign in a central region of half-width $b_+ = w - s_+$, before H_z reaches the value $-H_a + 2H_f$. The distribution of the magnetic induction thus is

$$B_z(x) = \begin{cases} -\mu_0 j_c |x| - \mu_0 (H_a - H_f) & \text{for } 0 \le |x| \le b_+ \\ \mu_0 j_c |x| + \mu_0 (H_z - H_f) & \text{for } b_+ \le |x| \le w, \end{cases} \tag{2.72}$$

consistent with the current density $j_y(x)$ adopting respective values j_c and $-j_c$ in the flux-reduced marginal parts of the slab and also values of reversed signs otherwise. The thermodynamic average magnetization, given by

$$\langle M_z \rangle = -(H_a + H_z)\left(1 - \frac{H_a + H_z}{4H_f}\right) + \frac{H_f}{2}, \tag{2.73}$$

shows a transition from a *paramagnetic* to a *diamagnetic* property of the slab, tending towards saturation as H_z approaches $-H_a + 2H_f$. Range $-H_a + 2H_f \le H_z \le H_a$: magnetic flux of the original sign forms in the interior and spreads across the superconductor slab, increasing the magnetic induction

$$B_z(x) = \mu_0 j_c |x| + \mu_0 (H_z - H_f) \quad \text{for } 0 \le |x| \le w \tag{2.74}$$

furthermore, yet sustaining a central minimum, until H_z reaches the maximum H_a, while the current density $j_y(x)$, at j_c and $-j_c$, occupies both halves of the slab. The thermodynamic average magnetization again persists in *diamagnetic* saturation,

$$\langle M_z \rangle = -\frac{H_f}{2}. \tag{2.75}$$

The evolution of the profile of the normalized magnetic induction B_z/B_f with $B_f = \mu_0 H_f$ during cycling of the applied magnetic field is shown in Fig. 2.5. This bears evidence of the *symmetry* of the magnetic induction with respect to the midplane $x = 0$ of the slab and, hence, of the *antisymmetry* of the distribution of the induced current density across the slab.

The corresponding variation of the normalized thermodynamic average magnetization $\langle M_z \rangle / M_s$ with $M_s = H_f/2$ is portrayed in Fig. 2.6. It thus appears, according

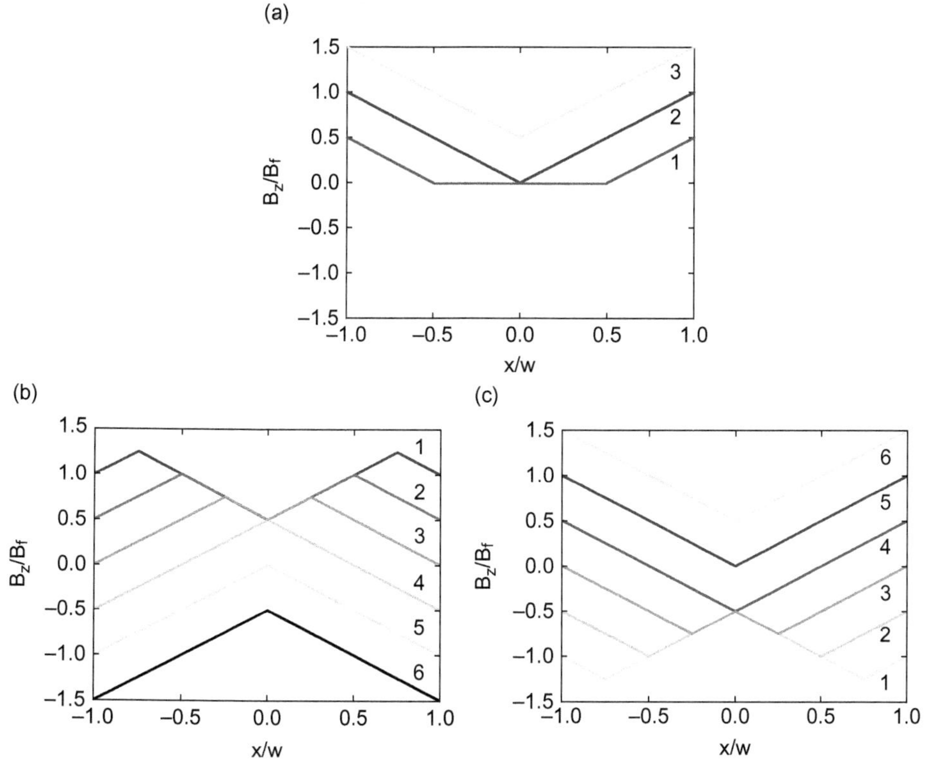

Fig. 2.5 Distribution of the normalized magnetic induction B_z/B_f as a function of the normalized coordinate x/w over the thickness of the superconductor slab for different values of the normalized applied magnetic field H_z/H_f during (a) the initial transient with increasing field, when $H_z/H_f = 0.5, 1, 1.5$, (b) a steady-state half-cycle with decreasing field, when $H_z/H_f = 1, 0.5, 0, -0.5, -1, -1.5$, and (c) a steady-state half-cycle with increasing field, when $H_z/H_f = -1, -0.5, 0, 0.5, 1, 1.5$, assuming $H_a/H_f = 1.5$.

to eqns. (2.71) and (2.75), that the hysteresis width of the magnetization loop admits a determination of the critical current density j_c from

$$wj_c = \langle M_z \rangle_{max} - \langle M_z \rangle_{min} \tag{2.76}$$

by employing magnetic measurements alone.

A recourse to the thermodynamic average magnetization also enables establishing the hysteretic AC loss, per unit area of the slab, U_{ac}, incurred during a steady-state cycle of the applied magnetic field,

$$U_{ac} = 2\mu_0 w \left| \oint_{\text{f.c.}} dH_z \langle M_z \rangle \right|, \tag{2.77}$$

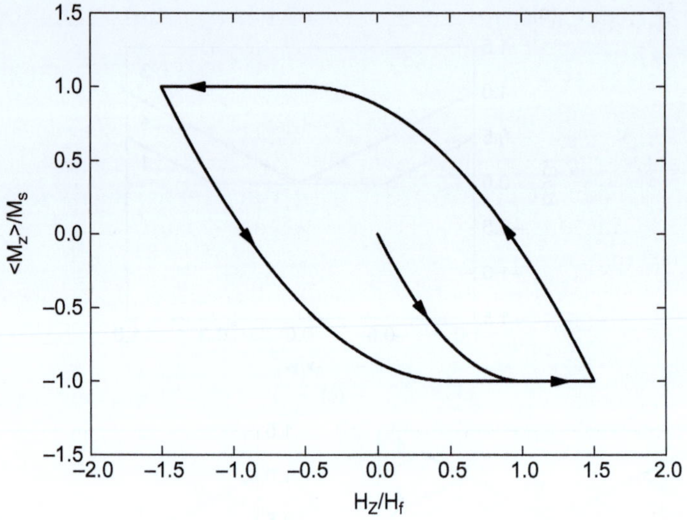

Fig. 2.6 Normalized thermodynamic average magnetization of the superconductor slab $\langle M_z \rangle / M_s$ as a function of the normalized applied magnetic field H_z/H_f during the initial transient and in steady state, assuming $H_a/H_f = 1.5$.

by integrating over a full cycle of the magnetization loop (Carr 2001). Hence, using the results, eqns. (2.69), (2.71), (2.73), and (2.75), in eqn. (2.77) yields the scaled universal form for $H_a \geq H_f$,

$$U_{ac}/H_f^2 = 4\mu_0 w \left(H_a/H_f - 2/3\right), \tag{2.78}$$

which reveals a linear dependence on H_a/H_f. Surely, if a steady-state cycle for $0 < H_a < H_f$, i.e. below saturation, were considered instead, the result

$$U_{ac}/H_f^2 = 4\mu_0 w \left(H_a/H_f\right)^3/3 \tag{2.79}$$

would be obtained. From $H_a = H_f$, the critical current density j_c appears to separate the linear and cubic regimes of the applied magnetic field.

We comment that the above analysis may be generalized in several respects. Thus, the field dependence of the critical current density j_c can be taken into account, if need be (Campbell and Evetts 1972); likewise, the simplifying premise of entry of magnetic flux at zero applied magnetic field can be relaxed and even a surface penetration field H_p allowed (Portis 1993). More sophisticated critical states, at last, with arbitrary orientations of the flow of the imposed transport current and the induced magnetic field or, respectively, the induced current and the applied magnetic field, could be discussed as in a number of works (Clem and Pérez-González 1984; Prigozhin 1996, 1997; Badía-Majós and López 2001; Brandt and Mikitik 2007; Badía-Majós, López, and Ruiz 2009; Badía-Majós and López 2014).

2.3 Interaction of superconductors with magnetic shields

One of the great assets of magnetically susceptible materials is their ability to guide magnetic fields. Thus, when used as elements of superconductor/paramagnet heterostructures, or even structures made up of superconductors alone, such materials can largely control the magnetic field around the superconducting constituents and also the magnetic induction inside, assuming certain criteria have been set. An obvious aim in exploiting this effect is the improvement of superconductor performance and properties, the current-carrying capability and the hysteretic AC loss naturally playing most significant parts. Successful realizations depend on a judicious choice of the permeability, the geometrical shape, and the proper positioning of the magnetic shields.

2.3.1 Examples of technological applications

A pragmatic statement says that applications of superconductors can be considered as falling into two realms identified by either physical size or the level of energy involved. The first category concerns bulk superconducting materials to form conductors with large cross sections. Such elements usually support a high current and often must operate in strong ambient magnetic fields. Speaking of high-power and high-field applications thus means employing an appropriate terminology. The second category is that of low-power electronic applications. This broad area of applications typically involves superconductors plated as thin films on dielectric substrates to form strip lines, wave guides, or resonant structures, apart from Josephson junctions and variants thereof. Important devices in their own right, Josephson junctions represent constitutive elements of superconducting quantum interference devices (SQUIDs) and also serve a variety of other applications including both analogue and digital circuitry. While the actual current flowing in an electronic conductor may be fairly small, the current density itself could be quite high (Doss 1989). The diversity of potential applications from within the above categories is immense and cannot be given an exhaustive description here, as the related literature impressively displays (Sheahen 1994; Weinstock 2000; Seidel 2015). We therefore restrict the following brief account to four select fields, emphasizing the role of magnetic shields in improving superconductor properties and performance too.

2.3.1.1 Electromagnetic power transmission

Resorting to high-temperature superconductors in components transmitting electromagnetic power for public utility applications seems advantageous over the use of conventional materials from an efficiency point of view, if suitable precautions can be met. These basically concern the problem of designing units in a form that enables providing cryogenic flow without significant thermal loss. The preference for superconductors has two reasons at least: their high current-carrying capability allows cables to transmit electromagnetic power of large strengths, permits transformers to benefit from strong magnetic fields, and makes fault-current limiters enjoy usability for long terms; their low dissipation of electromagnetic energy during reversals of magnetization tolerates curtailed requirements for the removal of Joule heat, when operating in an AC mode. Appropriate conditioning of magnetic fields can enhance the total transport current passing through a superconductor element and also reduce the hysteretic AC loss. Fundamental investigations on this topic and related subjects make up an extended share of the present text.

2.3.1.2 Magnetic energy storage

A device for storing energy in the magnetic field of a superconducting coil presents an attractive solution to the problem of load levelling in electrical grids. Being linked with the power grid via a rectifier/converter scheme, this device exploits the energy due to a persistent current flowing through the coil, which it stores during periods where demand is low, supplying it back to the grid when demand augments. Shielding the coil using a soft-magnetic core with a variable airgap can double the stored energy while keeping the critical current fixed at its original strength (Friedman et al. 1999). If the coil is wound with a high-temperature superconducting tape, guiding the magnetic field using soft-magnetic disks mounted at both ends of the coil can raise the performance of the storage device further still (Kondoh and Umeda 2002).

2.3.1.3 Macroscopic quantum interference

The phenomenon of quantization of the total magnetic flux encompassed by a supercurrent circulating around a pit, as discussed before, earlier becomes manifest if two Josephson junctions connected in parallel form a superconducting ring subject to an applied magnetic field and driven by an external current source. Such a system defines a basic SQUID capable of resolving minute changes of magnetic flux. Its theoretical analysis essentially follows that of a two-slit optical interference experiment. If placed in the airgap between two soft-magnetic rods concentrating the distribution of the magnetic field, a high-temperature superconductor SQUID can demonstrate an increase in sensitivity by up to two orders of magnitude, addressing magnetometer or gradiometer use (Bondarenko, Shablo, Pavlov, and Perepelkin 2002). A scanning microscope fitted with a magnetic sensor high-temperature superconductor SQUID, on the other hand, allows measurements of the local magnetic field near the surface of a sample at extreme sensitivity. If a magnetic flux guide made of a sharpened soft-magnetic wire is installed between the sample and the SQUID, the spatial resolution of such a microscope can improve by several times (Gudoshnikov et al. 2001).

2.3.1.4 Electromagnetic compatibility

The widespread use and large-scale integration of electronic devices for signal processing, communication, metrology, and other aims entails that circuits frequently operate in each other's close proximity. Electromagnetic interference thus presents a major problem to the designing engineer who must ensure products are made to work in the 'real world', with other equipment nearby, unaffected by external sources, and not polluting the environment by being a source of electromagnetic noise (Ott 2009). Suitably guiding the magnetic flux by either paramagnetic or superconducting shields can help to perform this task, a procedure routinely employed for protecting field-sensitive devices and personnel from unwanted interference with strong magnetic fields.

2.3.2 Illustration of the principle

Let us first look at the textbook problem of a magnetic sphere of radius R and permeability μ, surrounded by a vacuum of permeability μ_0 and placed in a uniform magnetic field $\mathbf{H}_\mathrm{a}$ parallel to the z-axis of a Cartesian coordinate system x, y, z with its centre at the origin of the sphere. We focus on the distributions of the magnetic field $\mathbf{H}$ therein and, respectively, the magnetic induction $\mathbf{B}$ in the entire space, proceeding from the homogeneous Maxwell equations

$$\nabla \times \mathbf{H} = 0 \quad \text{and} \quad \nabla \cdot \mathbf{B} = 0, \tag{2.80}$$

where $\mathbf{B} = \mu \mathbf{H}$ inside the sphere and $\mathbf{B} = \mu_0 \mathbf{H}$ around it. These equations suggest the introduction of a scalar magnetic potential ψ to yield the representation of the magnetic field,

$$\mathbf{H} = -\nabla \psi, \tag{2.81}$$

identically satisfying the first equation (2.80), whereas the second equation (2.80) entails the Laplace equation

$$\Delta \psi = 0. \tag{2.82}$$

Continuity of both the normal component of the magnetic induction B_r and the tangential component of the magnetic field H_ϑ across the surface of the sphere hence demands that the relations

$$\mu \left(\frac{\partial \psi}{\partial r} \right) \bigg|_{r=R-} = \mu_0 \left(\frac{\partial \psi}{\partial r} \right) \bigg|_{r=R+} \quad \text{and} \quad \frac{\partial \psi}{\partial \vartheta} \bigg|_{r=R-} = \frac{\partial \psi}{\partial \vartheta} \bigg|_{r=R+} \tag{2.83}$$

in spherical coordinates r, ϑ, ϕ hold; the asymptotic condition that $\mathbf{H}$ must tend to $\mathbf{H}_a$ far away from the sphere becomes

$$\psi(r, \vartheta) \to -H_a \, r \cos \vartheta \quad \text{as } r \to \infty. \tag{2.84}$$

The appropriate solution of eqn. (2.82) can be represented by the potential of a homogeneous magnetic field inside and the superposition with a magnetic dipole potential outside respectively,

$$\psi(r, \vartheta) = \begin{cases} -\left(\dfrac{3\mu_0}{\mu + 2\mu_0} \right) H_\mathrm{a} \, r \cos \vartheta & \text{for } 0 \le r \le R \\ -H_\mathrm{a} \, r \cos \vartheta + \left(\dfrac{\mu - \mu_0}{\mu + 2\mu_0} \right) H_\mathrm{a} \dfrac{R^3}{r^2} \cos \vartheta & \text{for } r > R \end{cases} \tag{2.85}$$

(Jackson 1998; Lehner 2010). Equation (2.81) thus furnishes the Cartesian components of the magnetic induction inside the sphere, $0 \le r \le R$,

$$B_x(r, \vartheta, \phi) = 0 \tag{2.86a}$$

$$B_y(r, \vartheta, \phi) = 0 \tag{2.86b}$$

$$B_z(r, \vartheta, \phi) = \left(\frac{3\mu}{\mu + 2\mu_0} \right) B_\mathrm{a}, \tag{2.86c}$$

and those of the magnetic field outside the sphere, $r > R$,

$$H_x(r, \vartheta, \phi) = 3 \left(\frac{\mu - \mu_0}{\mu + 2\mu_0} \right) H_\mathrm{a} \frac{R^3}{r^3} \sin\vartheta \cos\vartheta \cos\phi \tag{2.87a}$$

$$H_y(r, \vartheta, \phi) = 3 \left(\frac{\mu - \mu_0}{\mu + 2\mu_0} \right) H_\mathrm{a} \frac{R^3}{r^3} \sin\vartheta \cos\vartheta \sin\phi \tag{2.87b}$$

$$H_z(r, \vartheta, \phi) = H_a + \left(\frac{\mu - \mu_0}{\mu + 2\mu_0} \right) H_\mathrm{a} \frac{R^3}{r^3} \left(3\cos^2\vartheta - 1 \right), \tag{2.87c}$$

calling upon the magnetic induction of the applied magnetic field $B_\mathrm{a} = \mu_0 H_\mathrm{a}$.

It is instructive to portray the distribution of the magnetic induction **B** or the magnetic field **H** in the limiting cases of extreme magnetic behaviour of the sphere. If $\mu \to \infty$, representing a soft magnet in an idealized paramagnetic state, eqn. (2.86) reveals that $\mathbf{B} = 3\mathbf{B}_\mathrm{a}$ with $\mathbf{H} \equiv 0$ inside the sphere, displaying a homogeneous inductive field of maximum strength; eqn. (2.87), on the other hand, shows that the magnetic field adjacent to the polar regions of the sphere is enhanced, its lines meeting the sphere at normal incidence outside, the separatrices associated with the equatorial points as well as the lines beyond exempt, and therefore suffer refraction at the soft-magnet/vacuum interface (Fig. 2.7(a)). If $\mu = 0$, modelling a perfect diamagnet with complete expulsion of magnetic flux, eqn. (2.86) shows that $\mathbf{B} \equiv 0$ with $\mathbf{H} = 3\mathbf{H}_\mathrm{a}/2$ inside the sphere, reminiscent of a superconductor in the fully developed Meissner state, the flux-filled surface sheath exempt; eqn. (2.87), on the other hand, demonstrates that the magnetic field adjacent to the polar regions of the sphere is depressed, its lines passing the sphere outside, the closest ones tracing the sphere's circumference and opening up at the poles (Fig. 2.7(b)).The cases illustrate how the shape of magnetically susceptible bulk materials in conjunction with their permeability affects the distribution and strength of the magnetic induction and the magnetic field inside and around the sample, respectively, a fact which can be exploited for duly steering the magnetic flux.

Improvements in superconductor performance and properties due to the presence of magnetic shields can obviously be achieved in two respects: (i) by mitigating, or even averting, detrimental interactions which may occur in systems composed of several superconductor constituents, so that the unimpaired properties of the bare constituents come to the fore; and (ii) by improving properties of the superconductor constituents themselves through application of magnetic shields tailored to actual, given needs.

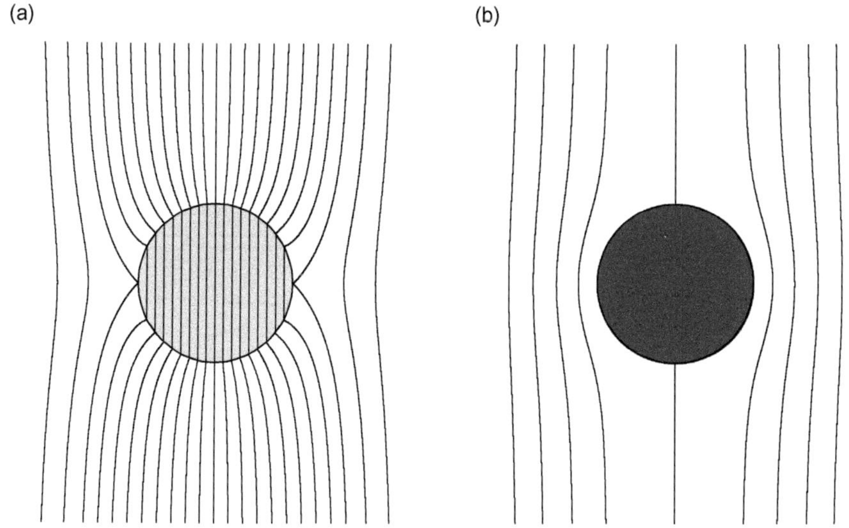

Fig. 2.7 Lines of the magnetic induction **B** inside, and lines of the magnetic field **H** around, a magnetic sphere placed in a uniform magnetic field $\mathbf{H}_a$ of fixed strength, when the permeability of the sphere (a) $\mu \to \infty$, as for a soft magnet in an idealized paramagnetic state, and (b) $\mu = 0$, as for a magnet in the perfectly diamagnetic state (meridional view).

2.3.3 Decoupling of superconductor filaments

It is the requirement of thermal stability and mechanical strength to favour the designs of superconducting cables or tapes with a multitude of thin current-carrying superconductor filaments embedded in a normal matrix supporting the composite. Straightforward partition of the superconductor constituent into small cross sections of many individual filaments though would entail feedback currents pervading the matrix and, hence, magnetically coupling the filaments. Such a phenomenon is by no means desirable, since it implies that the composite as a whole contrives the electric charge transport rather than the superconductor filaments alone, resulting in hysteretic AC losses being significantly enhanced along with thermal stability being greatly reduced.

A way out of this plight is to decouple the superconductor filaments by coating them with highly permeable soft-magnetic sheaths, a *passive*, but nonetheless efficient, form of utilizing magnetic shields proposed by Majoros, Glowacki, and Campbell (2000a,b) which works well for filaments with cross sections of elliptical or rectangular shapes. To explain the precept on the basis of a numerical test, let us consider a simplified model system made up of an infinitely extended and magnetically coated superconductor filament carrying a transport current in the normal state, encompassed by six like filaments in a current-free state, as depicted in Fig. 2.8. The effect of magnetic shielding on the distribution of the induced magnetic field **H** around this central filament becomes manifest when varying the

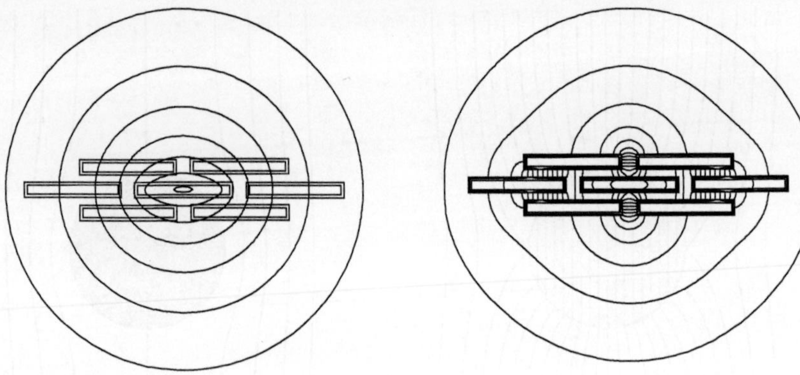

Fig. 2.8 Lines of the magnetic field **H** around a central current-carrying superconductor filament of rectangular cross section in the normal state, fitted with a magnetic sheath and encompassed by six like filaments in a current-free state, when the relative permeability of the sheath (a) $\mu/\mu_0 = 1$ and (b) $\mu/\mu_0 = 1000$.

permeability of the protecting magnetic sheaths. Thus, choosing the reference value $\mu/\mu_0 = 1$ first yields a distribution of the magnetic field that would prevail in the absence of magnetic shields: the magnetic field occupies all space, coupling the superconductor filaments because of the penetration of magnetic flux (Fig. 2.8(a)). Selecting the drastic value $\mu/\mu_0 = 1000$, on the other hand, leads to a distribution of the magnetic field that impressively demonstrates the action of the magnetic shields: the lines of the magnetic field tend to essentially pass through the magnetic sheaths, leaving the interior of the current-free filaments virtually unaffected by the influence of the current-carrying filament, i.e. devoid of magnetic flux (Fig. 2.8(b)). Interestingly enough, such a magnetic decoupling of superconductor filaments can also be helpful for reducing hysteretic AC losses of multifilamentary cables or tapes subject to applied magnetic fields.

2.3.4 Conditioning of magnetic fields

Like the need to minimize hysteretic AC loss, the quest to maximize current-carrying capability poses tremendous challenges regarding electromagnetic power transmission system designs. It thus seems feasible to start off with superconductors that primarily exhibit ample transport current support. The latter property is typical of high-temperature superconductors which have been carefully prepared as thin films or strips and therefore achieve the best values of the barrier-controlled total current I_b and the critical current I_c, rendering them particularly attractive in technological applications. The question then crops up: can the loss-free transport current carried by the superconducting constituents still be enhanced over and above I_b or I_c, the characteristics of isolated superconductors alone?

This may be the case indeed, Genenko, Usoskin, and Freyhardt (1999) assert, if superconductor strips are merged with magnetic shields to form superconductor/paramagnet (or even superconductor/diamagnet) heterostructures that offer suitably designed magnetic environments; an *active* and powerful way of tailoring currents through conditioning of magnetic fields. The idea relies on the dependence between the magnetic induction $\mathbf{B}$ and the density $\mathbf{j}$ of current flow in the superconductor domain D according to Ampère's law

$$\nabla \times \mathbf{B} = \mu_0 \mathbf{j}, \tag{2.88}$$

assuming static states or—approximately—quasistatic processes, as well as on the variation of the magnetic field $\mathbf{H}$ in the residual domains and the magnetic induction $\mathbf{B}$ in the entire space following the homogeneous Maxwell equations

$$\nabla \times \mathbf{H} = 0 \quad \text{and} \quad \nabla \cdot \mathbf{B} = 0. \tag{2.89}$$

If no, or only few, magnetic vortices are contained in the superconducting elements, the magnetic induction $\mathbf{B}$ within D may be determined by recourse to the generalized London theory, eqn. (2.20), for zero, or finite, vorticity $\mathbf{V}$; if, however, significant penetration of magnetic flux takes place, Bean's model of the critical state, eqn. (2.50), surely presents a more adequate approach. The current density $\mathbf{j}$ obtains from eqn. (2.88) in either case. Hence, the issue is, firstly, to identify distributions of the magnetic field $\mathbf{H}$ and the magnetic induction $\mathbf{B}$ as solutions of eqns. (2.88) and (2.89) which satisfy proper boundary conditions at the interfaces of the various constituents (save for possible asymptotic requirements far away) and concomitantly entail favourable profiles of the current density $\mathbf{j}$; secondly, to realize these distributions by a judicious choice of the permeability, geometrical shape, and proper positioning of the magnetic shields.

Such a task can be carried out most usefully for thin superconductor strips, the subject of Chapters 3 to 6, because of the prevailing local connection of essentially two-dimensional current flow with the tangential components of the magnetic field on the two surfaces of the strips. Yet, the method stated here reaches far beyond superconducting constituents of the named sort; it also applies, e.g., to superconductor filaments of cylindrical shapes dealt with in Chapter 7, and even superconductor bulks. Conditioning of magnetic fields works if the superconducting elements are in the Meissner state, with the current often limited to a thin surface sheath of large, otherwise flux-free specimens, or in the partially flux-filled critical state, with the current generally extending over the whole cross section of these constituents. An important merit of the concept of guiding the magnetic flux by magnetic shields, of course, is its validity regardless of the particular superconductor state and of the model which it describes, although the boundary conditions at the superconductor surfaces will naturally depend on that state.

2.3.4.1 Mathematical implementation

Solutions to the problem explained earlier are conveniently derived by employing the magnetic vector potential $\mathbf{A}$ introduced in eqn. (2.3) and adopting the Coulomb gauge $\nabla \cdot \mathbf{A} = 0$. Thus, for superconductor constituents, eqn. (2.88) yields a vector Poisson equation as the governing equation in the domain D,

$$\Delta \mathbf{A} = -\mu_0 \mathbf{j}, \tag{2.90}$$

which, by dint of eqn. (2.89), simplifies to the vector Laplace equation

$$\Delta \mathbf{A} = 0 \tag{2.91}$$

in the residual domains (Moon and Spencer 1988). An additional relationship between $\mathbf{j}$ and $\mathbf{A}$ delineating the superconductor constituents' state is often called for; this may be provided by the generalized London theory or Bean's model of the critical state. Boundary (and possibly asymptotic) conditions must be posed on $\mathbf{A}$.

There are a number of ready mathematical tools lending themselves to tackling eqns. (2.90) and (2.91): (i) the Green's function technique provides a representation of the field arising from an extended current source by calculating the effects of all elementary parts of source and integrating to give the desired result; (ii) the method of images often enables a construction of the Green's functions in the case of bounded domains; (iii) the singular integral equation technique provides a conducive trail to extracting the current distributions looked for in the superconductor domain; (iv) the conformal mapping technique can make problems of planar geometry amenable to analytical treatment even if intricate constituent boundaries persist; and (v) numerical methods such as the finite-element technique broaden further the spectrum of superconductor/paramagnet heterostructures accessible to theoretical studies by relaxing constraints which analytical investigations may demand. All of these tools are thoroughly explained in various classical and modern texts (Morse and Feshbach 1953; Tikhonov and Samarskii 1967; Nicolaides and Walkington 1996; Cook, Malkus, Plesha, and Witt 2002; Richards 2002; Cap 2003; Press, Teukolsky, Vetterling, and Flannery 2007; Riley, Hobson, and Bence 2010; Arfken, Weber, and Harris 2012; Pence and Wichman 2020).

We comment that, apart from this direct way of proceeding with the magnetostatic problem at hand, alternative routes to obtaining solutions in support of the search for appropriate shielding configurations still exist. One of these involves drawing up a corresponding electrostatic problem and resorting to magnetostatic-electrostatic analogues, the method established in Chapter 8. Accordingly, lines of the magnetic field $\mathbf{H}$ or the magnetic induction $\mathbf{B}$ of the original problem agree with equipotentials of the electric field $\mathbf{E}$ or, the electric displacement $\mathbf{D}$ respectively, of its electrostatic equivalent. Another route exploits the isomorphy of the pairs of fields $\mathbf{H}, \mathbf{E}$ and $\mathbf{B}, \mathbf{D}$ outside the domain occupied by the superconductor or its electrostatic equivalent, respectively, i.e. in the space where classical potential theory based on the Laplace equation applies. Last but not least, if the magnetic field $\mathbf{H}$ or the magnetic induction $\mathbf{B}$ in the space between a heterostructure's constituents only is addressed, taking advantage of links with the velocity field $\mathbf{v}$ of steady-state rotational flow of an incompressible fluid may sometimes help too. These observations elucidate that correct transcriptions of solutions known in areas like electrostatics, magnetostatics, or hydrodynamics can offer additional paths to finding solutions of the magnetostatic problem set out at first.

2.3.4.2 Practical realizations

Let us illustrate the concept of guiding magnetic fields with the basic example of an infinitely extended, thin type-II superconductor strip carrying a transport current in the flux-free Meissner state. If a magnetic environment is absent, such as the isolated superconductor strip shown in Fig. 2.9, the lines of the induced magnetic field **H** take on elliptical shapes whose focal points are fixed to the edges of the strip. The concentration of the lines of the magnetic field, and hence the increase in its strength, around the latter evokes peaks of the current density **j** at the edges of the strip. These, in turn, cause magnetic flux to penetrate the strip accompanied by a destruction of the Meissner state, if the total current I flowing through the strip exceeds the maximum value I_{b} that can be attained with some edge barrier mechanism acting to protect the Meissner state. Ideally therefore, conditioning of the magnetic field **H** by means of suitable magnetic environments should yield the magnetic field to weaken near the edges of the strip, and thus suppress the aforementioned peaks, but build up the current density **j** in the region beyond, while preserving the Meissner state. The possibility of such a control for open magnetic environments appears because the lines of the magnetic field can, on the one hand, be forced to intersect the surface of highly permeable bulk paramagnetic shields at angles that might be very steep; they can, on the other hand, be prompted to avoid the surface of completely impermeable bulk diamagnetic shields at all, facts which together with the geometrical shape of the shields make it possible to

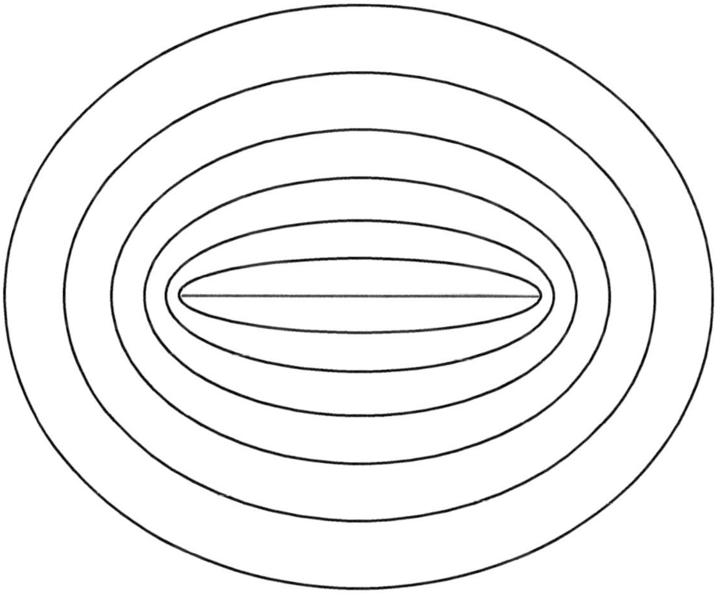

Fig. 2.9 Lines of the magnetic field **H** around an isolated current-carrying superconductor strip in the Meissner state (cross-sectional view).

locally guide the magnetic field (cf. Fig. 2.7). One- or two-sided magnetic shielding configurations may be addressed, and even closed magnetic environments envisaged too. Some cases in point will be given for paramagnetic shields with finite, or infinite, permeability μ and also for diamagnetic shields with zero permeability μ.

A straightforward way of detracting the lines of the magnetic field $\mathbf{H}$ from one edge of the superconductor strip using an open magnetic environment consists of placing the strip adjacent to a flat semi-infinite paramagnetic shield, as Fig. 2.10(a) depicts. The effect of an asymmetric weakening of the magnetic field becomes still more pronounced if the paramagnet assumes the form of a wedge-shaped cavity, as Fig. 2.10(b) reveals. Certainly, both configurations serve to depress the peaks of the current density $\mathbf{j}$ at the edge of the strip facing the magnetic shield; however, they also bring about a lowering of the current distribution in the contiguous region of the strip, which renders the maximum total current $I_{\max}$ achievable by them even smaller than the barrier-controlled total current I_{b} of the isolated strip.

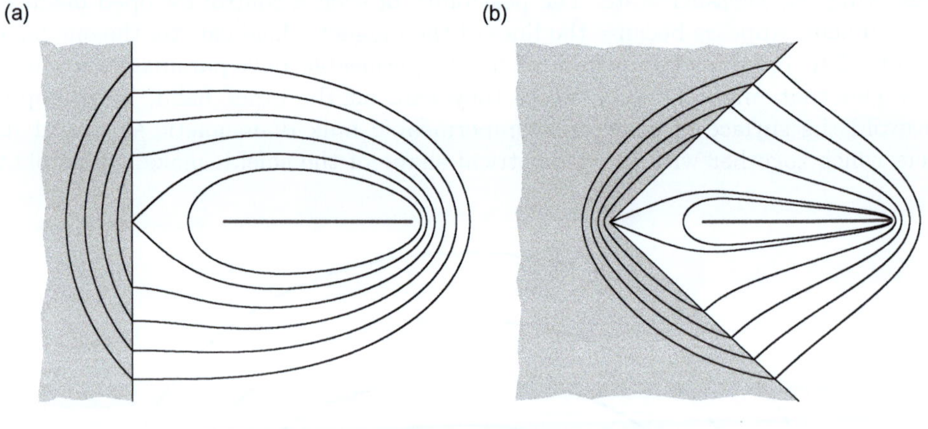

Fig. 2.10 Lines of the magnetic field $\mathbf{H}$ or, respectively, the magnetic induction $\mathbf{B}$ around a current-carrying superconductor strip in the Meissner state placed (a) adjacent to a flat semi-infinite paramagnet and (b) inside a wedge-shaped paramagnetic cavity, when the permeability of the magnets $\mu \to \infty$ (cross-sectional view).

These findings suggest trying out flat semi-infinite paramagnetic environments that provide shielding around both edges of the superconductor strip instead. Two principal orientations of such environments relative to the strip are at hand, viz. longitudinal and transverse, apt at symmetrically guiding the magnetic field, as Fig. 2.11 imparts. If the surfaces of the paramagnets run parallel to the strip, the lines of the magnetic field $\mathbf{H}$ outwardly depart from the strip over its entire width, producing a consistent *reduction* in the current density $\mathbf{j}$, and hence in the maximum loss-free current $I_{\max}$ which this configuration is able to sustain compared with the barrier-controlled total current I_{b} of the isolated strip (Fig. 2.11(a)).

If the surfaces of the paramagnets run perpendicular to the strip, the lines of the magnetic field $\mathbf{H}$ flatly escort the strip over its entire width, resulting in a consistent *increase* in the current density $\mathbf{j}$, and hence in the maximum loss-free current $I_{\max}$

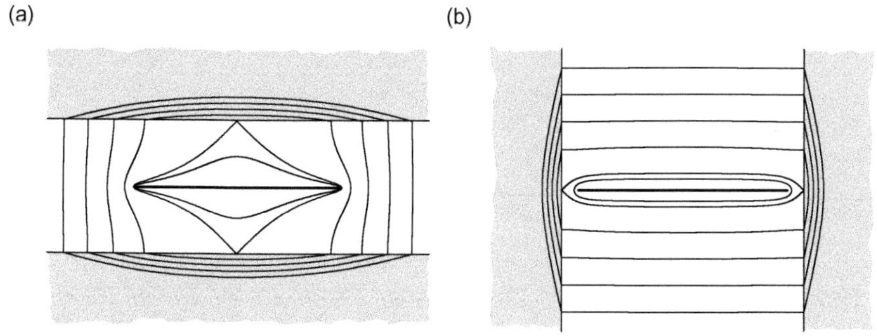

Fig. 2.11 Lines of the magnetic field **H** or the magnetic induction **B** around a current-carrying superconductor strip in the Meissner state between two flat semi-infinite paramagnets oriented (a) parallel to the strip and (b) perpendicular to the strip, when the relative permeability of the magnets $\mu/\mu_0 = 1000$ (cross-sectional view).

which this configuration is able to sustain compared with the barrier-controlled total current I_b of the isolated strip (Fig. 2.11(b)). The shielding effects can be substantial in either case, depending on the positioning of the magnetic shields.

A positive effect can also be derived if the superconductor strip is placed inside the closed environment of a cylindrical paramagnetic cavity, the configuration portrayed in Fig. 2.12, even though the contingency of guiding the magnetic field here is cut back to some extent. Limited by the constraint which the inner boundary of the magnetic shield sets, the lines of the magnetic field **H** about the strip adopt elliptical shapes tending to a circular shape, while their focal points move inward towards the centre of the strip. Accordingly, the current density **j** over the strip augments, yielding a rise in the maximum loss-free current $I_{\max}$ which may greatly exceed I_b, the value for the unshielded strip.

Results can be even more spectacular if the idealized paramagnets are replaced with perfect diamagnets while the geometry of the magnetic environment is maintained. We elucidate this point by taking the example of a superconductor strip placed between two *longitudinally* oriented flat semi-infinite *diamagnetic* shields, as in Fig. 2.13 by varying Fig. 2.11(a). The lines of the magnetic field **H** now appear compressed perpendicular to the plane of the strip and confined to the space outside the shields, affording an increase in the current density **j** over the greater part of the width of the strip, but relaxed otherwise, curtailing the current density peaks at the edges of the strip. This feature is reminiscent of the situation of a superconductor strip encompassed by two *transversely* oriented flat semi-infinite *paramagnetic* shields, Fig. 2.11(b). Like there, the maximum loss-free current $I_{\max}$ can be substantially enhanced compared with the barrier-controlled total current I_b of the isolated strip.

Detailed accounts of the theoretical analysis underlying the field and current distributions of the magnetic shielding configurations presented for the Meissner state, and supplemented with those for the critical state, can be found in Chapters 3 to 5.

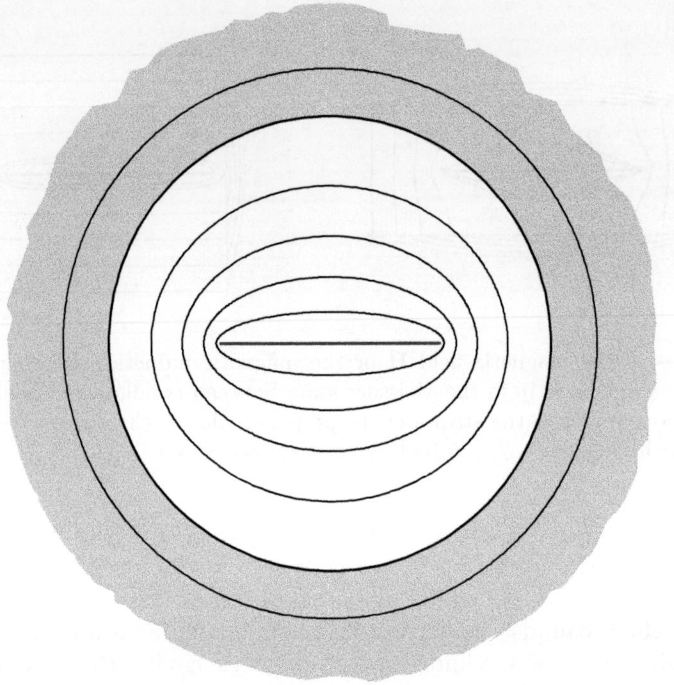

Fig. 2.12 Lines of the magnetic field **H** around a current-carrying superconductor strip in the Meissner state placed inside a cylindrical paramagnetic cavity with permeability $\mu \to \infty$ (cross-sectional view).

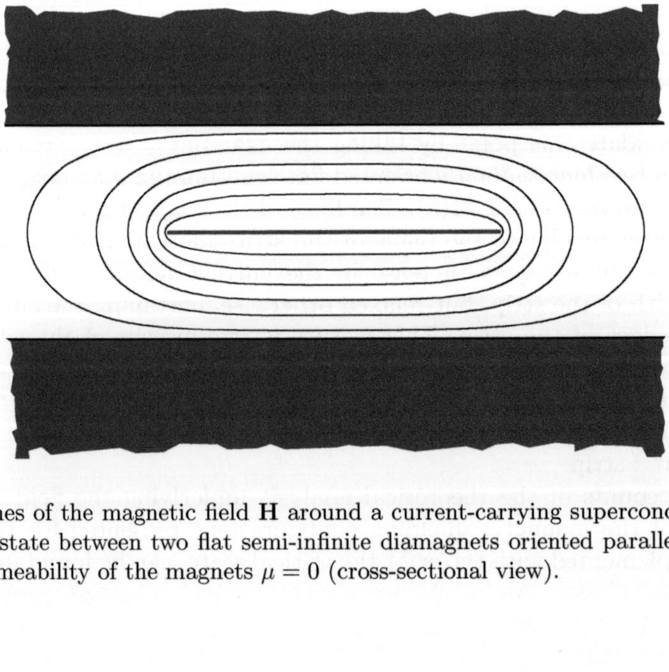

Fig. 2.13 Lines of the magnetic field **H** around a current-carrying superconductor strip in the Meissner state between two flat semi-infinite diamagnets oriented parallel to the strip, when the permeability of the magnets $\mu = 0$ (cross-sectional view).

3

Meissner state of a superconductor strip in ideal paramagnetic environments

Given the phenomenological concepts deployed thus far, the scene is set for turning to the electromagnetics of superconductor/paramagnet heterostructures, the theme of this book. We start by resuming the magnetic shielding configurations suggested at the end of Chapter 2: an infinitely extended, thin type-II superconductor strip with a transport current imposed in the flux-free Meissner state, surrounded by a number of different paramagnetic environments. Protection of this state against penetration by magnetic flux is achieved by invoking a distinct edge barrier in the profile of the sheet current beyond its critical limit due to pinning of magnetic flux, an assumption known to apply to single crystals of $Bi_2Sr_2Ca_2Cu_3O_{10}$ (Indenbom et al. 1994) as well as $YBa_2Cu_3O_7$ (Gardner et al. 1998) and, not least, to very thin MoGe strips whose edges show clear cuts (Plourde et al. 2000).

Our main interest here is to establish the principal effects of magnetic shielding on the distributions of the sheet current in, and hence of the magnetic field around, the superconductor strip. We therefore emphasize the limit of infinite permeability of the paramagnetic constituents, $\mu \to \infty$, illustrated by the example of Fig. 2.7(a) for an ideal spherical paramagnet with complete suppression of the magnetic field inside, apart from their infinite spatial extent. Magnetic environments considered involve open as well as closed geometries such as a single flat paramagnet, a wedge-shaped paramagnetic cavity, two flat paramagnets with principal orientations either longitudinal or transverse, and a cylindrical and a cusp-shaped paramagnetic cavity. The advantage thereby comes to light that readily amenable similarities with electrostatic or hydrodynamic problems are obtainable and analytical results exist. Fascinatingly enough, the integral equations governing the sheet current distributions in the presence of any of the open magnetic environments can be reduced to the generic equation for the unshielded strip. Another merit is the improvement in the geometrical shape of the paramagnetic constituents in regards to the distribution of the current that pervades the strip, through the implementation of the conformal mapping technique.

Electromagnetics of Superconductor/Paramagnet Heterostructures. Yuri Genenko and Hermann Rauh, Oxford University Press.
 DOI: 10.1093/9780191782855.003.0003

A split superconductor strip may serve for modelling a thin film with a grain boundary. This sort of defect can seriously curtail the utility of the film, i.e. reduce its current-carrying capability, by the action as a weaklink. Evidence stems from the fact that weaklinks in high-quality superconducting films represent locations which facilitate the penetration of magnetic vortices, thereby promoting the onset of electromagnetic loss; they often reveal properties Josephson junctions conventionally display (Tinkham and Lobb 1989; Rauh, Stoneham, and Choy 1992; Díaz, Méchin, Berghuis, and Evetts 1998a,b; Larbalestier, Gurevich, Feldman, and Polyanskii 2001; Rauh and Genenko 2004; Rauh, Genenko, and Rieck 2004). We therefore conclude by probing the influence of magnetic shielding on the distribution of the sheet current in a split superconductor strip, selecting the magnetic environment that performs best for the prototype configurations examined earlier.

3.1 Isolated superconductor strip: a reference

Let us consider an infinitely extended, thin type-II superconductor strip of width $2w$ and thickness d limited by the domain $-w \leq x \leq w$, $-d/2 \leq y \leq d/2$, the direction of translational invariance of this strip being parallel to the z-axis of a Cartesian coordinate system x, y, z. A rectangular cross section distinguishes the domain D occupied by the strip from the residual (vacuum) domain. The strip is supposed to reveal the flux-free Meissner state, carrying a transport current of magnitude I in the absence of an external magnetic field. This means the current density $\mathbf{j}$ only has a nonzero z-component depending on x and y. We understand that the inequalities $d, \lambda_\perp \ll w$ apply, where $\lambda_\perp = \lambda_L^2/d$ denotes the transverse penetration depth (Pearl 1964). By virtue of the restriction concerning the dimensions of the strip, spatial variations of the current and the induced magnetic field on a length scale less than d will be ignored and, for mathematical convenience, the strip will be regarded as infinitesimally thin, enabling the physical state of the strip to be characterized by the sheet current $J(x)$ (Brandt 1992). We therefore rely upon formally letting

$$j_z(x, y) \to J(x)\delta(y) \tag{3.1}$$

with the normalization

$$\int_{-w}^{w} dx' J(x') = I. \tag{3.2}$$

Although the concept that the superconductor strip should represent a planar singularity can only approximately hold, up to an error on the order h/w with h denoting the greater of the quantities d and $\lambda_\perp$ (Brandt and Indenbom 1993; Brandt and Mikitik 2000), this approach is important and justified by the fact that it often yields useful results for problems in connection with magnetic shields that otherwise might defy a theoretical analysis. It does, however, entail divergences of the sheet current at the edges of the strip and consequently necessitates introducing suitable cut-offs, for idealizing magnetic vortices with an infinitesimally thin core as proposed in section 2.1.3.

3.1.1 Field distribution

Under the premises stated above, eqns. (2.87) and (2.88) simplify to the Poisson equation for the single nontrivial component of the magnetic vector potential A_z,

$$\Delta A_z = \begin{cases} -\mu_0 J(x)\delta(y) & \text{in } -w \leq x \leq w \\ 0 & \text{otherwise.} \end{cases} \tag{3.3}$$

A similar equation for the scalar electric potential φ would obviously hold, with the inverse electric permittivity of a vacuum ε_0^{-1} and the surface charge density σ in place of μ_0 and J, if an electrically charged thin metal strip surrounded by an infinite vacuum domain would be considered instead. The solution of eqn. (3.3) can be obtained with the help of the Green's function technique by identifying the contribution $dI = dx'J(x')$ of a distributed current element at $(x = x', y = 0)$ and integrating over x' to yield the desired result in the vacuum space inclusive of the surface of the strip ∂D,

$$A_z(x, y) = \frac{\mu_0}{2\pi} \int_{-w}^{w} dx' J(x') \ln \frac{a}{\sqrt{(x - x')^2 + y^2}}, \tag{3.4}$$

introducing an arbitrary length a. Application of eqn. (2.3) thus gives the Cartesian components of the magnetic field $\mathbf{H}$,

$$H_x(x, y) = -\frac{1}{2\pi} \int_{-w}^{w} dx' J(x') \frac{y}{(x - x')^2 + y^2} \tag{3.5a}$$

$$H_y(x, y) = \frac{1}{2\pi} \int_{-w}^{w} dx' J(x') \frac{x - x'}{(x - x')^2 + y^2}, \tag{3.5b}$$

which, for the sheet current, eqn. (3.8), are reflected in Fig. 2.9.

3.1.2 Current distribution

The Meissner state of the superconductor strip implies absence of magnetic flux inside, and thus vanishing of the normal component of the magnetic field H_y across the width of the strip,

$$H_y(x) = 0 \quad \text{for } -w < x < w, y = 0. \tag{3.6}$$

Recalling eqn. (3.5b) hence furnishes the condition for this state

$$\int_{-w}^{w} dx' J(x') \frac{1}{x - x'} = 0 \quad \text{in } -w < x < w, \tag{3.7}$$

i.e. a singular integral equation of the Cauchy type (Mikhlin 1964; Muskhelishvili 2008). Its solution, normalized according to the requirement, eqn. (3.2), delineates the sheet current as

$$J(x) = \frac{I}{\pi} \frac{1}{\sqrt{(w-x)(x+w)}} \quad \text{for} \quad -w < x < w \tag{3.8}$$

(Rhoderick and Wilson 1962). Physically, since the magnetic field is bound to bypass the flux-free superconductor strip, the sheet current originates from the discontinuity of the tangential component of the magnetic field H_x across the thickness of the strip. The distribution of this current, therefore, is utterly contingent upon geometry and does not involve any characteristic of the superconductor material itself.

The form of the sheet current, eqn. (3.8), diverges at the edges of the superconductor strip, $x = \pm w$, suggesting destruction of the Meissner state by drawing magnetic flux inside, whereas in reality $J(x)$ saturates on the scale of h from the edges of the strip, adopting the maximum strength

$$J_{\max} \simeq \frac{I}{\pi\sqrt{2wh}}; \quad h \ll w \tag{3.9}$$

at $|x \pm w| = h$. This observation limits the validity of eqn. (3.8) to the range $-w + h \leq x \leq w - h$ and simultaneously recommends to also posit the value $J_{\max}$ from eqn. (3.9) in $|x \pm w| \leq h$. Hence, if we invoke protection of the strip against entry of magnetic flux owing to edge barriers typified by the sheet current value J_b, the virgin state of the strip is conserved as long as $J_{\max} < J_\mathrm{b}$ at both edges of the strip. The maximum loss-free total transport current possible, $I_{\max} = I_\mathrm{b}$, thus results for $J_{\max} = J_\mathrm{b}$. Figure 3.1 illustrates the variation of the normalized sheet current J/J_b over the width of the strip. A broad minimum of the symmetric

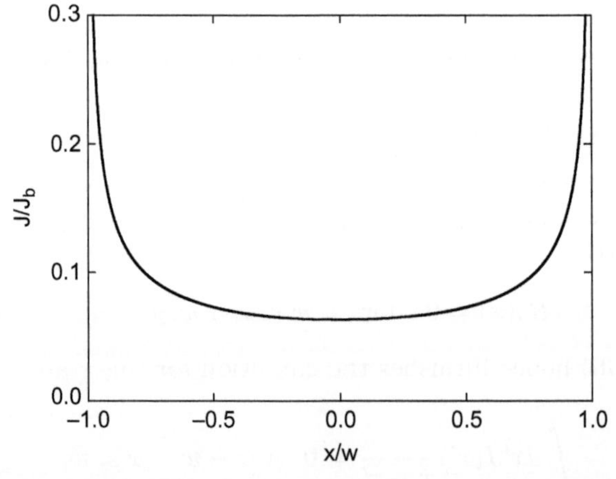

Fig. 3.1 Distribution of the normalized sheet current J/J_b as a function of the normalized coordinate x/w over the width of the isolated superconductor strip in the flux-free Meissner state, assuming $h/2w = 0.001$.

current profile around the central part, followed by a steep rise towards the edges, of the strip is evident.

3.1.3 Edge barrier considerations

For strips with a rectangular cross section, J_b is determined by the balance between the current-induced Lorentz-like force per unit length exerted upon magnetic vortices at the edges of the strip,

$$\mathbf{F}_\mathrm{L} = \pm\,(J_\mathrm{b}/d)\,\mathbf{e}_z \times (\Phi_0 \mathbf{e}_y)\,, \tag{3.10}$$

in analogy to eqn. (2.39), and the geometrical barrier force per unit length,

$$\mathbf{F}_\mathrm{b} = \pm\,(2E_\mathrm{V}/d)\,\mathbf{e}_x, \tag{3.11}$$

introducing the vortex self-energy per unit length E_V, where the plus/minus signs relate to the right and left edges, respectively, of the strip (Indenbom et al. 1994; Zeldov et al. 1994). The quantity E_V herein measures the bulk lower critical field H_c1 from $E_\mathrm{V} = H_\mathrm{c1}\Phi_0$ via

$$H_\mathrm{c1} \simeq \frac{\Phi_0}{4\pi\mu_0\lambda_\mathrm{L}^2}\ln\left(\frac{\lambda_L}{\xi_\mathrm{GL}}\right); \quad \xi_\mathrm{GL} \ll \lambda_\mathrm{L}, \tag{3.12}$$

as outlined in section 2.1.3.1. This field naturally governs the first entry of magnetic vortices from the edges of the strip. For a vanishing resultant total force per unit length,

$$\mathbf{F}_\mathrm{tot} = \mathbf{F}_\mathrm{L} + \mathbf{F}_\mathrm{b}, \tag{3.13}$$

eqns. (3.10) and (3.11) thus yield the link

$$J_\mathrm{b} = 2H_\mathrm{c1}. \tag{3.14}$$

Substituting $I_\mathrm{max} = 2w\bar{J}_\mathrm{max}$ for I in eqn. (3.9) when $J_\mathrm{max} = J_\mathrm{b}$ finally provides the maximum average sheet current

$$\bar{J}_\mathrm{max} \simeq \pi H_\mathrm{c1}\sqrt{\frac{2h}{w}}; \quad h \ll w. \tag{3.15}$$

Suppose now the premise $d \ll \lambda_\mathrm{L}$ holds. For thin films with clear-cut edges (Likharev 1971) or thick strips with carefully rounded edges (Benkraouda and Clem 1996), the penetration of magnetic flux then is governed by the Bean–Livingston barrier (Bean and Livingston 1964), which allows J_b to be as large as $d\,j_\mathrm{GL} \simeq 5H_\mathrm{c1}$, recalling the depairing critical current density j_GL. When edge defects of a size σ_d exist, this high value comes down to $J_\mathrm{b} \simeq H_\mathrm{c1}d/k$, where k denotes the greater of the quantities σ_d and ξ_GL (Kupriyanov and Likharev 1975). We tacitly assume protection of the Meissner state by the most robust geometrical edge barrier henceforth.

3.2 Superconductor strip adjacent to a flat semi-infinite paramagnet

Let us further develop the considerations of section 3.1, addressing an infinitely extended, thin current-carrying type-II superconductor strip in the Meissner state, placed at the distance c adjacent to a flat semi-infinite paramagnetic environment with permeability μ, as depicted in Fig. 3.2. This represents a simple example of the conditioning of magnetic fields by a one-sided open magnetic shield; its theoretical analysis draws upon the work of Genenko, Snezhko, and Freyhardt (2000).

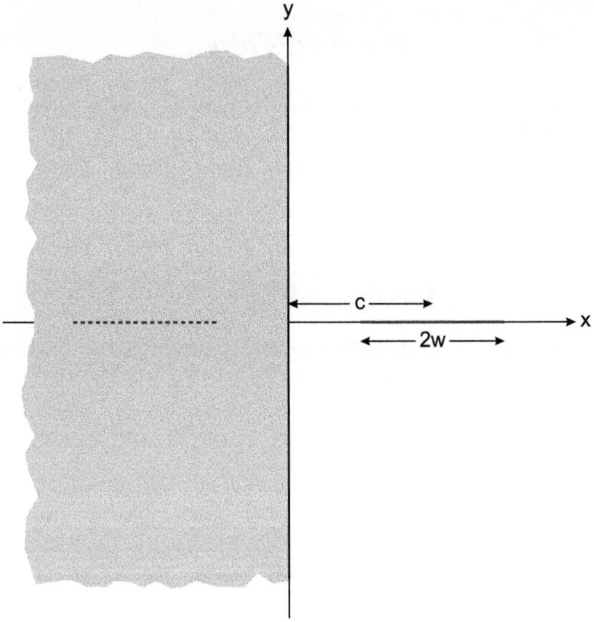

Fig. 3.2 Cross-sectional view of a superconductor strip (dark shading) adjacent to a flat semi-infinite paramagnet (light shading) extending infinitely in the z-direction of a Cartesian coordinate system x, y, z. The definition of the distance between the centre of the strip and the paramagnet c as well as the image of the strip (dotted line) are marked.

3.2.1 Field distribution

Here, eqns. (2.87) and (2.88) simplify to the Poisson equation for the single nontrivial component of the magnetic vector potential A_z,

$$\Delta A_z = \begin{cases} -\mu_0 J(x)\delta(y) & \text{in } c-w \le x \le c+w \\ 0 & \text{otherwise,} \end{cases} \tag{3.16}$$

supplemented with the boundary conditions

$$\lim_{x\to 0-} A_z(x,y) = \lim_{x\to 0+} A_z(x,y) \quad \text{and} \quad \frac{1}{\mu}\frac{\partial A_z}{\partial n}\bigg|_{x=0-} = \frac{1}{\mu_0}\frac{\partial A_z}{\partial n}\bigg|_{x=0+}. \tag{3.17}$$

The latter requirements exclude direct contact between the paramagnetic shield and the strip to avoid complications arising from the proximity effect (Schmidt 1997). A scheme similar to eqns. (3.16) and (3.17) obviously applies to the analogous problem of an electrically charged thin metal strip adjacent to a flat dielectric semi-space. The desired solution can be constructed by the method of images, calling in the image current strength $\kappa = (\mu - \mu_0)/(\mu + \mu_0)$, together with the Green's function technique. Identifying the contributions $dI = dx' J(x')$ of a distributed current element of unit weight at $(x = x', y = 0)$ as well as of an image current element of weight κ at $(x = -x', y = 0)$ and integrating over x' yields the convenient representation in the vacuum space, $x > 0$ inclusive of the surface of the strip ∂D,

$$A_z(x,y) = \frac{\mu_0}{2\pi}\int_{c-w}^{c+w} dx' J(x') \left\{ \ln \frac{c}{\sqrt{(x-x')^2 + y^2}} + \kappa \ln \frac{c}{\sqrt{(x+x')^2 + y^2}} \right\}. \tag{3.18}$$

Likewise, identifying the contribution $dI = dx' J(x')$ of an image current element of assigned weight $1 - \kappa$ at $(x = x', y = 0)$ and integrating over x' yields the representation in the paramagnet space, $x \leq 0$,

$$A_z(x,y) = \frac{\mu}{2\pi}\int_{c-w}^{c+w} dx' J(x')(1-\kappa) \ln \frac{c}{\sqrt{(x-x')^2 + y^2}}. \tag{3.19}$$

A fully analytic treatment is possible when considering the limit of infinite permeability of the paramagnetic shield, $\mu \to \infty$. We therefore turn to the representations in the vacuum space, $x > 0$ inclusive of ∂D,

$$\lim_{\mu\to\infty} A_z(x,y) = \frac{\mu_0}{2\pi}\int_{c-w}^{c+w} dx' J(x') \left\{ \ln \frac{c}{\sqrt{(x-x')^2 + y^2}} + \ln \frac{c}{\sqrt{(x+x')^2 + y^2}} \right\} \tag{3.20}$$

and in the paramagnet space, $x \leq 0$,

$$\lim_{\mu\to\infty} A_z(x,y) = \frac{\mu_0}{\pi}\int_{c-w}^{c+w} dx' J(x') \ln \frac{c}{\sqrt{(x-x')^2 + y^2}}. \tag{3.21}$$

Application of eqn. (2.3) to eqns. (3.20) and (3.21) thus gives the Cartesian components of the magnetic field $\mathbf{H}$ in the vacuum space, $x > 0$ inclusive of ∂D,

$$H_x(x,y) = -\frac{1}{2\pi}\int\limits_{c-w}^{c+w} dx'\,J(x')\left[\frac{y}{(x-x')^2+y^2}+\frac{y}{(x+x')^2+y^2}\right] \tag{3.22a}$$

$$H_y(x,y) = \frac{1}{2\pi}\int\limits_{c-w}^{c+w} dx'\,J(x')\left[\frac{x-x'}{(x-x')^2+y^2}+\frac{x+x'}{(x+x')^2+y^2}\right], \tag{3.22b}$$

and those of the magnetic induction **B** in the paramagnet space, $x \leq 0$,

$$B_x(x,y) = -\frac{\mu_0}{\pi}\int\limits_{c-w}^{c+w} dx'\,J(x')\frac{y}{(x-x')^2+y^2} \tag{3.23a}$$

$$B_y(x,y) = \frac{\mu_0}{\pi}\int\limits_{c-w}^{c+w} dx'\,J(x')\frac{x-x'}{(x-x')^2+y^2}. \tag{3.23b}$$

Both the magnetic field and the magnetic induction, for the sheet current, eqn. (3.27), are visualized in Fig. 2.10(a).

3.2.2 Current distribution

As noted earlier, the Meissner state of the superconductor strip requires the normal component of the magnetic field H_y to vanish across the width of the strip,

$$H_y(x) = 0 \quad \text{for } c-w < x < c+w, y = 0. \tag{3.24}$$

Remembering eqn. (3.22b) hence provides the condition for this state,

$$\int\limits_{c-w}^{c+w} dx'\,J(x')\frac{1}{x^2-x'^2} = 0 \quad \text{in } c-w < x < c+w. \tag{3.25}$$

The change of the variables x and x' to $u = x^2$ and $u' = x'^2$, respectively, together with the introduction of the substitutions $l_- = (c-w)^2$ and $l_+ = (c+w)^2$ transforms eqn. (3.25) into the singular integral equation of the Cauchy type

$$\int\limits_{l_-}^{l_+} du'\,S(u')\frac{1}{u-u'} = 0 \quad \text{in } l_- < u < l_+ \tag{3.26}$$

for the norm-conserving auxiliary function $S(u') = J(x')/2x'$. Its solution can be directly inferred from eqn. (3.8), representing the sheet current as

$$J(x) = \frac{2I}{\pi}\frac{x}{\sqrt{((c+w)^2-x^2)\,(x^2-(c-w)^2)}} \quad \text{for } c-w < x < c+w. \tag{3.27}$$

Although overstretching the model requirements, it is instructive to formally test eqn. (3.27) against the apparent situation of close strip-magnet contact, when $c \to w$. In this case, the sheet current and its image make up a continuous current sheet which occupies the range $-2w \le x \le 2w$. Applying the Meissner state condition, eqn. (3.24), to the right, real half of this sheet provides a current distribution given by the appropriately scaled shape of eqn. (3.8),

$$J(x) = \frac{2I}{\pi} \frac{1}{\sqrt{(2w - x)(x + 2w)}} \quad \text{for } 0 < x < 2w. \tag{3.28}$$

One can easily verify that, in the above limit, eqn. (3.27) exactly duplicates eqn. (3.28).

Reasoning in a similar vein to that for the isolated superconductor strip in the barrier-protected Meissner state of section 3.1 limits the validity of eqn. (3.27) to the range $c - w + h \le x \le c + w - h$ and simultaneously recommends imposing the maximum value $J_{\max} = J_{\mathrm{b}}$ in $c + w - h \le x \le c + w$. Substituting $I_{\max} = 2w\bar{J}_{\max}$ for I in eqn. (3.27) with eqn. (3.14) therefore yields the maximum average sheet current $\bar{J}_{\max}$, which may be approximated by the highly accurate simple form

$$\bar{J}_{\max} \simeq \pi H_{\mathrm{c1}} \sqrt{\frac{2h}{w}} \sqrt{\frac{c}{c + w}}; \quad h \ll w < c - h. \tag{3.29}$$

Comparing eqn. (3.29) with eqn. (3.15) shows that the maximum average sheet current $\bar{J}_{\max}$, and hence the maximum loss-free total current $I_{\max}$, in the presence of

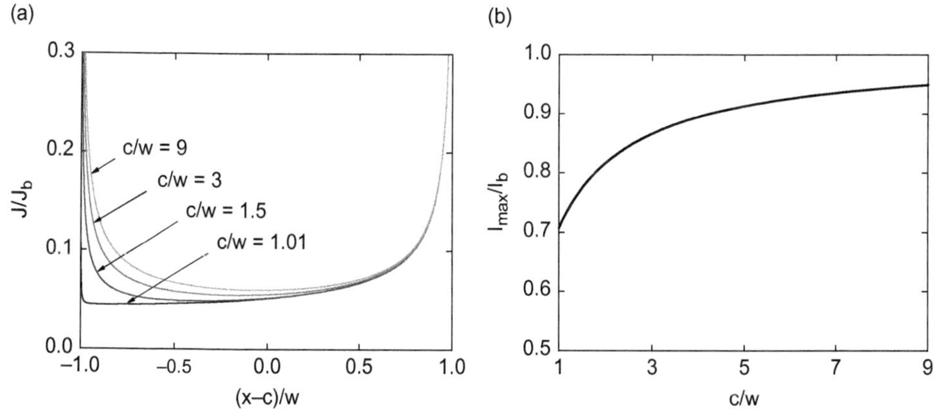

Fig. 3.3 (a) Distribution of the normalized sheet current J/J_{b} as a function of the normalized coordinate $(x - c)/w$ over the width of the magnetically shielded superconductor strip in the flux-free Meissner state for four different values of the normalized distance between the centre of the strip and the paramagnet c/w identified on the curves; (b) variation of the corresponding normalized maximum total transport current $I_{\max}/I_{\mathrm{b}}$ with the normalized distance between the centre of the strip and the paramagnet c/w, when $h/2w = 0.001$, referring to the flat semi-infinite paramagnet configuration of Fig. 3.2.

the flat paramagnetic shield are always *smaller* than the corresponding quantities for the unshielded strip, the strongest current depression $I_{\max}/I_{\mathrm{b}} \simeq 1/\sqrt{2}$ being attained if $c \to w$.

Figure 3.3(a) displays the variation of the normalized sheet current J/J_{b} over the width of the strip for a range of values of the normalized distance between the centre of the strip and the flat paramagnetic shield c/w. A general trait due to sheet current redistribution in the strip is the lowering of the current towards the left edge of the strip adjacent to the magnetic shield, which gets more pronounced as c/w abates, reaching saturation with a *finite* minimum in the limiting sheet current profile, eqn. (3.28). Figure 3.3(b) presents the variation of the corresponding normalized maximum total transport current $I_{\max}/I_{\mathrm{b}}$ with the normalized distance between the centre of the strip and the flat paramagnetic shield c/w. Starting from a distinct minimum at the smallest value of c/w, there is a monotonic rise followed by an approach towards unity, the value prevailing for the isolated strip, as c/w continues to grow.

3.3 Superconductor strip inside a wedge-shaped paramagnetic cavity

We now alter the geometry of section 3.2, considering an infinitely extended, thin current-carrying type-II superconductor strip in the Meissner state, placed at the distance c from the vertex of a wedge-shaped paramagnetic cavity with opening half-angle $\beta = \pi/4$ and infinite permeability, $\mu \to \infty$, as portrayed in Fig. 3.4. This represents a more elaborate example of the conditioning of magnetic fields by a one-sided open magnetic shield; its analysis that follows augments the work of Genenko, Snezhko, and Freyhardt (2000).

3.3.1 Field distribution

Here, eqns. (2.87) and (2.88) again simplify to the Poisson equation for the single nontrivial component of the magnetic vector potential A_z,

$$\Delta A_z = \begin{cases} -\mu_0 J(x)\delta(y) & \text{in } c - w \le x \le c + w \\ 0 & \text{otherwise,} \end{cases} \tag{3.30}$$

supplemented with the boundary conditions

$$\lim_{|y| \to x-} A_z(x,y) = \lim_{|y| \to x+} A_z(x,y) \quad \text{and} \quad \left.\frac{\partial A_z}{\partial n}\right|_{|y|=x+} = 0. \tag{3.31}$$

The latter requirements exclude direct contact between the paramagnetic cavity and the strip to avoid complications arising from the proximity effect (Schmidt 1997). A scheme similar to eqns. (3.30) and (3.31) obviously applies to the analogous problem of an electrically charged thin metal strip placed inside a wedge-shaped dielectric cavity. The desired solution can again be constructed by the method of images together with the Green's function technique. Identifying the contributions $dI = dx' J(x')$ of a distributed current element of unit weight at $(x = x', y = 0)$ as

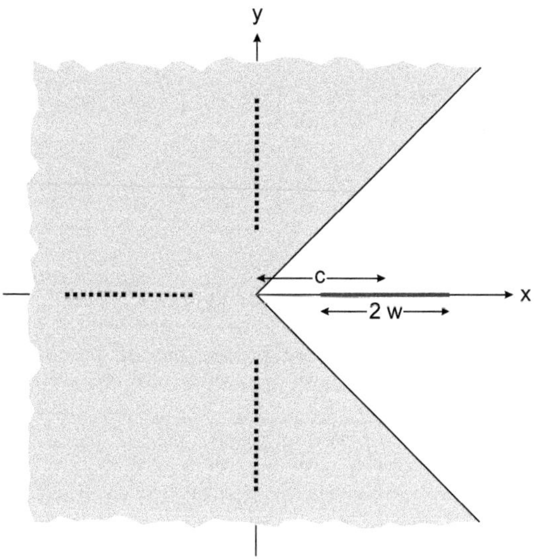

Fig. 3.4 Cross-sectional view of a superconductor strip (dark shading) inside a wedge-shaped paramagnetic cavity of opening half-angle $\beta = \pi/4$ (light shading) extending infinitely in the z-direction of a Cartesian coordinate system x, y, z. The definition of the distance between the centre of the strip and the vertex of the cavity c as well as the three images of the strip (dotted lines) are marked.

well as of image current elements of equal weight at $(x = -x', y = 0)$, $(x = 0, y = x')$, $(x = 0, y = -x')$ and integrating over x' yields the convenient representation in the vacuum space, $0 \leq |y| < x$ inclusive of the surface of the strip ∂D,

$$A_z(x,y) = \frac{\mu_0}{2\pi} \int_{c-w}^{c+w} dx' J(x') \left\{ \ln \frac{c}{\sqrt{(x-x')^2 + y^2}} + \ln \frac{c}{\sqrt{(x+x')^2 + y^2}} \right.$$
$$\left. + \ln \frac{c}{\sqrt{x^2 + (y-x')^2}} + \ln \frac{c}{\sqrt{x^2 + (y+x')^2}} \right\}. \tag{3.32}$$

Employing the proper Green's function together with the conformal mapping technique allows for an extension of this solution into the paramagnet space, $|y| \geq x$,

$$A_z(x,y) = \frac{\mu_0}{\pi^2} \int_{c-w}^{c+w} dx' J(x') \int_{-\infty}^{\infty} dp' \frac{q(x,y)}{(p(x,y) - p')^2 + q(x,y)^2}$$
$$\times \left\{ \ln \frac{c}{\sqrt{\left(p'^{3/2} - x'\right)^2 + p'^3}} + \ln \frac{c}{\sqrt{\left(p'^{3/2} + x'\right)^2 + p'^3}} \right\}. \tag{3.33}$$

The auxiliary functions herein read

$$p(x,y) = \left(\frac{x^2+y^2}{2}\right)^{1/3} \sin\left(\frac{2}{3}\phi(x,y)\right) \tag{3.34a}$$

$$q(x,y) = \left(\frac{x^2+y^2}{2}\right)^{1/3} \cos\left(\frac{2}{3}\phi(x,y)\right), \tag{3.34b}$$

where

$$\phi(x,y) = \pi\theta(x) - \arctan\frac{|y|}{x} \tag{3.35}$$

with Heaviside's step function θ. Evidently, because of

$$\lim_{|y|\to x-} \frac{q(x,y)}{(p(x,y)-p')^2 + q(x,y)^2} = \pi\delta\left(p(x,x) - p'\right) \tag{3.36}$$

and $p(x,x) = x^{2/3}$, continuity across the paramagnet/vacuum interface, as required by eqn. (3.31), exists. Application of eqn. (2.3) to eqns. (3.32) and (3.33) yields the Cartesian components of the magnetic field $\mathbf{H}$ in the vacuum space, $0 \le |y| < x$ inclusive of ∂D,

$$H_x(x,y) = -\frac{1}{2\pi}\int\limits_{c-w}^{c+w} dx'\,J(x')\left[\frac{y}{(x-x')^2+y^2} + \frac{y}{(x+x')^2+y^2}\right.$$
$$\left. + \frac{y-x'}{x^2+(y-x')^2} + \frac{y+x'}{x^2+(y+x')^2}\right] \tag{3.37a}$$

$$H_y(x,y) = \frac{1}{2\pi}\int\limits_{c-w}^{c+w} dx'\,J(x')\left[\frac{x-x'}{(x-x')^2+y^2} + \frac{x+x'}{(x+x')^2+y^2}\right.$$
$$\left. + \frac{x}{x^2+(y-x')^2} + \frac{x}{x^2+(y+x')^2}\right], \tag{3.37b}$$

and those of the magnetic induction $\mathbf{B}$ in the paramagnet space, $|y| \ge x$,

$$B_x(x,y) = \frac{\mu_0}{\pi^2}\int\limits_{c-w}^{c+w} dx'\,J(x')\frac{\partial}{\partial y}\int\limits_{-\infty}^{\infty} dp'\frac{q(x,y)}{(p(x,y)-p')^2+q(x,y)^2}$$
$$\times\left\{\ln\frac{c}{\sqrt{\left(p'^{3/2}-x'\right)^2+p'^3}} + \ln\frac{c}{\sqrt{\left(p'^{3/2}+x'\right)^2+p'^3}}\right\} \tag{3.38a}$$

$$B_y(x,y) = -\frac{\mu_0}{\pi^2}\int\limits_{c-w}^{c+w} dx'\,J(x')\frac{\partial}{\partial x}\int\limits_{-\infty}^{\infty} dp'\frac{q(x,y)}{(p(x,y)-p')^2+q(x,y)^2}$$
$$\times\left\{\ln\frac{c}{\sqrt{\left(p'^{3/2}-x'\right)^2+p'^3}} + \ln\frac{c}{\sqrt{\left(p'^{3/2}+x'\right)^2+p'^3}}\right\}. \tag{3.38b}$$

Both the magnetic field and the magnetic induction, for the sheet current, eqn. (3.42), are visualized in Fig. 2.10(b).

3.3.2 Current distribution

Naturally again the Meissner state of the superconductor strip requires the normal component of the magnetic field H_y to vanish across the width of the strip,

$$H_y(x) = 0 \text{ for } c - w < x < c + w, y = 0. \tag{3.39}$$

Recalling eqn. (3.37b) hence furnishes the condition for this state

$$\int_{c-w}^{c+w} dx' J(x') \frac{1}{x^4 - x'^4} = 0 \text{ in } c - w < x < c + w. \tag{3.40}$$

The change of the variables x and x' to $u = x^4$ and $u' = x'^4$, respectively, together with the introduction of the substitutions $l_- = (c-w)^4$ and $l_+ = (c+w)^4$ transforms eqn. (3.40) into the singular integral equation of the Cauchy type

$$\int_{l_-}^{l_+} du' S(u') \frac{1}{u - u'} = 0 \text{ in } l_- < u < l_+ \tag{3.41}$$

for the norm-conserving auxiliary function $S(u') = J(x')/4x'^3$. Its solution directly emerges from eqn. (3.8), establishing the sheet current as

$$J(x) = \frac{4I}{\pi} \frac{x^3}{\sqrt{((c+w)^4 - x^4)(x^4 - (c-w)^4)}} \quad \text{for } c - w < x < c + w. \tag{3.42}$$

Once more, though overstretching the model requirements, eqn. (3.42) may be formally tested against the apparent situation of close strip–magnet contact, when $c \to w$. In this case, the sheet current and its images make up two crossed continuous current sheets which occupy the ranges $-2w \leq x \leq 2w$ and $-2w \leq y \leq 2w$. Applying the Meissner state condition, eqn. (3.39), to the right, real half of the horizontal sheet provides a current distribution given by the appropriately scaled shape of eqn. (3.8),

$$J(x) = \frac{4I}{\pi} \frac{x}{\sqrt{((2w)^2 - x^2)(x^2 + (2w)^2)}} \quad \text{for } 0 < x < 2w. \tag{3.43}$$

One can easily verify that, in the above limit, eqn. (3.42) exactly reduces to eqn. (3.43).

Reasoning in a similar vein to that for the isolated superconductor strip in the barrier-protected Meissner state of section 3.1 limits the validity of eqn. (3.42) to the range $c - w + h \leq x \leq c + w - h$ and simultaneously recommends imposing the maximum value $J_{\text{max}} = J_{\text{b}}$ in $c + w - h \leq x \leq c + w$. Substituting $I_{\text{max}} = 2w\bar{J}_{\text{max}}$

for I in eqn. (3.42) with eqn. (3.14) therefore yields the maximum average sheet current $\bar{J}_{\max}$, which may be approximated by the highly accurate simple form

$$\bar{J}_{\max} \simeq \pi H_{c1}\sqrt{\frac{2h}{w}}\sqrt{\frac{c}{c+w}}\,\frac{\sqrt{c^2+w^2}}{c+w}; \quad h \ll w < c-h. \tag{3.44}$$

Comparing eqn. (3.44) with eqn. (3.29) shows that the maximum average sheet current $\bar{J}_{\max}$, and hence the maximum loss-free total current $I_{\max}$, in the presence of the wedge-shaped paramagnetic cavity is always *smaller* than the corresponding quantities for the flat paramagnetic shield, the strongest current depression $I_{\max}/I_b \simeq 1/2$ being achieved if $c \to w$.

Figure 3.5(a) displays the variation of the normalized sheet current J/J_b over the width of the strip for a range of values of the normalized distance between the centre of the strip and the wedge-shaped paramagnetic cavity c/w. A general trait due to sheet current redistribution in the strip is the lowering of the current towards the left edge of the strip adjacent to the magnetic shield, which gets more pronounced as c/w abates, reaching saturation with a *zero* minimum in the limiting sheet current profile, eqn. (3.43). Figure 3.5(b) presents the variation of the corresponding normalized maximum total transport current $I_{\max}/I_b$ with the normalized distance between the centre of the strip and the wedge-shaped paramagnetic cavity c/w. Starting from a distinct minimum at the smallest value of c/w, there is a monotonic rise followed by an approach towards unity, the value prevailing for the isolated strip, as c/w continues to grow.

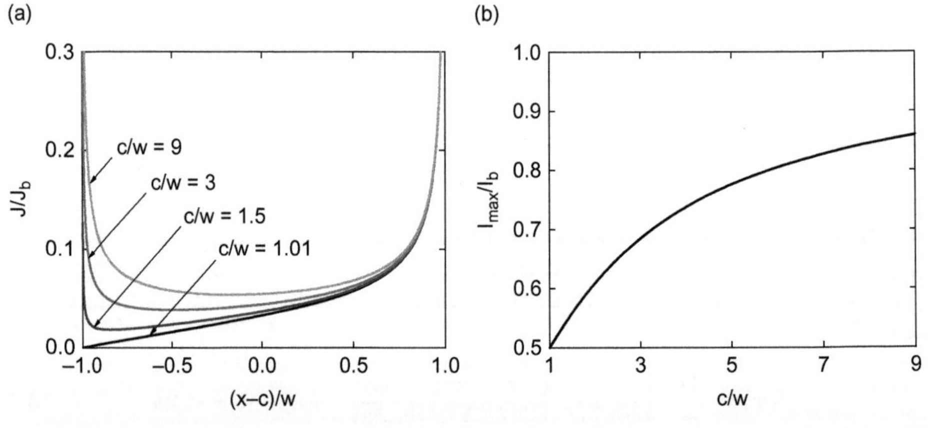

Fig. 3.5 (a) Distribution of the normalized sheet current J/J_b as a function of the normalized coordinate $(x-c)/w$ over the width of the magnetically shielded superconductor strip in the flux-free Meissner state for four different values of the normalized distance between the centre of the strip and the paramagnet c/w identified on the curves; (b) variation of the corresponding normalized maximum total transport current $I_{\max}/I_b$ with the normalized distance between the centre of the strip and the paramagnet c/w, when $h/2w = 0.001$, referring to the wedge-shaped paramagnet configuration of Fig. 3.4.

3.4 Superconductor strip between two flat semi-infinite paramagnets

We now go on to develop the considerations of section 3.1, addressing an infinitely extended, thin current-carrying type-II superconductor strip in the Meissner state, placed between two flat semi-infinite paramagnetic shields with permeability μ and principal orientations either longitudinal or transverse. These geometries represent examples of the conditioning of magnetic fields by two-sided open magnetic shields; their analysis that follows adds to the work of Genenko, Usoskin, and Freyhardt (1999).

3.4.1 Longitudinal shielding geometry

Let us first examine the case where the surfaces of the paramagnets run parallel to the strip, at the distance a from the plane of the strip, as illustrated in Fig. 3.6.

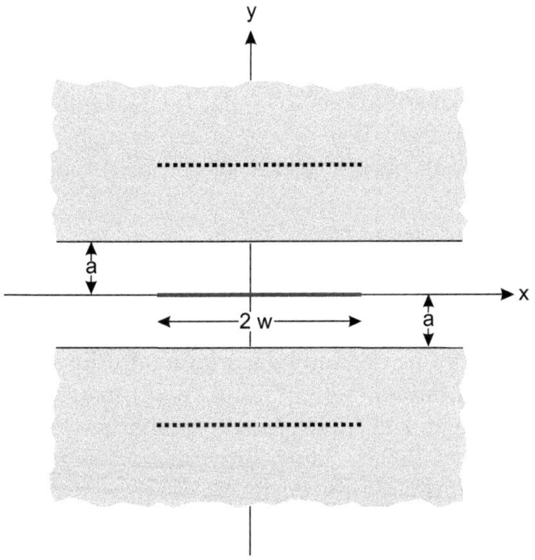

Fig. 3.6 Cross-sectional view of a superconductor strip (dark shading) between two flat semi-infinite paramagnets with longitudinal orientation (light shading) extending infinitely in the z-direction of a Cartesian coordinate system x, y, z. The definition of the distance between the plane of the strip and the paramagnets a as well as two of the images of the strip (dotted lines) are marked.

3.4.1.1 Field distribution

Clearly, eqns. (2.87) and (2.88) here simplify to the Poisson equation for the single nontrivial component of the magnetic vector potential A_z,

$$\Delta A_z = \begin{cases} -\mu_0 J(x)\delta(y) & \text{in } -w \le x \le w \\ 0 & \text{otherwise,} \end{cases} \tag{3.45}$$

supplemented with the boundary conditions

$$\lim_{|y|\to a-} A_z(x,y) = \lim_{|y|\to a+} A_z(x,y) \quad \text{and} \quad \left.\frac{1}{\mu_0}\frac{\partial A_z}{\partial n}\right|_{|y|=a-} = \left.\frac{1}{\mu}\frac{\partial A_z}{\partial n}\right|_{|y|=a+}. \tag{3.46}$$

Obviously, the latter demands rule out direct contact between the paramagnetic shields and the surfaces of the strip to avoid complications arising from the proximity effect (Schmidt 1997). The desired solution can again be constructed by the method of images, calling in their current strength $\kappa = (\mu - \mu_0)/(\mu + \mu_0)$, along with the Green's function technique. Identifying the contributions $dI = dx'J(x')$ of a distributed current element of unit weight at $(x = x', y = 0)$ as well as of image current elements of weight $\kappa^{|n|}$ at $(x = x', y = 2na)$ for $n = \pm 1, \pm 2, \pm 3, ...$ and integrating over x' yields the convenient representation in the vacuum space, $0 \le |y| < a$ inclusive of the surface of the strip ∂D,

$$A_z(x,y) = \frac{\mu_0}{2\pi}\sum_{n=-\infty}^{\infty}\kappa^{|n|}\int_{-w}^{w} dx'J(x')\ln\frac{a}{\sqrt{(x-x')^2+(y-2na)^2}}. \tag{3.47}$$

Likewise, identifying the contributions $dI = dx'J(x')$ of image current elements of weight κ^n at $(x = x', y = 2na)$ if $y \le -a$, or $(x = x', y = -2na)$ if $y \ge a$, for $n = 0, 1, 2, ...$ and integrating over x' yields the representation in the paramagnet space, $|y| \ge a$,

$$A_z(x,y) = \frac{\mu}{2\pi}(1-\kappa)\sum_{n=0}^{\infty}\kappa^n\int_{-w}^{w} dx'J(x')\ln\frac{a}{\sqrt{(x-x')^2+(|y|+2na)^2}}. \tag{3.48}$$

Compact expressions can be derived by summing up the partial derivatives with respect to x and y of A_z from eqns. (3.47) and (3.48) in terms of suitable integral forms of Lerch's transcendent Φ (Erdélyi, Magnus, Oberhettinger, and Tricomi 1953; Laurinčikas and Garunkštis 2002). Thus, integrating the total differential of A_z, eqn. (3.47), yields the exact representation in the vacuum space, $0 \le |y| < a$ inclusive of ∂D,

$$\begin{aligned} A_z(x,y) &= \frac{\mu_0}{2\pi}\int_{-w}^{w} dx'J(x')\ln\frac{a}{\sqrt{(x-x')^2+y^2}} \\ &+ \frac{\mu_0}{8\pi i}\int_{-w}^{w} dx'J(x')\int_C dt\,\frac{\csc(\pi t)\exp(-i\pi t)}{\kappa^t}\ln\left|\vartheta_+(x-x',y,t)\vartheta_-(x-x',y,t)\right|, \end{aligned} \tag{3.49}$$

where

$$\vartheta_\pm(x-x',y,t) = \left[((x-x')/a)^2 + (y/a \pm 2t)^2\right]/(1\pm 2t)^2, \tag{3.50}$$

C denoting the keyhole, or Hankel, contour, i.e. a path of integration that starts at infinity on the real t-axis, circles the origin in the counterclockwise direction, and returns to the start. The first term in eqn. (3.49) emulates the contribution of an isolated superconductor strip, recalling eqn. (3.4), whereas the second term in eqn. (3.49) reflects the influence of the paramagnetic shields. Likewise, integrating the total differential of A_z, eqn. (3.48), yields the exact representation in the paramagnet space, $|y| \geq a$,

$$A_z(x,y) = \frac{\mu}{2\pi}(1-\kappa)\int_{-w}^{w} dx' J(x') \int_0^\infty \frac{dt}{t} \frac{\cos(t(x-x')/2a)\exp(-t|y|/2a) - \exp(-t/2)}{1-\kappa\exp(-t)}. \tag{3.51}$$

Application of eqn. (2.3) to eqns. (3.49) and (3.51) thus gives the Cartesian components of the magnetic field $\mathbf{H}$ in the vacuum space, $0 \leq |y| < a$ inclusive of ∂D,

$$H_x(x,y) = -\frac{1}{2\pi}\int_{-w}^{w} dx' J(x') \frac{y}{(x-x')^2+y^2}$$

$$+\frac{1}{16\pi i a}\int_{-w}^{+w} dx' J(x') \int_C dt \frac{\csc(\pi t)\exp(-i\pi t)}{\kappa^t} \mathrm{Re}\left[\phi_+(x-x',y,t) + \phi_-(x-x',y,t)\right] \tag{3.52a}$$

$$H_y(x,y) = \frac{1}{2\pi}\int_{-w}^{w} dx' J(x') \frac{x-x'}{(x-x')^2+y^2}$$

$$+\frac{1}{16\pi i a}\int_{-w}^{+w} dx' J(x') \int_C dt \frac{\csc(\pi t)\exp(-i\pi t)}{\kappa^t} \mathrm{Im}\left[\phi_+(x-x',y,t) - \phi_-(x-x',y,t)\right], \tag{3.52b}$$

where

$$\phi_\pm(x-x',y,t) = 1/\left[(\pm i(x-x')+y)/2a + t\right] + 1/\left[(\pm i(x-x')+y)/2a - t\right], \tag{3.53}$$

and those of the magnetic induction $\mathbf{B}$ in the paramagnet space, $|y| \geq a$,

$$B_x(x,y) = -\mathrm{sgn}(y)\frac{\mu}{4\pi a}(1-\kappa)\int_{-w}^{w} dx' J(x') \int_0^\infty dt \frac{\cos(t(x-x')/2a)\exp(-t|y|/2a)}{1-\kappa\exp(-t)} \tag{3.54a}$$

$$B_y(x,y) = \frac{\mu}{4\pi a}(1-\kappa)\int_{-w}^{w} dx' J(x') \int_0^\infty dt \frac{\sin(t(x-x')/2a)\exp(-t|y|/2a)}{1-\kappa\exp(-t)}. \tag{3.54b}$$

A fully analytic treatment is possible when considering the limit of infinite permeability of the paramagnetic shields, $\mu \to \infty$. We therefore attend to the representation of the single component of the magnetic vector potential A_z in the vacuum space, $0 \leq |y| < a$ inclusive of ∂D,

$$\lim_{\mu \to \infty} A_z(x, y) = \frac{\mu_0}{2\pi} \int_{-w}^{w} dx' J(x') \ln \frac{1}{\sqrt{[\cosh(\pi(x - x')/a) - \cos(\pi y/a)]/2}} \tag{3.55}$$

together with the Cartesian components of the magnetic field $\mathbf{H}$,

$$H_x(x, y) = -\frac{1}{4a} \int_{-w}^{w} dx' J(x') \frac{\sin(\pi y/a)}{\cosh(\pi(x - x')/a) - \cos(\pi y/a)} \tag{3.56a}$$

$$H_y(x, y) = \frac{1}{4a} \int_{-w}^{w} dx' J(x') \frac{\sinh(\pi(x - x')/a)}{\cosh(\pi(x - x')/a) - \cos(\pi y/a)}, \tag{3.56b}$$

duplicating a result which can also directly be obtained by recourse to customary table sets (Prudnikov, Brychkov, and Marychev 1986). It is worth noting that in the above limit, A_z and $\partial A_z/\partial n$, whence B_x, suffer infinite jumps along the paramagnet/vacuum interface upon passing to the paramagnet space, $|y| \geq a$, behaviour demonstrating an extreme form of the boundary conditions, eqn. (3.46), whereas B_y remains confined. Both the magnetic field and the magnetic induction are visualized in Fig. 2.11(a) for a large but *finite* permeability μ in conjunction with the sheet current for $\mu \to \infty$, eqn. (3.60), thus disregarding its weak dependence on μ in the high permeability regime. Obviously, a congruence exists between the magnetic field $\mathbf{H}$ and the velocity field $\mathbf{v}$ of steady-state rotational flow of an incompressible fluid about a strip-like vortex core placed in a longitudinally oriented flat interspace, whose surfaces act as fluid sources or, respectively, sinks.

3.4.1.2 Current distribution

As noted already, the Meissner state of the superconductor strip requires the normal component of the magnetic field H_y to vanish across the width of the strip,

$$H_y(x) = 0 \quad \text{for} \quad -w < x < w, y = 0. \tag{3.57}$$

Recalling eqn. (3.56b) hence provides the condition for this state,

$$\int_{-w}^{w} dx' J(x') \coth(\pi(x - x')/2a) = 0 \quad \text{in} \quad -w < x < w. \tag{3.58}$$

The change of the variables x and x' to $u = \tanh(\pi x/2a)$ and $u' = \tanh(\pi x'/2a)$, respectively, together with the introduction of the substitution $q = \tanh(\pi w/2a)$ and the request for an even solution in x, transform eqn. (3.58) into the singular integral equation of the Cauchy type

$$\int_{-q}^{q} du' S(u') \frac{1}{u-u'} = 0 \quad \text{in} \quad -q < u < q \tag{3.59}$$

for the auxiliary function $S(u') = J(x')$. The functional form of its solution hence follows from eqn. (3.8); ensuring normalization, eqn. (3.2), finally yields the sheet current as

$$J(x) = \frac{I}{2a} \frac{\cosh(\pi x/2a)}{\sqrt{[\cosh(\pi w/a) - \cosh(\pi x/a)]/2}} \quad \text{for} \quad -w < x < w. \tag{3.60}$$

Although overstretching the model requirements, it is instructive to formally test eqn. (3.60) against the apparent situation of direct strip-magnet contact, when $a \to 0$. In this case, the sheet current and its images make up an effective current-carrying 'superconductor slab' which occupies the space between the planes $x = \pm w$. Current flow is limited to the surface layers of the slab beneath these planes, which is why $J(x) = 0$ for $-w < x < w$, in keeping with the Meissner state.

Reasoning in a similar vein to that for the isolated superconductor strip in the barrier-protected Meissner state of section 3.1 limits the validity of eqn. (3.60) to the range $-w+h \leq x \leq w-h$ and simultaneously recommends imposing the maximum value $J_{\max} = J_{\text{b}}$ in $|x \pm w| \leq h$. Substituting $I_{\max} = 2w\bar{J}_{\max}$ for I in eqn. (3.60) with eqn. (3.14) therefore yields the maximum average sheet current $\bar{J}_{\max}$, which may be approximated by the highly accurate simple form

$$\bar{J}_{\max} \simeq \pi H_{c1} \sqrt{\frac{2h}{w}} \sqrt{\frac{\tanh(\pi w/2a)}{\pi w/2a}}; \quad h \ll w, \; a > h. \tag{3.61}$$

Comparing eqn. (3.61) with eqn. (3.15) shows that the maximum average sheet current $\bar{J}_{\max}$, and hence the maximum loss-free total current $I_{\max}$, in the presence of the *longitudinally* oriented flat paramagnetic shields is always *smaller* than the corresponding quantities for the unshielded strip, virtually complete suppression being achieved if $a \to 0$.

Figure 3.7(a) displays the variation of the normalized sheet current J/J_{b} over the width of the strip for a range of values of the normalized distance between the plane of the strip and the longitudinally oriented flat paramagnetic shields a/w. A general trait due to sheet current redistribution in the strip is the depression of the symmetric current profile around the central part of the strip and the lowering of the total current itself, a phenomenon which gets more pronounced as a/w wanes, reaching saturation with a *zero* minimum in the limiting sheet current profile of eqn. (3.60). Figure 3.7(b) presents the variation of the corresponding normalized maximum total transport current $I_{\max}/I_{\text{b}}$ with the normalized distance between the plane of the strip and the longitudinally oriented flat paramagnetic shields a/w. Starting from zero minimum at the smallest value of a/w, there is a monotonic rise followed by an approach towards unity, the value prevailing for the isolated strip, as a/w continues to grow.

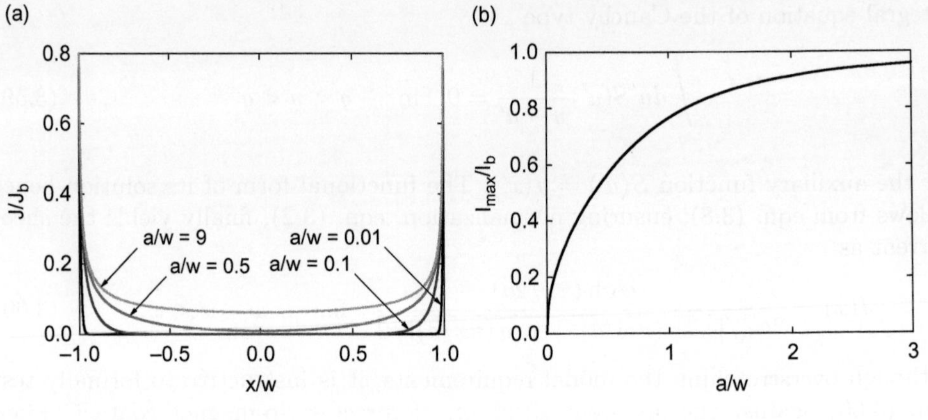

Fig. 3.7 (a) Distribution of the normalized sheet current J/J_b as a function of the normalized coordinate x/w over the width of the magnetically shielded superconductor strip in the flux-free Meissner state for four different values of the normalized distance between the plane of the strip and the paramagnets a/w identified on the curves; (b) variation of the corresponding normalized maximum total transport current $I_\mathrm{max}/I_\mathrm{b}$ with the normalized distance between the plane of the strip and the paramagnets a/w, when $h/2w = 0.001$, referring to the *longitudinal* paramagnet configuration of Fig. 3.6.

We comment that the allied problems of infinite, regular stacks of isolated, thin superconductor strips with an applied magnetic field (Mawatari 1996) or an imposed transport current (Müller 1997) have been addressed before.

3.4.2 Transverse shielding geometry

Let us now inspect the case where the surfaces of the paramagnets run perpendicular to the strip, at the distance c from the centre of the strip, as depicted in Fig. 3.8.

3.4.2.1 Field distribution

Clearly, eqns. (2.87) and (2.88) here again simplify to the Poisson equation for the single nontrivial component of the magnetic vector potential A_z,

$$\Delta A_z = \begin{cases} -\mu_0 J(x)\delta(y) & \text{in } -w \leq x \leq w \\ 0 & \text{otherwise,} \end{cases} \tag{3.62}$$

supplemented with the boundary conditions

$$\lim_{|x|\to c-} A_z(x,y) = \lim_{|x|\to c+} A_z(x,y) \quad \text{and} \quad \left.\frac{1}{\mu_0}\frac{\partial A_z}{\partial n}\right|_{|x|=c-} = \left.\frac{1}{\mu}\frac{\partial A_z}{\partial n}\right|_{|x|=c+} . \tag{3.63}$$

Obviously, the latter demands rule out direct contact between the paramagnetic shields and the surfaces of the strip to avoid complications arising from the proximity effect (Schmidt 1997). The desired solution can again be constructed by the method of images, calling in their current strength $\kappa = (\mu - \mu_0)/(\mu + \mu_0)$, along

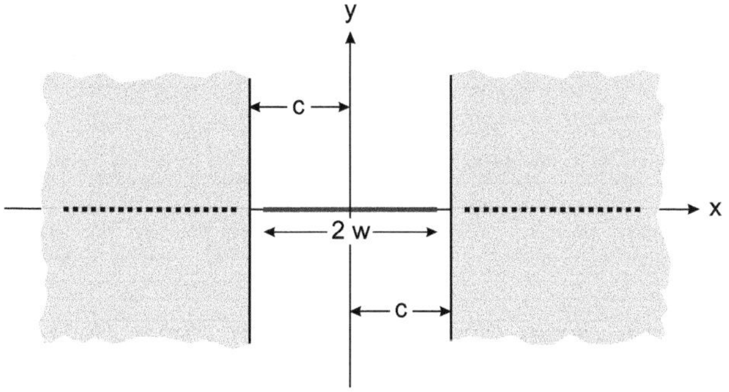

Fig. 3.8 Cross-sectional view of a superconductor strip (dark shading) between two flat semi-infinite paramagnets with transverse orientation (light shading) extending infinitely in the z-direction of a Cartesian coordinate system x, y, z. The definition of the distance between the centre of the strip and the paramagnets c as well as two of the images of the strip (dotted lines) are marked.

with the Green's function technique. Identifying the contributions $dI = dx'J(x')$ of a distributed current element of unit weight at $(x = x', y = 0)$ as well as of image current elements of weight $\kappa^{|n|}$ at $(x = x' + 2nc, y = 0)$ for $n = \pm1, \pm2, \pm3, ...$ and integrating over x' yields the convenient representation in the vacuum space, $0 \leq |x| < c$ inclusive of the surface of the strip ∂D,

$$A_z(x, y) = \frac{\mu_0}{2\pi} \sum_{n=-\infty}^{\infty} \kappa^{|n|} \int_{-w}^{w} dx' J(x') \ln \frac{c}{\sqrt{(x - x' - 2nc)^2 + y^2}}. \tag{3.64}$$

Likewise, identifying the contributions $dI = dx'J(x')$ of image current elements of weight κ^n at $(x = x' + 2nc, y = 0)$ if $x \leq -c$, or $(x = -x' - 2nc, y = 0)$ if $x \geq c$, for $n = 0, 1, 2, ...$ and integrating over x' yields the representation in the paramagnet space, $|x| \geq c$,

$$A_z(x, y) = \frac{\mu}{2\pi}(1 - \kappa) \sum_{n=0}^{\infty} \kappa^n \int_{-w}^{w} dx' J(x') \ln \frac{c}{\sqrt{(|x| + x' + 2nc)^2 + y^2}}. \tag{3.65}$$

Compact expressions can be derived by summing up the partial derivatives with respect to x and y of A_z from eqns. (3.64) and (3.65) in terms of suitable integral forms of Lerch's transcendent Φ (Erdélyi, Magnus, Oberhettinger, and Tricomi 1953; Laurinčikas and Garunkštis 2002). Thus, integrating the total differential of A_z, eqn. (3.64), yields the exact representation in the vacuum space, $0 \leq |x| < c$ inclusive of ∂D,

$$A_z(x,y) = \frac{\mu_0}{2\pi}\int_{-w}^{w} dx' J(x') \ln \frac{c}{\sqrt{(x-x')^2+y^2}}$$

$$+\frac{\mu_0}{8\pi i}\int_{-w}^{w} dx' J(x') \int_C dt \frac{\csc(\pi t)\exp(-i\pi t)}{\kappa^t} \ln\left|\vartheta_+(x-x',y,t)\vartheta_-(x-x',y,t)\right|, \quad (3.66)$$

where

$$\vartheta_\pm(x-x',y,t) = \left[((x-x')/c \pm 2t)^2 + (y/c)^2\right] / (1\pm 2t)^2, \quad (3.67)$$

C denoting the keyhole, or Hankel, contour, i.e. a path of integration that starts at infinity on the real t-axis, circles the origin in the counterclockwise direction, and returns to the start. The first term in eqn. (3.66) emulates the contribution of an isolated superconductor strip, recalling eqn. (3.4), whereas the second term reflects the influence of the paramagnetic shields. Likewise, integrating the total differential of A_z, eqn. (3.65), yields the exact representation in the paramagnet space, $|x| \geq c$,

$$A_z(x,y) = \frac{\mu}{2\pi}(1-\kappa)\int_{-w}^{w} dx' J(x') \int_0^{\infty} \frac{dt}{t} \frac{\exp(-t(|x|+x')/2c)\cos(ty/2c) - \exp(-t/2)}{1-\kappa\exp(-t)}. \quad (3.68)$$

Application of eqn. (2.3) to eqns. (3.66) and (3.68) thus gives the Cartesian components of the magnetic field $\mathbf{H}$ in the vacuum space, $0 \leq |x| < c$ inclusive of ∂D,

$$H_x(x,y) = -\frac{1}{2\pi}\int_{-w}^{w} dx' J(x') \frac{y}{(x-x')^2+y^2}$$

$$+\frac{1}{16\pi i c}\int_{-w}^{+w} dx' J(x') \int_C dt \frac{\csc(\pi t)\exp(-i\pi t)}{\kappa^t} \operatorname{Im}\left[\phi_+(x-x',y,t) - \phi_-(x-x',y,t)\right] \quad (3.69a)$$

$$H_y(x,y) = \frac{1}{2\pi}\int_{-w}^{w} dx' J(x') \frac{x-x'}{(x-x')^2+y^2}$$

$$+\frac{1}{16\pi i c}\int_{-w}^{+w} dx' J(x') \int_C dt \frac{\csc(\pi t)\exp(-i\pi t)}{\kappa^t} \operatorname{Re}\left[\phi_+(x-x',y,t) + \phi_-(x-x',y,t)\right], \quad (3.69b)$$

where

$$\phi_\pm(x-x',y,t) = 1/\left[(x-x'\pm iy)/2c + t\right] + 1/\left[(x-x'\pm iy)/2c - t\right], \quad (3.70)$$

and those of the magnetic induction $\mathbf{B}$ in the paramagnet space, $|x| \geq c$,

$$B_x(x,y) = -\frac{\mu}{4\pi c}(1-\kappa)\int_{-w}^{w} dx' J(x') \int_{0}^{\infty} dt \frac{\exp(-t(|x|+x')/2c)\sin(ty/2c)}{1-\kappa\exp(-t)} \qquad (3.71a)$$

$$B_y(x,y) = \operatorname{sgn}(x)\frac{\mu}{4\pi c}(1-\kappa)\int_{-w}^{w} dx' J(x') \int_{0}^{\infty} dt \frac{\exp(-t(|x|+x')/2c)\cos(ty/2c)}{1-\kappa\exp(-t)}. \qquad (3.71b)$$

A fully analytic treatment is possible when considering the limit of infinite permeability of the paramagnetic shields, $\mu \to \infty$. We therefore attend to the representation of the single component of the magnetic vector potential A_z in the vacuum space, $0 \leq |x| < c$ inclusive of ∂D,

$$\lim_{\mu\to\infty} A_z(x,y) = \frac{\mu_0}{2\pi}\int_{-w}^{w} dx' J(x') \ln \frac{1}{\sqrt{[\cosh(\pi y/c) - \cos(\pi(x-x')/c)]/2}} \qquad (3.72)$$

together with the Cartesian components of the magnetic field $\mathbf{H}$,

$$H_x(x,y) = -\frac{1}{4c}\int_{-w}^{w} dx' J(x') \frac{\sinh(\pi y/c)}{\cosh(\pi y/c) - \cos(\pi(x-x')/c)} \qquad (3.73a)$$

$$H_y(x,y) = \frac{1}{4c}\int_{-w}^{w} dx' J(x') \frac{\sin(\pi(x-x')/c)}{\cosh(\pi y/c) - \cos(\pi(x-x')/c)}, \qquad (3.73b)$$

duplicating a result which can also be directly obtained by recourse to customary table sets (Prudnikov, Brychkov, and Marychev 1986). It is worth noting that in the above limit, A_z and $\partial A_z/\partial n$, whence B_y, suffer infinite jumps along the paramagnet/vacuum interface upon passing into the paramagnet space, $|x| \geq c$, behaviour demonstrating an extreme form of the boundary conditions, eqn. (3.63), whereas B_x remains confined. Both the magnetic field and the magnetic induction are visualized in Fig. 2.11(b) for a large but *finite* permeability μ in conjunction with the sheet current for $\mu \to \infty$, eqn. (3.77), thus disregarding its weak dependence on μ in the high-permeability regime. Obviously, a congruence exists between the magnetic field $\mathbf{H}$ and the velocity field $\mathbf{v}$ of steady-state rotational flow of an incompressible fluid about a strip-like vortex core placed in a transversely oriented flat interspace, whose surfaces act as fluid sources or sinks.

3.4.2.2 Current distribution

As noted already, the Meissner state of the superconductor strip requires the normal component of the magnetic field H_y to vanish across the width of the strip,

$$H_y(x) = 0 \quad \text{for} \quad -w < x < w, y = 0. \qquad (3.74)$$

Recalling eqn. (3.73b) hence furnishes the condition for this state

$$\int_{-w}^{w} dx'\, J(x') \cot\left(\pi(x - x')/2c\right) = 0 \quad \text{in} \; -w < x < w. \tag{3.75}$$

The change of the variables x and x' to $v = \tan(\pi x/2c)$ and $v' = \tan(\pi x'/2c)$, respectively, together with the introduction of the substitution $s = \tan(\pi w/2c)$ and the request for an even solution in x, transform eqn. (3.75) into the singular integral equation of the Cauchy type

$$\int_{-s}^{s} dv'\, S(v') \frac{1}{v - v'} = 0 \quad \text{in} \; -s < v < s \tag{3.76}$$

for the auxiliary function $S(u') = J(x')$. The functional form of its solution hence follows from eqn. (3.8); ensuring normalization, eqn. (3.2), finally yields the sheet current as

$$J(x) = \frac{I}{2c} \frac{\cos(\pi x/2c)}{\sqrt{[\cos(\pi x/c) - \cos(\pi w/c)]/2}} \quad \text{for} \; -w < x < w. \tag{3.77}$$

Once more, though overstretching the model requirements, eqn. (3.77) may be formally tested against the apparent situation of direct strip–magnet contact, when $c \to w$. Here, the sheet current and its images make up an effective current-carrying 'superconductor sheet' which occupies the plane $y = 0$. The direction of the magnetic field is parallel to the surfaces of this sheet, consistent with a uniform sheet current distribution in the strip, $J(x) = I/2w$ for $-w < x < w$, as demanded by the Meissner state.

Reasoning in a similar vein to that for the isolated superconductor strip in the barrier-protected Meissner state of section 3.1 limits the validity of eqn. (3.77) to the range $-w + h \le x \le w - h$ and simultaneously recommends imposing the maximum value $J_{\text{max}} = J_{\text{b}}$ in $|x \pm w| \le h$. Substituting $I_{\text{max}} = 2w\bar{J}_{\text{max}}$ for I in eqn. (3.77) with eqn. (3.14) therefore yields the maximum average sheet current $\bar{J}_{\text{max}}$, which may be approximated by the highly accurate simple form

$$\bar{J}_{\text{max}} \simeq \pi H_{c1} \sqrt{\frac{2h}{w}} \sqrt{\frac{\tan(\pi w/2c)}{\pi w/2c}}; \quad h \ll w < c - h. \tag{3.78}$$

Comparing eqn. (3.78) with eqn. (3.15) shows that the maximum average sheet current $\bar{J}_{\text{max}}$, and hence the maximum loss-free total current I_{max}, in the presence of the *transversely* oriented flat paramagnetic shields is always *larger* than the respective quantities for the unshielded strip, the strongest current enhancement $I_{\text{max}}/I_{\text{b}} \simeq (2/\pi)\sqrt{w/h}$ being achieved if $c \to w$.

Figure 3.9(a) displays the variation of the normalized sheet current J/J_{b} over the width of the strip for a range of values of the normalized distance between the

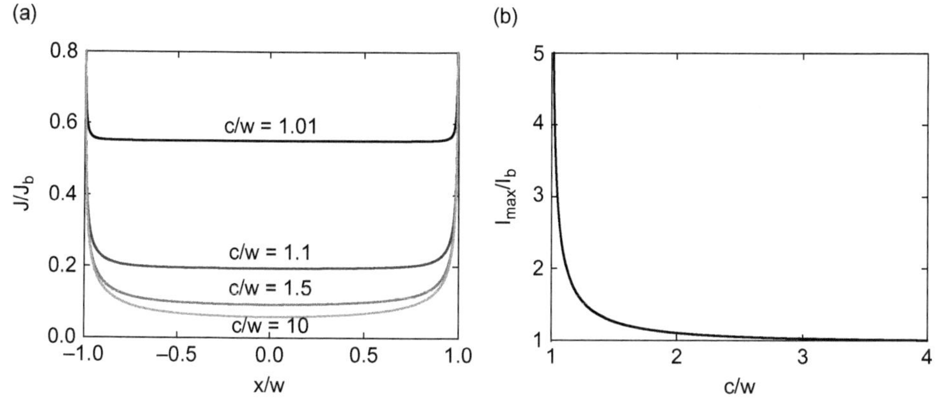

Fig. 3.9 (a) Distribution of the normalized sheet current J/J_b as a function of the normalized coordinate x/w over the width of the magnetically shielded superconductor strip in the flux-free Meissner state for four different values of the normalized distance between the centre of the strip and the paramagnets c/w identified on the curves; (b) variation of the corresponding normalized maximum total transport current $I_\mathrm{max}/I_\mathrm{b}$ with the normalized distance between the centre of the strip and the paramagnets c/w, when $h/2w = 0.001$, referring to the *transverse* paramagnet configuration of Fig. 3.8.

centre of the strip and the transversely oriented flat paramagnetic shields c/w. A general trait due to sheet current redistribution in the strip is the steady evening out of the symmetric current profile around the central part of the strip and the increase of the total current itself, a phenomenon which gets more pronounced as c/w wanes, reaching saturation with a *uniform* maximum in the limiting sheet current profile of eqn. (3.77). Figure 3.9(b) presents the variation of the corresponding normalized maximum total transport current $I_\mathrm{max}/I_\mathrm{b}$ with the normalized distance between the centre of the strip and the transversely oriented flat paramagnetic shields c/w. Starting from a distinct peak at the smallest value of c/w, there is a monotonic drop followed by an approach towards unity, the value prevailing for the isolated strip, as c/w continues to grow.

We comment that the allied problems of infinite, regular arrays of isolated, thin superconductor strips with an applied magnetic field (Mawatari 1996) or an imposed transport current (Mûller 1997) have been addressed before.

3.4.3 Possible experimental confirmation

The enhancement of the maximum total transport current I_max above the barrier-controlled total current I_b of an isolated superconductor strip predicted in connection with the transverse shielding geometry of Fig. 3.8 offers an important starting point for probing the assumptions of the edge-barrier model versus experiment. The aim here is to ensure that the mechanism bringing about loss-free current transport is limited by the entry of magnetic vortices from the edges, rather than from the surfaces, of the strip. Thin single-crystal films of superconductors like $Bi_2Sr_2Ca_2Cu_3O_{10}$ (Indenbom et al. 1994) or $YBa_2Cu_3O_7$ (Gardner et al. 1998) seem to typify preferable materials constituents regarding such a test. Alternatively, thin type-II superconductor strips with irradiation-enhanced edge pinning of magnetic vortices (Schuster et al. 1994) could be studied too.

Let us concentrate on an epitaxial $YBa_2Cu_3O_7$ film or strip, with its crystallographic *a*–*b*-planes parallel to the surfaces of the strip. The geometrical dimensions shall be $w = 2.5\,mm$ and $d = 50\,\text{nm}$; the material characteristics in respect of these planes shall be $\lambda_L = 0.5\,\mu m$ (whence $\lambda_\perp = 5\,\mu m$) and $\xi_{GL} = 3.6\,\text{nm}$ at the liquid nitrogen temperature $T = 77$ K (Fiory, Hebard, Mankievich, and Howard 1988; Kunchur, Christen, and Phillips 1993). Equation (3.12) then yields the value $H_{c1} \simeq 2.6 \times 10^3\,A/m$ for the lower critical field governing the first entry of magnetic vortices from the *edges* of the strip, whereas eqn. (3.14) suggests the value $J_b/d \simeq 1.0 \times 10^{11}\,A/m^2$ for the critical current density at the edges of the strip, which itself is close to the depairing critical current density $j_{GL} \simeq 1.1 \times 10^{11}\,A/m^2$ from eqn. (2.31). If, as an example, we consider a gap $a = c - w$ of $0.2\,mm$ between the edges of the strip and the surfaces of the magnetic shields, the maximum average sheet current density, from eqn. (3.78), becomes $\bar{J}_{max}/d \simeq 2.5 \times 10^{10}\,A/m^2$, possibly exceeding the bulk pinning critical current density representative of wide $YBa_2Cu_3O_7$ films.

High Meissner currents sustained by the shielding arrangement under consideration might give rise to the penetration of magnetic vortices from the *surfaces* of the strip; a phenomenon which could severely restrict the loss-free current transport in anisotropic single crystals possessing low critical fields H_{c1}^* for the first entry of horizontal magnetic vortices. Such an assertion, however, does not apply to thin strips having $d \ll \lambda_L$, where

$$H_{c1}^* \simeq \frac{2\Phi_0}{\pi\mu_0\gamma_a d^2} \ln\left(\frac{\gamma_a d}{\xi_{GL}}\right); \quad \xi_{GL} \ll d \ll \lambda_L, \tag{3.79}$$

with the anisotropy parameter γ_a relating the values of λ_L, or ξ_{GL}, parallel or orthogonal, respectively, to the *a*–*b*-planes of the strip (Abrikosov 1988; Blatter, Geshkenbein, and Larkin 1992; Brandt 1995b). Look at the extreme situation of direct strip–magnet contact, when a uniform sheet current attributed to the depairing critical current density j_{GL} permeates the strip. This would produce a magnetic field on the surfaces of the strip smaller than H_{c1}^*, as long as $d\,j_{GL}/2 \le H_{c1}^*$ or, consulting eqns. (2.31) and (3.79),

$$\gamma_a d^3 \le 12\sqrt{3}\,\lambda_L^2 \xi_{GL} \ln\left(\frac{\gamma_a d}{\xi_{GL}}\right). \tag{3.80}$$

The value $\gamma_a \simeq 6$ for $YBa_2Cu_3O_7$ single crystals hence furnishes $d_e \simeq 260\,\text{nm}$ as an upper bound on the thickness of the strip, along with $H_{c1}^* \simeq 1.5 \times 10^4\,A/m$, the corresponding lower bound on the critical field for the entry of magnetic vortices from the surfaces of the strip. It should be noted, however, for overall consistency with the above premise on the thickness of the strip, that in actual fact d should be well below d_e. Reverting to $d = 50$ nm thus renders the even higher value $H^*_{c1} \simeq 3.1 \times 10^5\ A/m$ as the corresponding lower bound on the critical field for the penetration of horizontal magnetic vortices into the strip.

These numerical estimates clearly demonstrate the hierarchy $H_{c1} \ll H^*_{c1}$, in the case of thin $YBa_2Cu_3O_7$ single crystals at least, and therefore support the assumption of magnetic vortices entering from the edges of the strip.

3.5 Superconductor strip inside a cylindrical paramagnetic cavity

We go ahead to develop the considerations of section 3.1, looking at an infinitely extended, thin current-carrying type-II superconductor strip in the Meissner state, now placed inside a cylindrical paramagnetic cavity with radius a and permeability μ, as illustrated in Fig. 3.10. This represents a sophisticated example of the conditioning of magnetic fields by a two-sided closed magnetic shield; its subsequent analysis partly adheres to the works of Genenko, Snezhko, and Freyhardt (2000) as well as Genenko and Rauh (2006).

3.5.1 Field distribution

Here, eqns. (2.87) and (2.88) simiplify to the Poisson equation for the single nontrivial component of the magnetic vector potential A_z,

$$\Delta A_z = \begin{cases} -\mu_0 J(x)\delta(y) & \text{in } -w \le x \le w \\ 0 & \text{otherwise,} \end{cases} \tag{3.81}$$

supplemented with the boundary conditions

$$\lim_{r \to a-} A_z(x,y) = \lim_{r \to a+} A_z(x,y) \quad \text{and} \quad \frac{1}{\mu_0}\frac{\partial A_z}{\partial n}\bigg|_{r=a-} = \frac{1}{\mu}\frac{\partial A_z}{\partial n}\bigg|_{r=a+}, \tag{3.82}$$

introducing cylindrical coordinates r, ϕ, z adapted to the paramagnetic cavity. The latter demands exclude direct contact between the cavity and the edges of the strip to avoid complications arising from the proximity effect (Schmidt 1997). The desired solution can again be constructed by the method of images, calling in their associated current strength $\kappa = (\mu - \mu_0)/(\mu + \mu_0)$, together with the Green's function technique. Identifying the contributions $dI = dx' J(x')$ of a distributed current element of unit weight at $(x = x', y = 0)$, the point marked M in Fig. 3.10, as well as of an image current element of weight κ at $(x = a^2/x', y = 0)$, the point marked N in Fig. 3.10, and integrating over x' yields the representation in the vacuum domain,

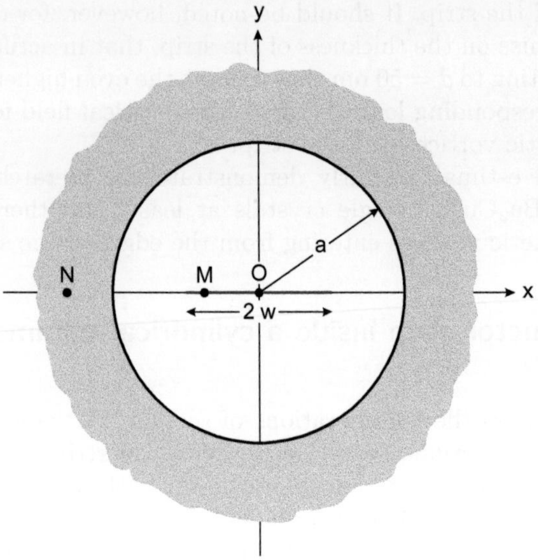

Fig. 3.10 Cross-sectional view of a superconductor strip (dark shading) inside a cylindrical paramagnetic cavity (light shading) extending infinitely in the z-direction of a Cartesian coordinate system x, y, z. The definition of the radius of the cavity a as well as a current element of the strip (M) and two images (N, O) are marked.

$0 \leq r < a$ inclusive of the surface of the strip ∂D,

$$A_z(x, y) = \frac{\mu_0}{2\pi} \int_{-w}^{w} dx' J(x') \left\{ \ln \frac{a}{\sqrt{(x-x')^2 + y^2}} + \kappa \ln \frac{a}{\sqrt{(x - a^2/x')^2 + y^2}} \right\}. \tag{3.83}$$

Likewise, identifying the contributions $dI = dx' J(x')$ of a distributed current element of assigned weight $1 - \kappa$ at $(x = x', y = 0)$, the point marked M in Fig. 3.10, as well as of an image current element of assigned weight κ at $(x = 0, y = 0)$, the point marked O in Fig. 3.10, and integrating over x' yields the representation in the paramagnet space, $r \geq a$,

$$A_z(x, y) = \frac{\mu}{2\pi} \int_{-w}^{w} dx' J(x') \left\{ (1-\kappa) \ln \frac{b}{\sqrt{(x-x')^2 + y^2}} + \kappa \ln \frac{b}{\sqrt{x^2 + y^2}} \right\} \tag{3.84}$$

with b determined from

$$I \ln(b/a) = (\mu_0/\mu)\, \kappa \int_{-w}^{w} dx' J(x') \ln |x'/a|, \tag{3.85}$$

upon observing the normalization, eqn. (3.2).

A fully analytic treatment is possible in the limit of infinite permeability of the paramagnetic shield, $\mu \to \infty$. We therefore turn to considering the representations in the vacuum domain, $0 \leq r < a$ inclusive of ∂D,

$$\lim_{\mu\to\infty} \frac{1}{\mu_0} A_z(x,y) = \frac{1}{2\pi} \int_{-w}^{w} dx' J(x') \left\{ \ln \frac{a}{\sqrt{(x-x')^2+y^2}} + \ln \frac{a}{\sqrt{(x-a^2/x')^2+y^2}} \right\} \tag{3.86}$$

and in the paramagnet space, $r \geq a$,

$$\lim_{\mu\to\infty} \frac{1}{\mu} A_z(x,y) = \frac{I}{2\pi} \ln \frac{a}{\sqrt{x^2+y^2}} \tag{3.87}$$

instead, revealing a complete separation of the component of the magnetic vector potential A_z in both domains. Application of eqn. (2.3) to eqns. (3.86) and (3.87) thus gives the Cartesian components of the magnetic field $\mathbf{H}$ in the vacuum domain, $0 \leq r < a$ inclusive of ∂D,

$$H_x(x,y) = -\frac{1}{2\pi} \int_{-w}^{w} dx' J(x') \left[\frac{y}{(x-x')^2+y^2} + \frac{y}{(x-a^2/x')^2+y^2} \right] \tag{3.88a}$$

$$H_y(x,y) = \frac{1}{2\pi} \int_{-w}^{w} dx' J(x') \left[\frac{x-x'}{(x-x')^2+y^2} + \frac{x-a^2/x'}{(x-a^2/x')^2+y^2} \right] \tag{3.88b}$$

and in the paramagnet space, $r \geq a$,

$$H_x(x,y) = -\frac{I}{2\pi} \left(\frac{y}{x^2+y^2} \right) \tag{3.89a}$$

$$H_y(x,y) = \frac{I}{2\pi} \left(\frac{x}{x^2+y^2} \right) \tag{3.89b}$$

which, for the sheet current, eqn. (3.98), are reflected in Fig. 2.12. Apparently, a reciprocity exists between the magnetic field $\mathbf{H}$ and the electric field $\mathbf{E}$ around an electrically charged thin metal strip placed inside the cylindrical cavity of a metallic conductor of infinite extent, i.e. an electrical transmission line.

3.5.2 Current distribution

The Meissner state of the superconductor strip implies the absence of magnetic flux inside, and thus the vanishing of the normal component of the magnetic field H_y across the width of the strip,

$$H_y(x) = 0 \quad \text{for} \quad -w < x < w, y = 0. \tag{3.90}$$

Recalling eqn. (3.88b) hence furnishes the condition for this state

$$\int_{-w}^{w} dx' J(x') Q(x, x') = 0 \quad \text{in} \quad -w < x < w, \tag{3.91}$$

i.e. a homogeneous Fredholm integral equation with the singular two-part kernel

$$Q(x, x') = \frac{1}{x - x'} + \frac{1}{x - a^2/x'}. \tag{3.92}$$

The expedient change of the variables x to $u = x(a^2 + w^2)/(a^2 + x^2)$ and x' to $u' = x'(a^2 + w^2)/(a^2 + x'^2)$ transforms eqn. (3.91) into the Cauchy-like integral equation

$$\int_{-w}^{w} du' S(u') \frac{1}{u - u'} = -P(u) \quad \text{in} \quad -w < u < w \tag{3.93}$$

for the auxiliary function $S(u') = J(x')/(du'/dx')$ subject to the normalization

$$\int_{-w}^{w} du' S(u') = I \tag{3.94}$$

in line with eqn. (3.2), introducing the singular two-part inhomogeneity

$$P(u) = \frac{I}{u} \left(\frac{1}{\sqrt{1 - mu^2/w^2}} - 1 \right) \tag{3.95}$$

with the parameter $m = \left(2aw/(a^2 + w^2)\right)^2$. The general solution of eqn. (3.93) can be expressed as

$$S(u) = \frac{1}{\pi^2} \int_{-w}^{w} du' \sqrt{\frac{w^2 - u'^2}{w^2 - u^2}}\, P(u') \frac{1}{u - u'} + \frac{N}{\pi\sqrt{w^2 - u^2}} \quad \text{in} \quad -w < u < w, \tag{3.96}$$

the term labelled with a free constant N on the right-hand side of eqn. (3.96) representing a solution of the homogeneous equation to eqn. (3.93) (Mikhlin 1964). Evaluation of the integral in eqn. (3.96) with eqn. (3.95) yields

$$S(u) = \frac{2I}{\pi^2} \frac{1}{\sqrt{w^2 - u^2}} \left[K(m) + \left(\frac{w^2}{u^2} - 1 \right) \Pi \left(\frac{w^2}{u^2} \middle| m \right) \right] + \frac{N - I}{\pi\sqrt{w^2 - u^2}}$$
$$\text{in} \quad -w < u < w, \tag{3.97}$$

where K and Π denote complete elliptic integrals of the first and third kinds, respectively (Abramowitz and Stegun 1972). Since in the limit of the radius of the paramagnetic cavity tending to infinity, $a \to \infty$, the sheet current J under consideration, and hence the auxiliary function S given by eqn. (3.97), must recover the solution appropriate for the isolated superconductor strip, eqn. (3.8), the choice $N = I$ is evident. Thus, when reverting to the original quantities J and x, use of eqn. (3.97) delineates the sheet current in the form

$$J(x) = \frac{2I}{\pi^2}\left(\frac{a^2 - x^2}{a^2 + x^2}\right)\frac{a^2 + w^2}{\sqrt{q(x)^2 - p(x)^2}}\left[K(m) + \left(\frac{q(x)^2}{p(x)^2} - 1\right)\Pi\left(\frac{q(x)^2}{p(x)^2}\middle| m\right)\right]$$

$$\text{for } -w < x < w, \tag{3.98}$$

where

$$p(x) = x\left(a^2 + w^2\right) \quad \text{and} \quad q(x) = w\left(a^2 + x^2\right). \tag{3.99}$$

Again reasoning similar to that for the isolated superconductor strip in the barrier-protected Meissner state of section 3.1 limits the validity of eqn. (3.98) to the range $-w+h \le x \le w-h$ and simultaneously recommends imposing the maximum value $J_{\mathrm{max}} = J_{\mathrm{b}}$ in $|x \pm w| \le h$.

Figure 3.11(a) illustrates the variation of the normalized sheet current J/J_{b} over the width of the strip for a range of values of the normalized radius of the cavity a/w. A general trait due to sheet current redistribution in the strip is the steady *warping* of the symmetric current profile around the central part of the strip and the increase in the total current itself, a phenomenon which gets more pronounced as a/w wanes. Figure 3.11(b) presents the variation of the corresponding maximum total transport current $I_{\mathrm{max}}/I_{\mathrm{b}}$ with the normalized radius of the cavity a/w. Starting from a distinct peak at the smallest value of a/w, there is a sharp monotonic drop followed by a gradual approach towards unity, the value prevailing for the isolated strip, as a/w continues to grow.

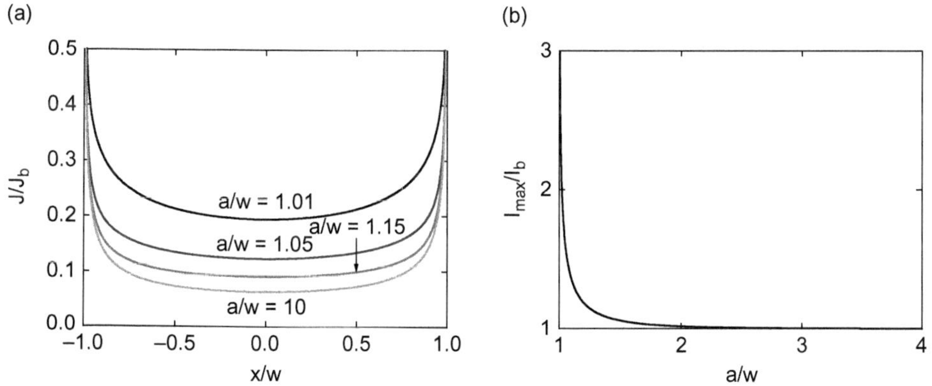

Fig. 3.11 (a) Distribution of the normalized sheet current J/J_{b} as a function of the normalized coordinate x/w over the width of the magnetically shielded superconductor strip in the flux-free Meissner state for four different values of the normalized radius of the cavity a/w identified on the curves. (b) Variation of the corresponding normalized maximum total transport current $I_{\mathrm{max}}/I_{\mathrm{b}}$ with the normalized radius of the cavity a/w, when $h/2w = 0.001$, referring to the cylindrical paramagnet configuration of Fig. 3.10.

The latter behaviour testifies to two antagonistic effects at work in the closed geometry of the cylindrical magnetic shield: elements of the cavity near the edges of the strip, on the one hand, favour an *enhancement* of both the sheet current and the maximum total transport current, like the flat semi-infinite paramagnets in the transverse shielding configuration of Fig. 3.8; elements of the cavity around the centre the strip, on the other hand, reinforce a *depression* of both the sheet current and the maximum total transport current, like the flat semi-infinite paramagnets in the longitudinal shielding configuration of Fig. 3.6. It is the interplay of both traits that brings about the unveiled results. These amount to the flat semi-infinite paramagnets in the transverse shielding configuration of Fig. 3.8 rendering the sheet current profile $J/J_{\rm b}$ more *homogeneous* and increasing the maximum total transport current $I_{\rm max}/I_{\rm b}$ to a large extent, whereas the cylindrical magnetic shield of Fig. 3.10 makes the sheet current profile $J/J_{\rm b}$ more *inhomogeneous* and augments the maximum total transport current $I_{\rm max}/I_{\rm b}$ to a lesser extent.

3.6 Improved shaping of the paramagnetic environment around a superconductor strip

The results of the preceding sections obtained for a thin current-carrying type-II superconductor strip in the Meissner state surrounded by various paramagnetic environments impart at least two desirable traits: (i) a suppression of the peaks of the sheet current J near the edge of the strip facing the magnetic shield, which is particularly effective in the case of the one-sided shielding of a wedge-shaped magnetic environment, Fig. 3.4; and (ii) an increase in the maximum total transport current $I_{\rm max}$ above the barrier-controlled total current $I_{\rm b}$ of the isolated strip in the case of the two-sided shielding of transversely oriented flat magnetic environments, Fig. 3.8. These findings suggest pondering whether a still further increase in $I_{\rm max}$ can be achieved by uniting the assets of both types of environments, possibly with even more sophisticated shapes. Developing a positive answer in the end, we mainly build on the work of Genenko, Snezhko, and Freyhardt (2000).

3.6.1 Conformal mapping technique

Let us start by resorting to the conformal mapping technique discussed previously (cf. section 2.3.4.1), a tool whose importance for scientists and engineers derives from its benefit to the solution of planar problems in electrostatics, magnetostatics, hydrodynamics, and other fields.

3.6.1.1 Introductory remarks

A conformal transformation distinguishes itself by the following three properties: firstly, continuous lines map into continuous lines; secondly, the angle between two intersecting curves is preserved; and thirdly, a solution of the two-dimensional Laplace equation remains a solution of the two-dimensional Laplace equation which, if a complete set of boundary conditions is fulfilled, turns out to be unique (Tikhonov and Samarskii 1967). Referring, e.g., to the electrostatics of metallic conductors, this means that lines of the electric field **E** (which always intersect such a conductor's surface rectangularly) will remain orthogonal to the mapped surface after a conformal transformation has been applied. The latter statement, in its proper magnetostatic reinterpretation, suggests that the lines of the magnetic field **H** (which intersect the surface of a paramagnet of infinite permeability, $\mu \to \infty$, rectangularly) will remain orthogonal to the mapped surface after a conformal transformation has been applied. In that limit, therefore, the technique of conformal mapping can also be used for calculating the distributions of the magnetic field as well as the electrical current in magnetostatic analogues.

The angles γ and γ_0 which the lines of the magnetic field **H** make with the normal to the boundary surface between a paramagnet of arbitrary permeability μ and a vacuum are connected by the law of refraction

$$\tan\gamma/\tan\gamma_0 = \mu/\mu_0 \tag{3.100}$$

(Landau, Lifshitz, and Pitaevskii 1995). If $\mu \to \infty$, eqn. (3.100) may be satisfied by two alternatives: (i) $\gamma_0 \to 0$ while $0 < \gamma < \pi/2$, supporting the electrostatic analogue, with the pattern of the electric field **E** near a conductor's surface in a vacuum; and (ii) $\gamma \to \pi/2$ while $0 < \gamma_0 < \pi/2$, resulting in another pattern of the magnetic field **H**. Both alternatives can be realized for different geometries; however, merely one option applies to a specific case since the underlying solution of Laplace's equation is unique.

One can easily prove that option (i) defies the situation of a transport current I flowing through a superconductor strip placed inside a *closed* paramagnetic cavity with permeability $\mu \to \infty$, the example of section 3.5. Indeed, if the lines of the magnetic field **H** were perpendicular to the paramagnet/vacuum interface ∂S of the cylindrical cavity throughout, the contour integral about ∂S would yield

$$\oint_{\partial S} \mathrm{d}\mathbf{l}\cdot\mathbf{H} = 0, \tag{3.101}$$

rather than amount to I, as Ampère's law, eqn. (2.49), demands. It is apparent, for the same reason too, that the condition $\partial A_z/\partial n = 0$ on the single nontrivial component of the magnetic vector potential A_z, which governs the problems considered here, cannot be imposed at the internal boundary ∂S of any closed paramagnetic cavity, even if $\mu \to \infty$. Clearly in this instance, option (ii) holds true instead, and the lines of the magnetic field **H** run along the contour ∂S, as shown in Fig. 2.12. This is why we confine ourselves to *open* paramagnetic configurations in the limit $\mu \to \infty$, like the examples of magnetic extremes of sections 3.2 to 3.4, where option (i) applies (see Fig. 2.10). One can easily see that the contour integral about any path ∂S encompassing the strip then is

$$\oint_{\partial S} \mathrm{d}\mathbf{l}\cdot\mathbf{H} = I, \tag{3.102}$$

in accordance with Stokes' theorem and Ampère's law, eqn. (2.49).

3.6.1.2 *Mathematical implementation*

A conformal transformation effects a change of coordinates from the complex variable $s = x + iy$ to the complex variable $\sigma = \xi + i\eta$ by means of some analytic function $\sigma = f(s)$ with a nonzero derivative df/ds. This maps areas of the s-plane onto areas of the σ-plane, implying $x = x(\xi, \eta)$ and $y = y(\xi, \eta)$. The properties of a conformal map duly interpreted in terms of A_z admit the following two statements to be made: firstly, given that the component of the original magnetic vector potential, $A_z(x, y)$, satisfies the Laplace equation $\Delta A_z = 0$ around the superconductor strip with the requirement of continuity of A_z and the condition $\partial A_z/\partial n = 0$ at the original paramagnet/vacuum interface ∂S, the transformed component of the magnetic vector potential, $\hat{A}_z(\xi, \eta)$, defined by

$$\hat{A}_z(\xi, \eta) = A_z[x(\xi, \eta), y(\xi, \eta)], \tag{3.103}$$

satisfies the Laplace equation $\Delta \hat{A}_z = 0$ around the superconductor strip with the requirement of continuity of $\hat{A}_z$ and the condition $\partial \hat{A}_z/\partial n = 0$ at the transformed paramagnet/vacuum interface $\partial \hat{S}$; and secondly, because of the preservation of the angle between two intersecting curves, lines of the magnetic field $\mathbf{H}$ parallel to the superconductor strip change into lines of the transformed magnetic field $\hat{\mathbf{H}}$ parallel to the superconductor strip. The latter fact notably implies that conformal mapping also preserves the Meissner state. Thus, any conformal transformation which maintains the width and the position of the superconductor strip yields another permissible distribution of the sheet current for the strip in this state, corresponding to the geometry of the transformed paramagnetic environment. While the components of the transformed magnetic field $\hat{\mathbf{H}}$ in the vacuum domain inclusive of ∂D emerge from

$$\mu_0 \hat{H}_\xi(\xi, \eta) = \frac{\partial \hat{A}_z}{\partial \eta} \quad \text{and} \quad \mu_0 \hat{H}_\eta(\xi, \eta) = -\frac{\partial \hat{A}_z}{\partial \xi}, \tag{3.104}$$

observing eqn. (2.3), the distribution of the new sheet current $\hat{J}$, which is related to the discontinuity of the tangential component of the transformed magnetic field $\hat{H}_\xi$ across the thickness of the strip, comes out as

$$\hat{J}(\xi) = J[x(\xi, \eta)] \left. \frac{\partial y}{\partial \eta} \right|_{\eta=0}, \tag{3.105}$$

making use of the particulars defining the map. Suffice it to add that $\hat{J}$ preserves normalization, eqn. (3.2), with ξ' in place of x'.

3.6.2 Elementary realizations

To demonstrate the application of the conformal mapping technique for calculating the distribution of the sheet current due to an improved shape of a paramagnetic environment of infinite permeability, $\mu \to \infty$, we proceed in two steps, commencing with basic one-sided shielding geometries, which primarily are of theoretical use, before continuing in section 3.7 with intricate two-sided shielding geometries, which should serve experimental interests too.

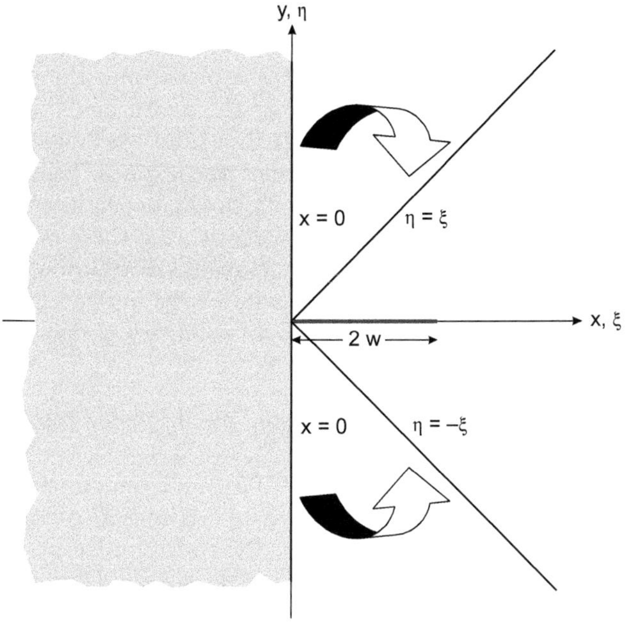

Fig. 3.12 Schematic view of the conformal mapping of the left half-domain $x \leq 0$ of the s-plane, occupied by a flat semi-infinite paramagnet (light shading), onto the three-quarter domain $|\eta| \geq \xi$ of the σ-plane, occupied by a wedge-shaped paramagnetic cavity of opening half-angle $\beta = \pi/4$, with a superconductor strip (dark shading) placed in front. The width and the position of the strip are fixed.

We first recall the superconductor strip placed adjacent to a flat semi-infinite paramagnet or inside a wedge-shaped paramagnetic cavity of opening half-angle $\beta = \pi/4$, problems dealt with in sections 3.2 and 3.3 which here, however, are understood as conformal mapping tasks. Simpler than Figs. 3.2 and 3.4 suggest, we thereby assume direct contact between the strip and the respective magnetic environment, letting $c \to w$. Figure 3.12 illustrates the transformation to be addressed. Mathematically, mapping of the left half-domain $x \leq 0$ of the s-plane onto the three-quarter domain $|\eta| \geq \xi$ of the σ-plane, while preserving the width and the position of the strip, is described by the analytic function $f(s) = (2ws)^{1/2}$, save for a cut on the real axis, $x < 0, y = 0$. This generates the explicit transformation relationship $x + iy = (\xi + i\eta)^2/2w$ from which the dependence $x(\xi, \eta)$ and the derivative $\partial y/\partial \eta$ at $\eta = 0$ can be obtained. Substituting these quantities into the right-hand side of eqn. (3.105) for the sheet current distribution in the Meissner state due to the flat magnetic environment, eqn. (3.28), yields the sheet current distribution in that state appropriate to the rectangular wedge-shaped magnetic cavity,

$$\hat{J}(\xi) = \frac{2I}{\pi w} \frac{\xi/2w}{\sqrt{1 - (\xi/2w)^4}} \quad \text{for } 0 < \xi < 2w, \tag{3.106}$$

in formal agreement with eqn. (3.43).

The presentation above can easily be extended to the case of a wedge-shaped paramagnetic cavity with an opening half-angle $\beta = p\pi/2$, where $0 < p \leq 1$. Then, mapping of the left half-domain $x \leq 0$ of the s-plane onto the sectorial domain $|\eta| \geq \xi \tan\beta$ of the σ-plane, while maintaining the width and the position of the strip, is procured by the analytic function $f(s) = (2w)^{1-p}s^p$, again save for a cut on the real axis, $x < 0, y = 0$. This furnishes the explicit transformation relationship $x + iy = (\xi + i\eta)^{1/p}(2w)^{1-1/p}$ from which the dependence $x(\xi, \eta)$ and the derivative $\partial y/\partial\eta$ at $\eta = 0$ can be obtained. Substituting these quantities into the right-hand side of eqn. (3.105) for the sheet current distribution in the Meissner state due to the flat magnetic environment, eqn. (3.28), now yields the sheet current distribution in that state appropriate to an arbitrary wedge-shaped magnetic cavity,

$$\hat{J}(\xi) = \frac{I}{p\pi w}\frac{(\xi/2w)^{1/p-1}}{\sqrt{1-(\xi/2w)^{2/p}}} \quad \text{for } 0 < \xi < 2w, \tag{3.107}$$

generalizing the result for $p = 1/2$, eqn. (3.106), and reproducing the result for $p = 1$, eqn. (3.28). Accordingly, sheet current redistribution causes a suppression of the current in a region from the left edge of the strip adjacent to the magnetic shield, which gets broader and more pronounced as p, and hence β, abates. We comment that, although preferable in view of this physical effect, the fabrication of a wedge-shaped magnetic cavity with a sharp opening half-angle, $\beta \ll \pi/2$, certainly poses a challenge to the experimentalist.

3.7 Superconductor strip between two curved semi-infinite paramagnets

We now return to the superconductor strip placed between two transversely oriented flat semi-infinite paramagnetic shields, the problem dealt with in section 3.4.2, extending it to curved semi-infinite paramagnetic shields by means of the conformal mapping technique. As Fig. 3.8 suggests, we thereby allow the strip to be detached from the magnetic environment, assuming $c > w$. The following presentation again draws upon the work of Genenko, Snezhko, and Freyhardt (2000).

3.7.1 Conformal transformation

Let us map the flat half-domains $|x| \geq c$ of the s-plane under consideration onto some cusp-shaped half-domains of the σ-plane, with opening half-angle of their vertices $\beta = p\pi/2$, where $0 < p \leq 1$, by employing the function

$$f_w(s) = c_w\left[(1+s/c)^p - (1-s/c)^p\right] \quad \text{with } c_w = w/\left[(1+w/c)^p - (1-w/c)^p\right]. \tag{3.108}$$

This is analytic in the whole s-plane, apart from two cuts on the real axis, $x < -c, y = 0$ and $x > c, y = 0$. Application of eqn. (3.108) transforms the points $(x = \pm c, y = 0)$ of the s-plane into the points $(\xi = \pm f_w(c), \eta = 0)$ of the σ-plane, while keeping the width and the position of the strip fixed. The initial gap between the edges of the strip and the surfaces of the magnetic shields, $a = c - w$, thus changes to the distance between the edges of the strip and the vertices of the magnetic shields, $a_w = f_w(c) - w$. Evidently, the latter distance a_w and the opening half-angle β measured via p represent the geometrical characteristics of the cusp-shaped magnetic cavity; they can be installed so as to control the distribution of the sheet current in the strip by a judicious choice of the parameters w, c, and p.

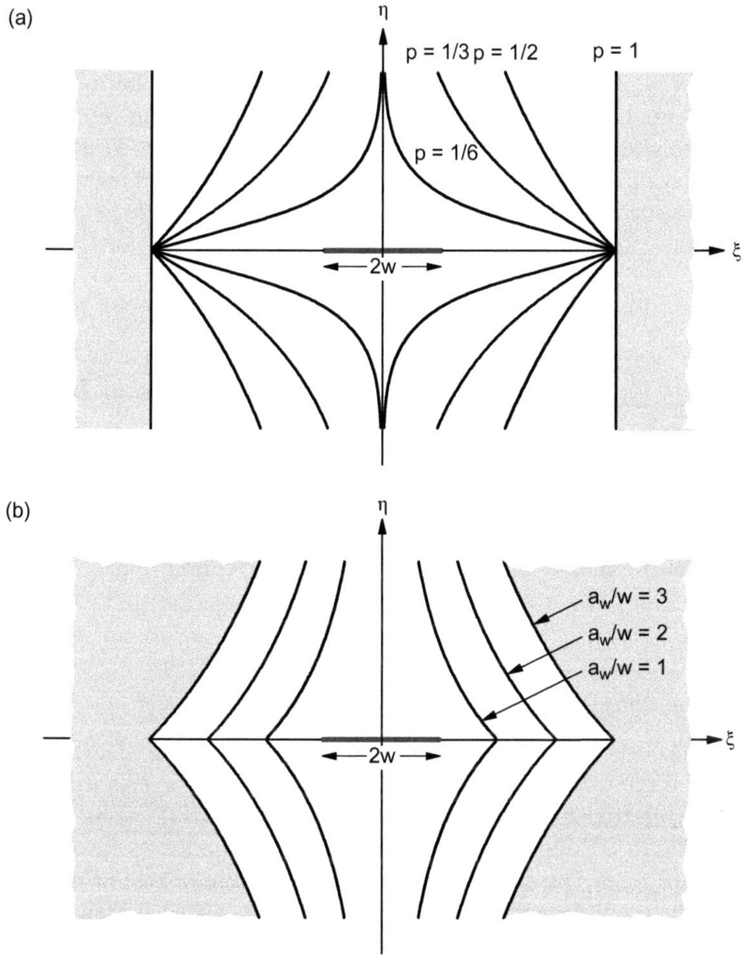

Fig. 3.13 Cross-sectional view of the boundaries of the semi-spaces in the σ-plane, occupied by a cusp-shaped paramagnetic cavity with a superconductor strip placed inside, for (a) the normalized distance between the edges of the strip and the vertices of the magnetic shields $a/w = 3$ as well as four different values of the parameter of the conformal map p identified on the curves, and for (b) the parameter of the conformal map $p = 1/2$ as well as three different values of the normalized distance between the edges of the strip and the vertices of the magnetic shields a/w identified on the curves. The width and the position of the strip are fixed.

The variation of the shape of the magnetic cavity with a_w and p is revealed in Fig. 3.13. Accordingly, the cavity consists of two separate parts, which are elongated in the direction normal to the plane of the strip, forming a pair of constrictions and approaching each other asymptotically far away from the strip. For a_w held fixed, the area encompassed by the cavity shrinks if p, and hence β, wanes (Fig. 3.13(a)). The same is true if a_w wanes, for p, and hence β, held fixed (Fig. 3.13(b)).

The asymptotic decrease to zero of the constrictions means that the two parts of the cusp-shaped magnetic cavity meet at infinity. However, a topological difference to the closed cavity of Fig. 3.10 persists, with the lines of the magnetic field $\mathbf{H}$ there running along the paramagnet/vacuum interface ∂S, while here the lines of the transformed magnetic field $\hat{\mathbf{H}}$ are always perpendicular to the respective interface $\partial\hat{S}$, thus ensuring that Ampère's law, eqn. (2.49), is obeyed. This can be seen in the following way. The asymptotic behaviour of the directional element $\delta\xi(\eta)$ of the contour $\partial\hat{S}$ bridging the constrictions between the two parts of the magnetic cavity, according to eqn. (3.108), comes out as

$$\delta\xi(\eta) \to -2p\eta/\left(|\eta|/2c_w \sin\beta\right)^{1/p} \quad \text{as } \eta \to \pm\infty. \tag{3.109}$$

By contrast, the asymptotic behaviour of the relevant component of the transformed magnetic field $\hat{H}_\xi(\xi,\eta)$, derived from eqn. (3.104) using eqns. (3.73), (3.103), and (3.108), appears as

$$\hat{H}_\xi(\xi,\eta) \to -\left(I/4p\eta\right)\left(|\eta|/2c_w \sin\beta\right)^{1/p} \quad \text{as } \eta \to \pm\infty. \tag{3.110}$$

In the contour integral of $\hat{\mathbf{H}}$ about $\partial\hat{S}$, closed across the said constrictions, only two contributions of the latter, $\delta\xi(\eta)\hat{H}_\xi(\xi,\eta) \to I/2$, remain, because of the orthogonality of $\hat{\mathbf{H}}$ along $\partial\hat{S}$, together making up I, in agreement with the result, eqn. (3.102).

3.7.2 Current distribution

The rectangular cusp-shaped magnetic cavity depicted in Fig. 3.13(b) may serve as an example for further scrutiny. Thus, eqn. (3.108) with $p = 1/2$ affords an implicit transformation relationship from which the dependence $x(\xi,\eta)$ and the derivative $\partial y/\partial\eta$ at $\eta = 0$ can be deduced. Substituting these quantities into the right-hand side of eqn. (3.105) for the sheet current distribution in the Meissner state due to the flat magnetic environment, eqn. (3.77), yields the sheet current distribution in that state appropriate to the rectangular cusp-shaped magnetic cavity,

$$\hat{J}(\xi) = \frac{I}{2c_w}\frac{d\vartheta_w}{d\xi}\frac{\cos\left(\pi\vartheta_w(\xi)/2c_w\right)}{\sqrt{\left[\cos\left(\pi\vartheta_w(\xi)/c_w\right) - \cos\left(\pi\vartheta_w(w)/c_w\right)\right]/2}} \quad \text{for } -w < \xi < w, \tag{3.111}$$

where

$$\vartheta_w(\xi) = \xi\left(1 - \xi^2/4c_w^2\right)^{1/2}. \tag{3.112}$$

Once more, though overstretching the model requirements as on the occasion of flat paramagnetic shields, eqn. (3.111) may be formally checked against the situation of direct strip–magnet contact, when $c \to w$. Here, the magnetic field concentrates around the centre, and falls off towards the edges, of the strip, entailing a cap-like sheet current distribution over the width of the strip as expected from eqn. (3.106), viz. $\hat{J}(\xi) = (I/2c_w)(d\vartheta_w/d\xi)$ for $-w < \xi < w$. This elucidates a modulation of the Meissner state given by the derivative of the auxiliary function, eqn. (3.112), which ultimately governs the current-carrying capability of the strip, a genuine asset of curved paramagnetic shields. Again reasoning similar to that for the magnetically shielded superconductor strip in the barrier-protected Meissner state of section 3.4.2 limits the validity of eqn. (3.111) to the range $-w+h \leq \xi \leq w-h$ and simultaneously recommends imposing the maximum value $J_{\max} = J_{\mathrm{b}}$ in $|\xi \pm w| \leq h$.

Figure 3.14(a) displays the variation of the normalized sheet current $\hat{J}/J_{\mathrm{b}}$ over the width of the strip for a range of values of the normalized distance between the edges of the strip and the vertices of the magnetic shields a_w/w. A general trait due to sheet current redistribution in the strip is the warping of the symmetric current profile around the central part of the strip and the increase in the total current itself, a phenomenon which gets more pronounced as a_w/w wanes, reaching saturation with a *nonuniform* maximum in the limiting sheet current profile of eqn. (3.111), when the magnetic cavity touches the strip.

Additionally, a comparison with the situation of flat paramagnetic shields, Fig. 3.9(a), reveals that, for a_w/w or, equivalently, c/w kept fixed, the sheet current throughout the width of the strip is consistently *higher* in the case of curved paramagnetic shields. The appearance of a central maximum in particular allows for sheet

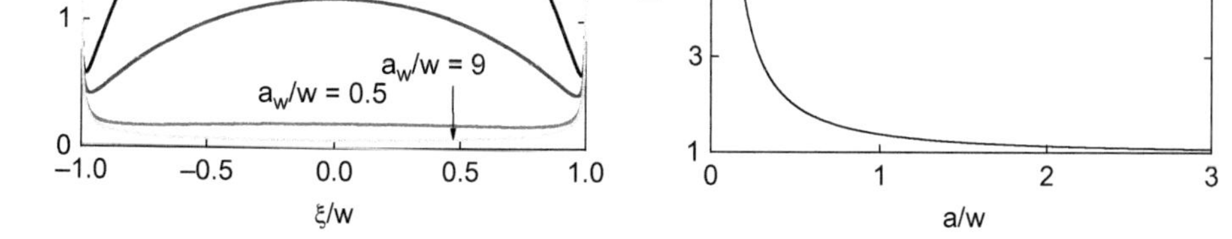

Fig. 3.14 (a) Distribution of the normalized sheet current $\hat{J}/J_{\mathrm{b}}$ as a function of the normalized coordinate ξ/w over the width of the magnetically shielded superconductor strip in the flux-free Meissner state for four different values of the normalized distance between the edges of the strip and the vertices of the magnetic shields a_w/w identified on the curves; (b) variation of the corresponding normalized maximum total transport current $I_{\max}/I_{\mathrm{b}}$ with the normalized distance between the edges of the strip and the vertices of the magnetic shields a_w/w, when $h/2w = 0.001$, referring to the rectangular cusp-shaped paramagnet configuration of Fig. 3.13(b).

currents exceeding the barrier value at the edges of the strip, $\hat{J}/J_b > 1$, separated by transitional zones of sheet current with $\hat{J}/J_b < 1$ in the marginal parts of the strip. Figure 3.14(b) presents the variation of the corresponding normalized maximum total transport current I_{max}/I_b with the normalized distance between the edges of the strip and the vertices of the magnetic shields a_w/w. Starting from a pronounced maximum at the smallest value of a_w/w, there is a sharp monotonic drop followed by a gradual approach towards unity, the value prevailing for the isolated strip, as a_w/w continues to grow. A comparison with the situation of flat paramagnetic shields, Fig. 3.9(b), accordingly shows that the maximum total transport current for fixed a_w/w or c/w is always *larger* in the case of curved paramagnetic shields.

Finally, it appears from Figs. 3.9 and 3.14 on the one hand, and from Fig. 3.11 on the other, that the distribution of the sheet current and the maximum total transport current too can be more effectively controlled by the open paramagnetic shields of Figs. 3.8 and 3.13, rather than by the closed paramagnetic cavity of Fig. 3.10. This reflects the enhanced possibility of locally guiding the magnetic field with the former type of environments as compared with the latter, a fact already noted in section 2.3.4.2.

It should be emphasized, however, that a realization of the predicted enhancement of the total transport current is, in actual fact, conditional upon two limitations at least. Firstly, high total transport currents associated with strong tangential components of the magnetic field around the central part of the strip might give rise to penetration of magnetic flux from the *surfaces* of the strip, and hence could severely restrict the loss-free current transport in type-II superconductors possessing small lower critical fields, as discussed in section 3.4.3. Secondly, large sheet currents might exceed the constraint which the depairing critical current density sets, given the thickness of the strip, and therefore could destroy superconductivity, of course.

3.8 Magnetically shielded superconductor strip with a grain boundary defect

We return to the issue of section 3.4.2 and ask what changes the sheet current J and the maximum total transport current I_{max} will undergo if the superconductor strip (which hitherto has been tacitly assumed uncut) contains a grain boundary defect. Since an overall rigorous analysis of this complex problem is beyond the present scope, we restrict ourselves to a qualitative estimate. The following account furthers, and simultaneously augments, the work of Genenko (2002a).

3.8.1 Theoretical approach

Let us therefore again proceed from an infinitely extended, thin current-carrying type-II superconductor strip in the Meissner state, placed between two transversely oriented flat semi-infinite paramagnetic shields with infinite permeability, $\mu \to \infty$, at the distance c from the centre of the strip, the magnetic shielding geometry portrayed in Fig. 3.15. The strip is now thought to be split into two parts by a

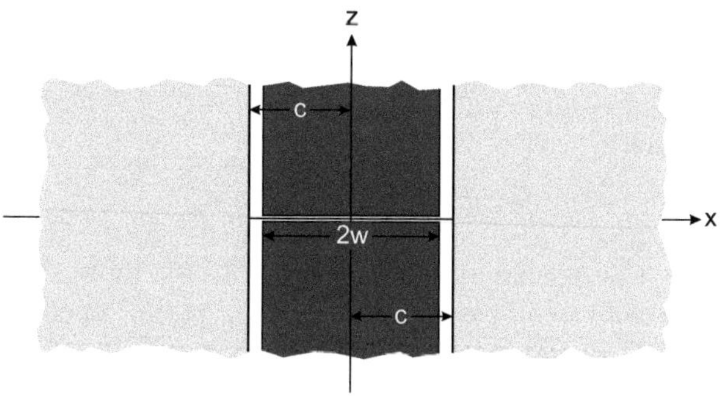

Fig. 3.15 Longitudinal-section view of a split superconductor strip (dark shading) between two flat semi-infinite paramagnets with transverse orientation (light shading) extending infinitely in the z-direction of a Cartesian coordinate system x, y, z. The discontinuity in the strip as well as the definition of the distance between the centre of the strip and the paramagnets c are marked.

grain boundary delineated by the cross-sectional plane $z = 0$, breaking the translational symmetry in the z-direction of the Cartesian coordinate system x, y, z. We understand that this defect acts like a weak-link which can be modelled as an ideal, 'butt-joint' Josephson junction with an infinitesimally thin tunnel layer in between. We furthermore assume that transport current flow along the strip subject to the paramagnetic environment and tunnelling supercurrent flow through the junction itself can be treated independently, an important simplification disregarding mutual interaction effects.

Given the premises stated earlier, the magnetic field due to the imposed transport current, $\mathbf{H}$, is delineated by eqn. (3.73). The magnetic field induced by the tunnelling supercurrent, $\mathbf{H}_s$, adds to this, making up the total magnetic field,

$$\mathbf{H}_{\text{tot}} = \mathbf{H} + \mathbf{H}_s. \tag{3.113}$$

An equation for $\mathbf{H}_s$ is easily derived when neglecting edge effects arising from the finiteness of the width of the strip. The formal substitution regarding the phase θ of the superfluid wavefunction Ψ,

$$\nabla \times \nabla\theta \rightarrow \left(\frac{d\varphi}{dx}\right)\delta(z)\mathbf{e}_y, \tag{3.114}$$

with the phase difference across the junction

$$\varphi(x) = \theta(x,z)\big|_{z=0+} - \theta(x,z)\big|_{z=0-} \tag{3.115}$$

then holds, the said geometrical and physical conditions implied (Kuzovlev and Lomtev 1997; Kogan et al. 2001). Thus, by taking the curl of the fundamental relation for the tunnelling supercurrent density $\mathbf{j}_s$, eqn. (2.10), employing Ampère's law, eqn. (2.19), observing the derivation, eqn. (2.3), and making use of the definition of the quantum of magnetic flux Φ_0, eqn. (2.17), the generalized London equation

$$\mathbf{H}_\mathrm{s} + \lambda_\mathrm{L}^2 \nabla \times (\nabla \times \mathbf{H}_\mathrm{s}) = \mathbf{V}_0 \tag{3.116}$$

with the scaled vorticity

$$\mathbf{V}_0 = \frac{\Phi_0}{2\pi\mu_0} \left(\frac{d\varphi}{dx} \right) \delta(z) \mathbf{e}_y \tag{3.117}$$

obtains. The required particular integral of eqn. (3.116) must ensure continuity of the tangential component of the magnetic field $H_{\mathrm{s},x}$ at the superconductor/vacuum interface and vanishing of the magnetic field $\mathbf{H}_\mathrm{s}$ as $z \to \pm\infty$.

Recourse to Ampère's law, eqn. (2.19), therefore allows the z-component of the tunnelling supercurrent density $j_{\mathrm{s},z}$ in the cross-sectional plane $z = 0$ to be expressed by the nonlocal form

$$j_{\mathrm{s},z}(x,z)\big|_{z=0} = j_\mathrm{c} \left(\frac{\lambda_\mathrm{J}^2}{\pi\lambda_\mathrm{L}} \right) \frac{d}{dx} \int_{-w}^{w} dx' \left(\frac{d\varphi}{dx'} \right) Q(x,x') \tag{3.118}$$

with the Josephson critical current density j_c and the Josephson penetration depth λ_J defined by

$$\lambda_\mathrm{J} = \sqrt{\frac{\Phi_0}{4\pi\mu_0\lambda_\mathrm{L} j_\mathrm{c}}}, \tag{3.119}$$

in addition to the derivative $d\varphi/dx'$ as the source bringing about excitations in the junction described by the kernel $Q(x,x')$ (Ivanchenko and Soboleva 1990a,b; Aliev, Silin, and Uryupin 1992). It is due to the latter fact that the current density distribution, eqn. (3.118), does not participate in current transport along the strip,

$$\int_{-w}^{w} dx\, j_{\mathrm{s},z}(x,z)\big|_{z=0} = 0. \tag{3.120}$$

If $\lambda_\mathrm{L} \ll d \ll w$, corresponding to the limit of a split superconductor bulk, the kernel in eqn. (3.118) is well approximated by

$$Q(x,x') \simeq K_0 \left(\frac{|x-x'|}{\lambda_\mathrm{L}} \right), \tag{3.121}$$

ad hoc invoking translational symmetry in the x-direction and thence again neglecting edge effects (Aliev, Silin, and Uryupin 1992; Kuzovlev and Lomtev 1997; Rauh, Genenko, and Rieck 2004). If, in addition, $\lambda_\mathrm{L} \ll \lambda_\mathrm{J}$, as for a junction with low j_c, the extended characteristic, eqn. (3.121), may be replaced with sufficient accuracy by its confined extreme, $Q(x,x') \to \pi\lambda_\mathrm{L}\delta(x-x')$, while preserving its strength. This admits the local current–phase relationship

$$j_{\mathrm{s},z}(x,z)\big|_{z=0} \simeq j_c \lambda_\mathrm{J}^2 \left(\frac{d^2\varphi}{dx^2} \right) \tag{3.122}$$

which, from Ampère's law, eqn. (2.19), in turn yields the local estimate for the

y-component of the magnetic field due to the tunnelling supercurrent

$$H_{s,y}(x,z)\big|_{z=0} \simeq j_c\lambda_J^2\left(\frac{d\varphi}{dx}\right). \tag{3.123}$$

The tunnelling supercurrent density, eqn. (3.122), together with the contribution from the transport current, eqn. (3.77), make up the total current density in the cross-sectional plane $z = 0$ of the strip,

$$j_{\text{tot}}(x) = j_{s,z}(x,z)\big|_{z=0} + J(x)/d. \tag{3.124}$$

However, only the second term on the right-hand side of eqn. (3.124) accounts for the total transport current I permeating the junction, in line with eqns. (3.2) and (3.120). Balancing j_{tot} from eqn. (3.124) against the Josephson tunnel current density

$$j_J(x) = j_c \sin\varphi(x) \tag{3.125}$$

thus constitutes the governing equation

$$\lambda_J^2\left(\frac{d^2\varphi}{dx^2}\right) = \sin\varphi(x) - J(x)/J_c \quad \text{for} \quad -w < x < w, \tag{3.126}$$

with $J_c = d\,j_c$, i.e. a generalized Sine–Gordon equation for a static state. This unites aspects of tunnelling supercurrent flow through the junction and transport current flow along the strip in a magnetically shielded environment. It shows that the Josephson penetration depth λ_J measures the spatial changes of the phase difference $\varphi(x)$, which in turn is driven by the sheet current $J(x)$. Proper boundary conditions supplementing eqn. (3.126) demand

$$\left.\frac{d\varphi}{dx}\right|_{x=\pm w} = 0, \tag{3.127}$$

consistent with eqns. (3.120) and (3.122); they imply that $d\varphi/dx = 0$ at $x = 0$ from the symmetry property of $J(x)$. Solutions for $\varphi(x)$ determine both the supercurrent density, eqn. (3.122), and the Josephson tunnel current density, eqn. (3.125), in the cross-sectional plane $z = 0$ of the strip.

3.8.2 Practical realizations

The methods for treating the problem stated in eqns. (3.126) and (3.127) can be classified according to whether a finite transport current or its zero limit persists. While an analytical solution obtains in the latter case, recourse to numerical techniques must be made otherwise.

3.8.2.1 Solution with a transport current absent

Here, $J(x) \equiv 0$ throughout the width of the strip, rendering eqn. (3.126) a homogeneous substitute,

$$\lambda_J^2 \left(\frac{d^2\varphi}{dx^2}\right) = \sin\varphi(x) \quad \text{for} \quad -w < x < w. \tag{3.128}$$

Upon applying the methods of separation of variables, quadrature, and inversion to eqn. (3.128) under consideration of eqn. (3.127), we get

$$\varphi(x) = 2\arcsin\left[\chi(x)\cos\left(\varphi_0/2\right)\right] + (2n+1)\pi; \quad n = 0, \pm 1, \pm 2..., \tag{3.129}$$

where

$$\chi(x) = \begin{cases} \mathrm{sn}[(x+w)/\lambda_J + K(m)|m] & \text{for} \quad -w \le x \le 0 \\ \mathrm{sn}[(w-x)/\lambda_J + K(m)|m] & \text{for } 0 \le x \le w, \end{cases} \tag{3.130}$$

introducing the Jacobian elliptic function sn and the complete elliptic integral of the first kind K with the parameter $m = \cos^2(\varphi_0/2)$ (Abramowitz and Stegun 1972). The latter quantity, and hence the value φ_0 of the phase difference $\varphi(x)$ at $x = 0$, is determined by the condition

$$K(m) = w/2\lambda_{\mathrm{J}}. \tag{3.131}$$

It obviously suffices to only consider the range $0 \le \varphi_0 \le 2\pi$ without any loss of generality. The solution, eqn. (3.129), thus appears as two infinite sets of pairs of opposite phase difference kinks, having convex shapes for $0 \le \varphi_0 < \pi$ and concave shapes for $\pi < \varphi_0 \le 2\pi$. We focus on the principal branches $n = 0$. Finally, since

$$\lim_{m \to 0} K(m) = \pi/2, \tag{3.132}$$

an upper bound $\lambda_{\mathrm{J,max}} = w/\pi$ to the Josephson penetration depth λ_{J} for the existence of *nontrivial* solutions of eqn. (3.128) with eqn. (3.127) in the limit of a vanishing transport current exists: if $\lambda_{\mathrm{J}} \ge \lambda_{\mathrm{J,max}}$, we find $\varphi(x) \equiv \pi$, apart from integer multiples of 2π, and hence $j_{\mathrm{J}}(x) \equiv 0$ in the junction over the whole width of the strip. A lower bound to the Josephson penetration depth λ_{J} arises from the model requirement $\lambda_{\mathrm{L}} \ll \lambda_{\mathrm{J}}$ and also from the fact that the Josephson critical current density j_{c} in eqn. (3.119) will, by all means, be less than the Ginzburg–Landau depairing critical current density j_{GL}.

3.8.2.2 Solution with a transport current present

If $J(x) \neq 0$ across the width of the strip, eqn. (3.126) is suitably decomposed into the set of differential equations

$$\left.\begin{aligned} \lambda_{\mathrm{J}}\left(\frac{d\varphi}{dx}\right) &= \psi(x) \\ \lambda_{\mathrm{J}}\left(\frac{d\psi}{dx}\right) &= \sin\varphi(x) - J(x)/J_{\mathrm{c}} \end{aligned}\right\} \quad \text{for} \quad -w < x < w, \tag{3.133}$$

preparing the use of numerical techniques. The profile of the sheet current $J(x)$ herein, as given by eqn. (3.77), is again limited to the range $-w+h \leq x \leq w-h$ like before, its maximum J_{max} at $|x \pm w| = h$ also being posited in $|x \pm w| \leq h$, establishing a relationship between J_{max} and I. Since the maximum total transport current I_{max} is limited by the presence of the grain boundary, it is conveniently measured in terms of the Josephson critical current $I_{\mathrm{c}} = 2wJ_{\mathrm{c}}$. The above equations together with the requirement $\psi(x) = 0$ at $x = \pm w$ define a two-point boundary value problem analogous to that of section 3.8.2.1, whose solution $\varphi(x)$ again consists of two infinite sets of pairs of opposite phase difference kinks distinguished by integer multiples of 2π, adopting convex shapes for the total transport current I but concave shapes for a total transport current $-I$. This can be verified by methods developed for initial value problems first, in principle at least (Press, Teukolsky, Vetterling, and Flannery 2007). The procedure runs accordingly: (i) given a fixed value of I, start the computation with a trial value φ_0 of $\varphi(x)$ at $x = 0$, where $\psi(x) = 0$, selected from the range $0 \leq \varphi_0 \leq 2\pi$ without any loss of generality; (ii) continue by forward integration up to $x = w$; (iii) iterate and adjust φ_0 by interpolation so as to satisfy the boundary condition $\psi(x) = 0$ at $x = w$; and (iv) extend the solution $\varphi(x)$ eventually found within the range $0 \leq x \leq w$ into the range $-w \leq x \leq 0$ by reflection about $x = 0$. Investigations of the strip's marginal Meissner state carrying the maximum total transport current I_{max} in the absence of Josephson vortices therefore best progress from $I = 0$ and successively work their way towards the maximum value of I for which a permissible solution $\varphi(x)$ still exists.

3.8.3 Numerical results

In order to elucidate the various processes involved and to facilitate understanding how they interact, we reveal the predictions of our model in three consecutive steps. Let us first inspect the zero limit of both the total transport current I and the induced magnetic field $\mathbf{H}$, in which case the problem reduces to the Josephson tunnelling effect, where the tunnelling supercurrent density $j_{\mathrm{s},z}$, eqn. (3.122), tallies with the Josephson tunnel current density j_{J}, eqn. (3.125). Figure 3.16 presents the variation of three key quantities over the width of the strip for a fixed value of the normalized Josephson penetration depth λ_{J}/w. The normalized phase difference φ/π, focusing on one convex pair of opposite phase difference kinks of point symmetry centred at $x/w = \pm 1/2$, is shown in Fig. 3.16(a). This involves a pair of Josephson vortices of the same symmetry positioned alike, as the normalized tunnelling supercurrent density $j_{\mathrm{s},z}/j_{\mathrm{c}}$ displayed in Fig. 3.16(b) reveals. The associated normalized magnetic field $H_{\mathrm{s},y}/j_{\mathrm{c}}\lambda_{\mathrm{J}}$ with alternating signs depicted in Fig. 3.16(c) corroborates the latter fact and makes clear that the total magnetic flux through the junction disappears.

We next allow for a finite total transport current I, and hence for a finite induced magnetic field $\mathbf{H}$, but first *omit* the effect of the paramagnetic shields, addressing the situation of an isolated split superconductor strip in the marginal Meissner state. The induced magnetic field here is given by eqn. (3.5), with the sheet current defined by eqn. (3.8). Part of Fig. 3.17 presents the variation of three key quantities

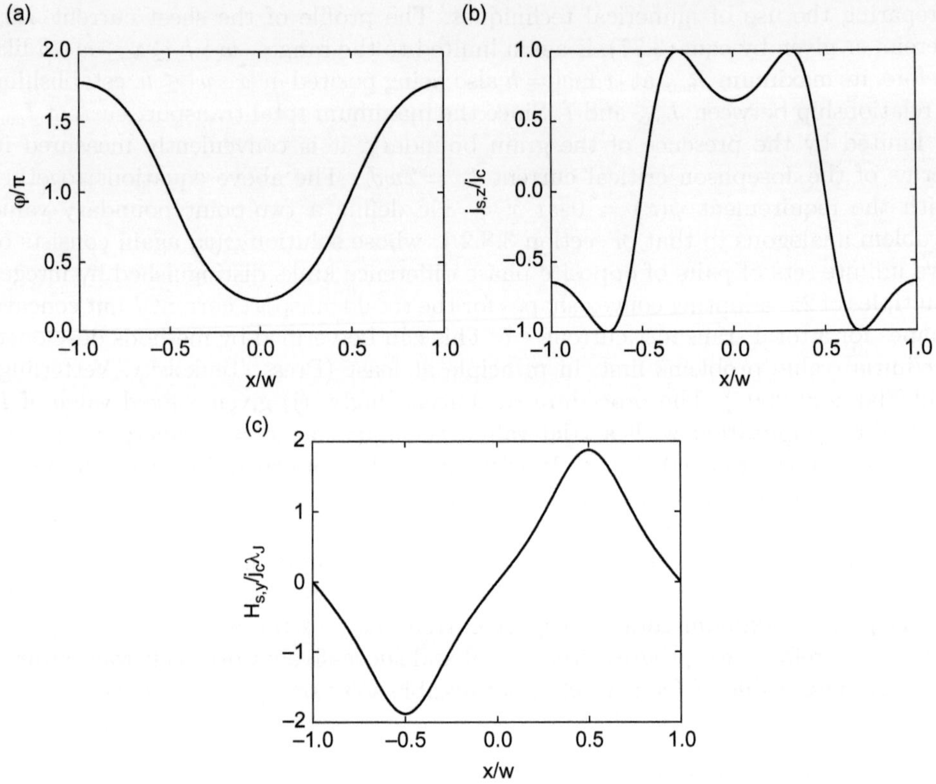

Fig. 3.16 Dependence of (a) the normalized phase difference φ/π, (b) the normalized tunnelling supercurrent density $j_{s,z}/j_c$, and (c) the associated normalized magnetic field $H_{s,y}/j_c\lambda_J$ on the normalized coordinate x/w over the width of the strip in the absence of a transport current, for the normalized Josephson penetration depth $\lambda_J/w = 0.2$.

over the width of the strip for the value of the normalized Josephson penetration depth λ_J/w from before.

The normalized phase difference φ/π confined to one convex pair of opposite phase difference kinks of lost point symmetry, with turning points displaced towards the edges of the strip, is shown in Fig. 3.17(a). This implies a pair of asymmetric Josephson vortices positioned accordingly, as the zero-crossings of the normalized tunnelling supercurrent density $j_{s,z}/j_c$ displayed in Fig. 3.17(b) reveal. The normalized total current density j_{tot}/j_c, which includes the contribution of the sheet current too, is depicted in Fig. 3.17(c). It makes up for the negative overcritical shares that the tunnelling supercurrent density entails, producing an overall subcritical current density in the direction of transport current flow. Its symmetric profile appears strongly nonuniform, reminiscent of the Meissner state of an isolated thin superconductor film (Rhoderick and Wilson 1962), with critical peaks of the current density displaced from the edges of the strip. The latter fact obviously

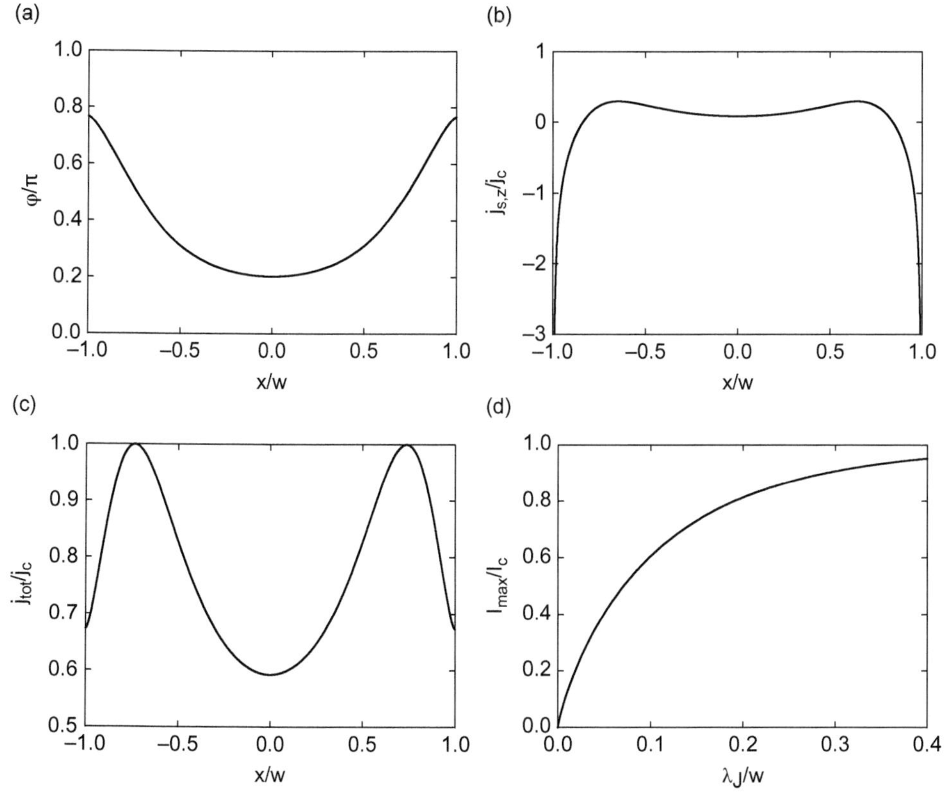

Fig. 3.17 Dependence of (a) the normalized phase difference φ/π, (b) the normalized tunnelling supercurrent density $j_{s,z}/j_c$, and (c) the normalized total current density j_{tot}/j_c on the normalized coordinate x/w over the width of an isolated current-carrying superconductor strip in the marginal Meissner state, for the normalized Josephson penetration depth $\lambda_J/w = 0.2$; (d) variation of the normalized maximum total transport current I_{max}/I_c with the normalized Josephson penetration depth λ_J/w, when $h/2w = 0.001$.

derives from the grain boundary defect, as a comparison with the profile of Fig. 3.1 for an uncut strip suggests. Figure 3.17(d) unfolds the variation of the corresponding normalized maximum total transport current I_{max}/I_c with the normalized Josephson penetration depth λ_J/w. Starting from zero minimum at the smallest value of λ_J/w, there is a monotonic rise followed by an approach towards unity due to the increased current filling of the junction, as λ_J/w continues to grow.

Lastly, we allow for a finite total transport current I, and hence for a finite induced magnetic field $\mathbf{H}$, and also *include* the effect of the paramagnetic shields, addressing the originally proposed situation of a magnetically shielded split superconductor strip in the marginal Meissner state. The induced magnetic field therefore is given by eqn. (3.73), with the sheet current defined by eqn. (3.77). Figure 3.18(a) displays the variation of the normalized total current density j_{tot}/j_c

over the width of the strip for a range of values of the normalized distance between the centre of the strip and the transversely oriented flat paramagnetic shields c/w, likewise adopting the value of the normalized Josephson penetration depth λ_J/w from before. A general trait due to sheet current redistribution in the strip with the grain boundary defect is the continuous evening out of the symmetric current profile between its critical peaks, which again are displaced from the edges of the strip, and the increase in the current itself, a phenomenon which gets more pronounced as c/w wanes. Although parallels in tendency to the situation of an uncut strip exist according to the evidence from Fig. 3.9(a), the stronger inhomogeneity of the current profile also here must be ascribed to the grain boundary defect. Magnetic shielding obviously mitigates this trait, as a comparison with Fig. 3.17(c) affirms. Figure 3.18(b) presents the variation of the corresponding normalized maximum total transport current I_{max}/I_c with the normalized distance between the centre of the strip and the transversely oriented flat paramagnetic shields c/w. Starting from unity at the smallest value of c/w, there is a slight monotonic fall towards the limit prevailing for the isolated split superconductor strip, as c/w continues to grow. A comparison with Fig. 3.9(b) for an uncut strip suggests that the grain boundary tends to smooth out the magnetic shielding effect; it also lowers the maximum loss-free current that can be carried by the strip. Thus, I_{max} could be reduced by eight orders of magnitude as compared to the situation of an uncut strip in the case of the numerical example chosen for λ_J/w, given the values of the geometrical and physical parameters quoted in section 3.4.3.

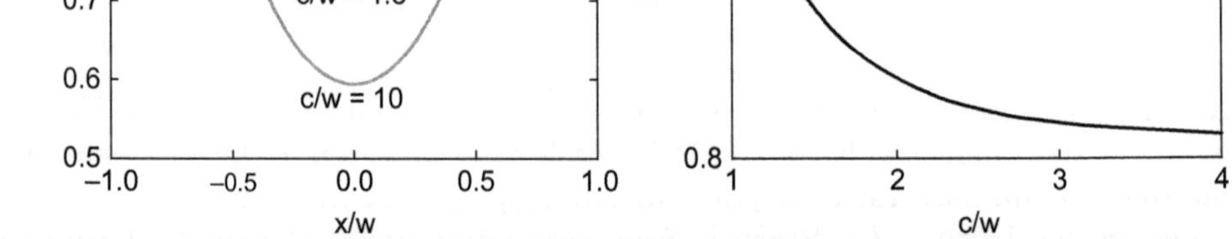

Fig. 3.18 (a) Distribution of the normalized total current density j_{tot}/j_c as a function of the normalized coordinate x/w over the width of the magnetically shielded current-carrying superconductor strip in the marginal Meissner state for three different values of the normalized distance between the centre of the strip and the paramagnets c/w identified on the curves; (b) variation of the corresponding normalized maximum total transport current I_{max}/I_c with the normalized distance between the centre of the strip and the paramagnets c/w, for the normalized Josephson penetration depth $\lambda_J/w = 0.2$, when $h/2w = 0.001$, referring to the transverse paramagnet configuration of Fig. 3.15.

We end by commenting that the two cardinal effects resulting from the grain boundary in the magnetically shielded superconductor strip—an enhancement of the inhomogeneity of the normalized total current density $j_{\rm tot}/j_{\rm c}$ and a reduction in the maximum total transport current $I_{\rm max}$—become more pronounced as the Josephson critical current density $j_{\rm c}$ wanes, and hence the normalized Josephson penetration depth $\lambda_{\rm J}/w$ augments, understanding that the magnetic shielding geometry delineated by the ratio c/w is fixed.

4
Critical state of a superconductor strip in ideal paramagnetic environments

Whereas proceeding from the flux-free Meissner state represents a useful approach for capturing the principal effects of magnetic shielding on the distributions of the sheet current in, and the magnetic field around, an infinitely extended, thin current-carrying type-II superconductor strip, the objective of Chapter 3, practice nevertheless shows that for small magnitudes of the total transport current only does the said state actually exist. This limitation is due to the peaks of the sheet current near the edges of the strip which, although if suppressed by the action of the magnetic shields, are still distinct. High values of the current, in turn, entail a strong Lorentz-like force exerted upon fluctuation-generated magnetic vortices, tending to drag them towards the centre, against an opposing barrier force at the edges of the strip (cf. section 3.1.3). The latter force, however, is typically weak except in special circumstances (cf. section 3.4.3). Having entered the strip, the vortices interact with each other and with pinning centres inside the strip to form metastable structures of magnetic flux, depending on the external conditions that actually prevail.

Such structures are often well described within the frame of Bean's model of the critical state introduced in section 2.2. As discussed there, we employ the underlying idea in its elementary form, i.e. assume a field-independent critical current density j_c. Duly adapted to the geometry of a thin superconductor strip, this concept yields predictions regarding the distributions of the current density and the magnetic field significantly different from those in a superconductor bulk, a fact most clearly demonstrated by the partially flux-filled critical state. Thus, for an infinitely extended slab, the two nonzero components of these quantities are related to each other locally; the current density and the magnetic field, hence, coexist in the flux-penetrated parts, but disappear in the flux-free region of the slab (Bean 1962; Campbell and Evetts 1972). In contrast, for an infinitely extended thin strip, with nonzero tangential components of the magnetic field on its surfaces and essentially two-dimensional shielding current flow across its entire width, only the normal

Electromagnetics of Superconductor/Paramagnet Heterostructures. Yuri Genenko and Hermann Rauh, Oxford University Press.
 DOI: 10.1093/9780191782855.003.0004

component of the magnetic field on the strip's surfaces along the flux-free region disappears (Norris 1970; Brandt and Indenbom 1993; Zeldov, Clem, McElfresh, and Darwin 1994). Direct magneto-optical observations of the structures of magnetic flux in wide superconductor strips and the profiles of the sheet current extracted thereof chiefly support these claims by theory and experiment (Brandt 1995b; Wijngaarden, Spoelder, Surdeanu, and Griessen 1996; Jooss, Warthmann, Forkl, and Kronmüller 1998; Gaevski et al. 1999).

With magnetic flux and current flow pervading substantial parts of a superconductor constituent, the critical state admits loss-free transport currents of much higher strengths than those achievable in the Meissner state. Oscillating transport currents, on the other hand, invariably give rise to the phenomenon of hysteretic AC loss. Whilst curtailing the dissipation of electromagnetic energy still is a major issue for superconductor applications at large scales (Gömöry et al. 2006; Malozemoff et al. 2008; Clem 2009; Grilli and Kario 2016), a promising way for flat superconductors such as thin superconductor strips consists of exploiting the shielding effect of magnetically susceptible environments, thereby controlling the distributions of the transport current and the magnetic field (Majoros, Glowacki, and Campbell 2000a,b; Genenko 2002c; Genenko and Rauh 2007; Aladyshkin, Silhanek, Gillijns, and Moshchalkov 2009; Genenko, Rauh, Krüger, and Narayanan 2009; Genenko and Rauh 2010; Genenko, Rauh, and Krüger 2011; Krüger et al. 2013). Here, we focus on some of the environments examined in Chapter 3, again referring to the limit of infinite permeability of the paramagnetic constituents, $\mu \to \infty$, the integral equations governing the sheet current distributions in the Meissner state providing the base for construction of the critical states. As an important result, curved paramagnetic shields may even generate overcritical states in a superconductor strip, with considerably increased critical currents and significantly reduced hysteretic AC loss.

4.1 Isolated superconductor strip: a reference

Let us start by revisiting the infinitely extended, thin type-II superconductor strip of width $2w$ and thickness d limited by the domain $-w \leq x \leq w$, $-d/2 \leq y \leq d/2$, the direction of translational invariance of this strip being parallel to the z-axis of a Cartesian coordinate system x, y, z, with geometrical dimensions and material characteristics satisfying the conditions assumed previously (cf. section 3.1). The strip now is supposed to reveal the partially flux-filled critical state, carrying a transport current of magnitude I in the absence of an external magnetic field. A convenient simplification, for critical sheet currents $J_c = d\,j_c$ much larger than the sheet current value J_b typifying the edge barriers of the strip, is to model the magnetic flux as entering the strip at transport current $I \to 0$. We thus understand that said state has been established by progressing from the flux-free virgin state up to the state with a finite value of I below the critical current $I_c = 2wJ_c$, i.e. the maximum loss-free transport current attainable by saturating the strip with magnetic flux. If using, *mutatis mutandis*, the predictions of Bean's model of the critical state of a superconductor slab derived in section 2.2.2.1 as a reference for characterizing the partially flux-filled critical state of the thin superconductor strip, the following picture ensues: magnetic flux penetrating from both edges fills the margins of the strip, exempting a central region of width $2b$, where the normal component of the magnetic field H_y disappears,

$$H_y(x) = 0 \ \text{ for } \ 0 \leq |x| \leq b,\ y = 0, \tag{4.1}$$

while the sheet current J adopts its critical value in the margins of the strip,

$$J(x) = J_c \ \text{ for } \ b \leq |x| < w. \tag{4.2}$$

Naturally, b is controlled by the transport current I.

4.1.1 Current and field distributions

The distributions of the sheet current and the magnetic field across the width of the superconductor strip in the partially flux-filled critical state are suitably obtained by calling upon eqn. (3.7) to delineate the flux-free Meissner state,

$$\int_{-w}^{w} dx' J(x') \frac{1}{x - x'} = 0 \quad \text{in} \quad -w < x < w. \tag{4.3}$$

Since the sheet current $J(x)$ within the flux-free region of the strip, $0 \leq |x| \leq b$, can be envisaged as arising from the effect of current filaments at $x = \pm x'$ for $b \leq x' < w$ within the flux-filled margins of the strip, $b \leq |x| < w$, the ansatz

$$J(x) \to \begin{cases} \chi(x, x') & \text{for } \ 0 \leq |x| \leq b \\ \delta(x - x') + \delta(x + x') & \text{for } \ b \leq |x| < w \end{cases} \tag{4.4}$$

introducing a linear response function, $\chi(x, x')$, suggests itself. Substitution of eqn. (4.4) into eqn. (4.3) yields the Cauchy-like integral equation

$$\int_{-b}^{b} dx'' \chi(x'', x') \frac{1}{x - x''} = -\gamma(x, x') \ \text{ in } \ 0 \leq |x| \leq b \tag{4.5}$$

with the singular two-part inhomogeneity

$$\gamma(x, x') = \frac{1}{x - x'} + \frac{1}{x + x'}. \tag{4.6}$$

Its general solution expresses the required function in the form

$$\chi(x, x') = \frac{1}{\pi^2} \int_{-b}^{b} dx'' \sqrt{\frac{b^2 - x''^2}{b^2 - x^2}}\, \gamma(x'', x') \frac{1}{x - x''} + \frac{\alpha}{\pi\sqrt{b^2 - x^2}} \quad \text{for } \ 0 \leq |x| \leq b, \tag{4.7}$$

the term labelled with a free constant α on the right-hand side of eqn. (4.7) satisfying the homogeneous equation to eqn. (4.5) (Mikhlin 1964). Evaluating the integral in eqn. (4.7) with eqn. (4.6) gives

$$\chi(x,x') = \frac{1}{\pi}\sqrt{\frac{x'^2-b^2}{b^2-x^2}}\left(\frac{1}{x-x'} - \frac{1}{x+x'}\right) + \frac{\alpha+2}{\pi\sqrt{b^2-x^2}} \quad \text{for } 0 \le |x| \le b. \tag{4.8}$$

The above approach, upon observing the demand, eqn. (4.2), admits the representation of the distribution of the sheet current $J(x)$ within the flux-free region of the strip,

$$J(x) = J_\text{c}\int_b^w dx'\,\chi(x,x') \quad \text{for } 0 \le |x| \le b. \tag{4.9}$$

Thus, using eqn. (4.8) in eqn. (4.9) and ensuring regularity as well as continuity of $J(x)$ at $x = \pm b$ through the choice

$$\alpha + 2 = 2\sqrt{w+b}/\sqrt{w-b} \tag{4.10}$$

furnishes the distribution of the sheet current across the entire width of the strip,

$$J(x) = \begin{cases} \dfrac{2J_\text{c}}{\pi}\arctan\sqrt{\dfrac{w^2-b^2}{b^2-x^2}} & \text{for } 0 \le |x| \le b \\ J_\text{c} & \text{for } b \le |x| < w, \end{cases} \tag{4.11}$$

duplicating a result previously obtained by means of the conformal mapping technique (Norris 1970). This, by reference to eqn. (3.2), allows for a straightforward evaluation of the total transport current

$$I = 2J_\text{c}(w-b) + \int_{-b}^{b} dx'\,J(x') \tag{4.12}$$

in terms of b. On the other hand, substituting eqn. (4.11) into eqn. (4.12) and solving for b yields the half-width of the flux-free region in terms of I,

$$b = w\sqrt{1 - I^2/I_\text{c}^2}. \tag{4.13}$$

The corresponding distribution of the normal component of the magnetic field across the entire width of the strip, which can be calculated from eqn. (3.5b) with eqns. (4.1) and (4.11), finally reads

$$H_y(x) = \begin{cases} 0 & \text{for } 0 \le |x| \le b, y = 0 \\ H_\text{c}\,\text{sgn}(x)\text{arctanh}\sqrt{\dfrac{x^2-b^2}{w^2-b^2}} & \text{for } b \le |x| < w, y = 0, \end{cases} \tag{4.14}$$

where $H_\text{c} = J_\text{c}/\pi$ denotes a characteristic magnetic field.

Figure 4.1(a) displays the variation of the normalized sheet current J/J_c over the width of the strip for a fixed value of the normalized total transport current I/I_c. A characteristic trait due to shielding current flow is the dip of the symmetric current

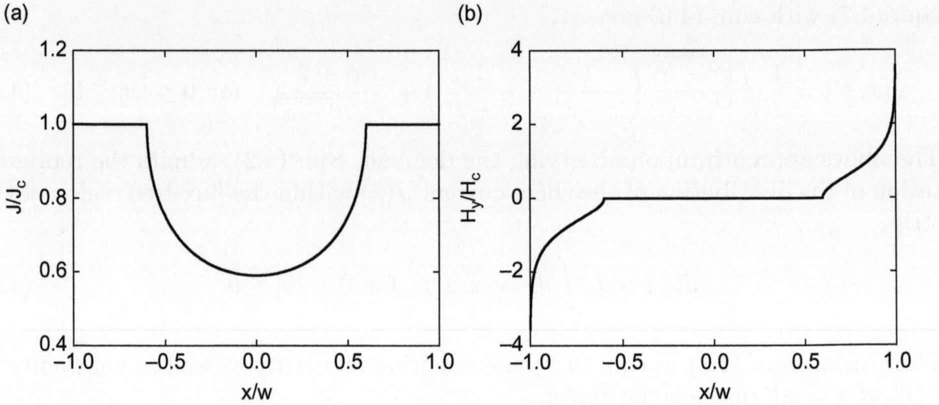

Fig. 4.1 Distribution of (a) the normalized sheet current J/J_c and (b) the normalized magnetic field H_y/H_c as functions of the normalized coordinate x/w over the width of the isolated superconductor strip in the partially flux-filled critical state, when the normalized total transport current $I/I_c = 0.8$.

profile within the central flux-free region of the strip between the maxima of critical sheet current adopted in the flux-filled margins of the strip. Figure 4.1(b) depicts the variation of the corresponding normalized magnetic field H_y/H_c over the width of the strip. Prominent aspects here are the sharp peaks of the antisymmetric field profile near the edges of the strip, growing out of the central region of the strip, at vanishing magnetic field.

4.1.2 Hysteretic AC loss

We assume that, like in the static situation considered previously, the scenario of entry and exit of magnetic flux in the partially flux-filled critical state of the strip, with an oscillating transport current imposed, is again controlled by the critical sheet current J_c. The distribution of the magnetic field across the width of the strip, emerging during a gradual increase in the transport current from the virgin state of the strip up to the state associated with the current's maximum value, changes sign as the current alternates between I and $-I$. Accordingly, by resorting to a quasistatic approach, the hysteretic AC loss, per unit length of the strip, U_{ac}, incurred during a steady-state cycle of the transport current, is determined by the normal component of the magnetic field H_y for a given value of I,

$$U_{ac} = 8\mu_0 J_c \int_b^w dx \int_b^x dx'\, H_y(x') \tag{4.15}$$

(Norris 1970; Brandt and Indenbom 1993). Substituting eqn. (4.14) into eqn. (4.15), changing the variables x and x' to $u = x/w$ and $u' = x'/w$, respectively, as well as introducing the substitution $l = b/w$ creates the form

$$U_{\mathrm{ac}} = U_0 \int_l^1 du \int_l^u du' \operatorname{arctanh}\sqrt{\frac{u'^2 - l^2}{1 - l^2}}, \tag{4.16}$$

where $U_0 = 2\mu_0 I_c^2/\pi$. Evaluation of eqn. (4.16) with eqn. (4.13) finally yields the explicit result

$$U_{\mathrm{ac}} = U_0 f\left(I/I_c\right) \quad \text{with } f(i) = (1-i)\ln(1-i) + (1+i)\ln(1+i) - i^2. \tag{4.17}$$

We note that, for small transport current strengths, eqn. (4.17) reveals the quartic law

$$U_{\mathrm{ac}} \simeq \mu_0 \left(I^4/I_c^2\right)/3\pi; \quad 0 \leq I \ll I_c, \tag{4.18}$$

but assigns for the maximum transport current strength

$$U_{\mathrm{ac}} \simeq 3\mu_0 I_c^2/4\pi; \quad I = I_c, \tag{4.19}$$

i.e. a quadratic change with the critical quantity, as established in the case of a superconductor slab, eqn. (2.59).

Figure 4.2(a) shows the variation of the normalized half-width of the flux-free region b/w with the normalized transport current I/I_c. Accordingly, penetration of magnetic flux is seen to hardly proceed at first, as I/I_c augments, but to increase rapidly, as the transition from the partially flux-filled critical state to the full critical state pends, a fact manifest in the steep fall of b/w. Figure 4.2(b) portrays the variation of the corresponding normalized hysteretic AC loss U_{ac}/U_0 with the normalized transport current I/I_c. Evidently, the power dependence, eqn. (4.18),

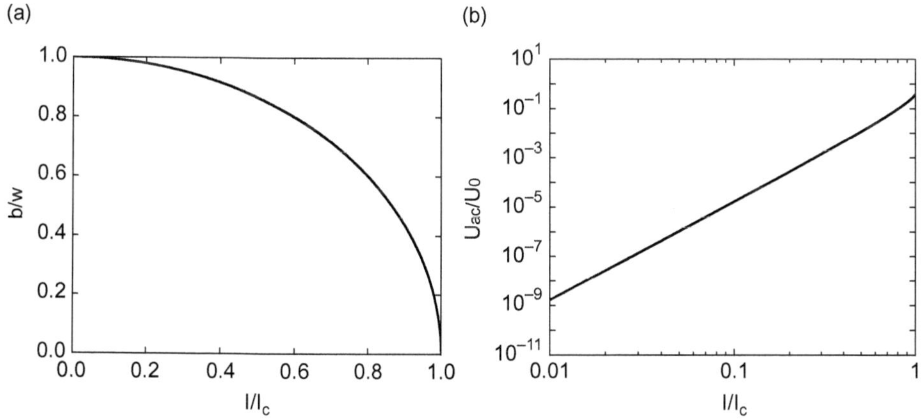

Fig. 4.2 Dependence of (a) the normalized half-width of the flux-free region b/w and (b) the normalized hysteretic AC loss U_{ac}/U_0 on I/I_c, the normalized total transport current imposed on the isolated superconductor strip in the partially flux-filled critical state.

approximates the increase in U_{ac}/U_0 well even for values of I/I_{c} midway to unity.

4.2 Superconductor strip between two flat semi-infinite paramagnets

We further develop the considerations of section 4.1, addressing an infinitely extended, thin current-carrying type-II superconductor strip in the partially flux-filled critical state, now placed between two flat semi-infinite paramagnetic shields with infinite permeability, $\mu \to \infty$, and principal orientations either longitudinal or transverse, the geometries studied earlier (cf. section 3.4). The following analysis supplements the works of Genenko, Snezhko, and Freyhardt (2000) as well as Genenko and Rauh (2010).

4.2.1 Longitudinal shielding geometry

Let us first examine the case where the surfaces of the paramagnets run parallel to the strip, at the distance a from the plane of the strip, as illustrated in Fig. 3.6. The symmetry of this configuration with respect to the midplane $x = 0$ of the strip entails that the characteristics, eqns. (4.1) and (4.2), still apply.

4.2.1.1 Current and field distributions

In order to obtain the distributions of the sheet current and the magnetic field across the width of the superconductor strip in the partially flux-filled critical state, we again proceed by resorting to eqn. (3.58) characterizing flux-free Meissner state,

$$\int_{-w}^{w} dx' J(x') \coth\left(\pi(x - x')/2a\right) = 0 \quad \text{in} \quad -w < x < w. \tag{4.20}$$

Since, like in section 4.1, the sheet current $J(x)$ within the flux-free region of the strip, $0 \leq |x| \leq b$, can be envisaged as originating from the effect of current filaments at $x = \pm x'$ for $b \leq x' < w$ within the flux-filled margins of the strip, $b \leq |x| \leq w$, the ansatz

$$J(x) \to \begin{cases} \chi(x, x') & \text{for } 0 \leq |x| \leq b \\ \delta(x - x') + \delta(x + x') & \text{for } b \leq |x| < w \end{cases} \tag{4.21}$$

introducing a linear response function, $\chi(x, x')$, in eqn. (4.20) suggests itself. The change of the variables x and x' to $u = \tanh\left(\pi x/2a\right)$ and $u' = \tanh\left(\pi x'/2a\right)$, respectively, together with the substitutions $p = \tanh\left(\pi b/2a\right)$ and $q = \tanh\left(\pi w/2a\right)$ as well as the request for an even solution in x thus transform eqn. (4.20) into the Cauchy-like integral equation

$$\int_{-p}^{p} du'' \zeta(u'', u') \frac{1}{u - u''} = -\gamma(u, u') \quad \text{in } 0 \leq |u| \leq p \tag{4.22}$$

for the auxiliary function $\zeta(u'', u') = \chi(x'', x')$, with the singular two-part inhomogeneity

$$\gamma(u, u') = \frac{\pi}{2a}\left(1 - u'^2\right)\left(\frac{1}{u - u'} + \frac{1}{u + u'}\right). \tag{4.23}$$

Owing to the formal analogy of eqns. (4.22), (4.23) with eqns. (4.5), (4.6) derived for the isolated superconductor strip, the required solution can be directly transcribed from eqn. (4.8) to read

$$\zeta(u, u') = \frac{1 - u'^2}{2a}\sqrt{\frac{u'^2 - p^2}{p^2 - u^2}}\left(\frac{1}{u - u'} - \frac{1}{u + u'}\right) + \frac{1 - u'^2}{2a}\frac{\alpha + 2}{\sqrt{p^2 - u^2}} \quad \text{for } 0 \le |u| \le p \tag{4.24}$$

with a free constant α. This admits the representation of the distribution of the sheet current $J(x)$ within the flux-free region of the strip,

$$J(u) = J_c\left(\frac{2a}{\pi}\right)\int_p^q du'\zeta(u(x), u')\frac{1}{1 - u'^2} \quad \text{for } 0 \le |x| \le b. \tag{4.25}$$

Thus, using eqn. (4.24) in eqn. (4.25) and ensuring regularity as well as continuity of $J(x)$ at $x = \pm b$ through the choice

$$\alpha + 2 = 2\sqrt{q + p}/\sqrt{q - p} \tag{4.26}$$

furnishes the distribution of the sheet current across the entire width of the magnetically shielded strip,

$$J(x) = \begin{cases} \dfrac{2J_c}{\pi}\arctan\sqrt{\dfrac{q^2 - p^2}{p^2 - u(x)^2}} & \text{for } 0 \le |x| \le b \\ J_c & \text{for } b \le |x| < w. \end{cases} \tag{4.27}$$

Substituting eqn. (4.27) into eqn. (4.12) and solving for b yields the half-width of the flux-free region in terms of I,

$$b = \frac{2a}{\pi}\operatorname{arctanh} p \quad \text{with } p = \sqrt{1 - \cosh^2\left(\frac{\pi w}{2a}\frac{I}{I_c}\right)\operatorname{sech}^2\left(\frac{\pi w}{2a}\right)}. \tag{4.28}$$

Likewise, eqn. (4.14) can be used to establish the corresponding distribution of the normal component of the magnetic field across the width of the magnetically shielded strip,

$$H_y(x) = \begin{cases} 0 & \text{for } 0 \le |x| \le b, y = 0 \\ H_c \operatorname{sgn}(x)\operatorname{arctanh}\sqrt{\dfrac{u(x)^2 - p^2}{r^2 - p^2}} & \text{for } b \le |x| < w, y = 0, \end{cases} \tag{4.29}$$

where $H_c = J_c/\pi$.

Figure 4.3(a) displays the variation of the normalized sheet current J/J_c over the width of the strip for a range of values of the normalized distance between the

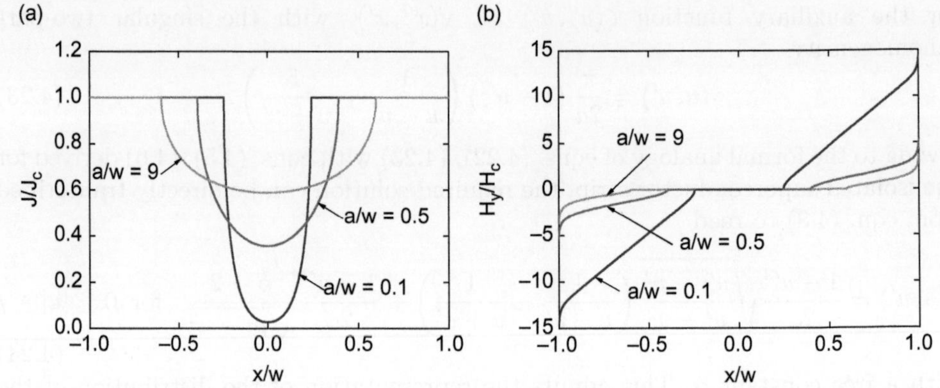

Fig. 4.3 Distribution of (a) the normalized sheet current J/J_c and (b) the normalized magnetic field H_y/H_c as functions of the normalized coordinate x/w over the width of the magnetically shielded superconductor strip in the partially flux-filled critical state for three different values of the normalized distance between the plane of the strip and the paramagnets a/w identified on the curves, when the normalized total transport current $I/I_c = 0.8$, referring to the *longitudinal* paramagnet configuration of Fig. 3.6.

plane of the strip and the longitudinally oriented flat paramagnetic shields a/w, adopting a fixed value of the normalized total transport current I/I_c. A general trait due to sheet current redistribution is the *deepening* and *narrowing* of the symmetric current profile within the central flux-free region of the strip between the maxima of critical sheet current adopted in the flux-filled margins of the strip; a phenomenon which gets more pronounced as a/w abates. Figure 4.3(b) depicts the variation of the corresponding normalized magnetic field H_y/H_c over the width of the strip for the said range of values of the normalized distance between the plane of the strip and the longitudinally oriented flat paramagnetic shields a/w and the same value of the normalized total transport current I/I_c. Evidently, whereas at large a/w the field profile virtually duplicates that of an isolated strip, it *augments* in strength while leaving a *reduced* central flux-free zone, as a/w wanes.

We note that the distributions of both the sheet current and the magnetic field in the limit of direct strip–magnet contact are reminiscent of those of an isolated superconductor slab, for which the strip together with its images effectively assembles into a regular stack of flat unshielded superconducting films, as remarked on previously (cf. section 3.4.1.2).

4.2.1.2 Hysteretic AC loss

The dynamics of magnetic flux in the partially flux-filled critical state of the magnetically shielded strip, with an oscillating transport current imposed, is basically the same as for an unshielded strip. Hence, eqn. (4.15) determines the electromagnetic energy dissipated during a steady-state cycle of the transport current in the presence of paramagnetic shields as well. Upon substituting the normal component of the magnetic field, eqn. (4.29), the hysteretic AC loss per unit length of the strip, U_{ac}, for the *longitudinal* shielding geometry at hand thus can be put as

$$U_{\mathrm{ac}} = U_0 \left(\frac{2a}{\pi w}\right)^2 \int_p^q \frac{du}{1-u^2} \int_p^u \frac{du'}{1-u'^2} \operatorname{arctanh}\sqrt{\frac{u'^2-p^2}{q^2-p^2}}, \tag{4.30}$$

where $U_0 = 2\mu_0 I_{\mathrm{c}}^2/\pi$. In the limit of the paramagnets situated close to the strip, eqn. (4.30) may be approximated by the highly accurate simple expression

$$U_{\mathrm{ac}} \simeq U_0 \left[\frac{\pi w}{12a}\left(\frac{l}{w}\right)^3 + \frac{\ln 2}{2}\left(\frac{l}{w}\right)^2\right]; \quad h < a \ll w, \tag{4.31}$$

which introduces the flux penetration depth $l = w - b$, with b related to p via eqn. (4.28). An essentially equivalent representation unfolding the dependence on the current I, albeit confined to the range $a/w \le I/I_{\mathrm{c}} \le 1 - a/w$, emerges from eqn. (4.31) when making use of the approximations

$$p \simeq 1 - \frac{1}{2}\exp\left[-\frac{\pi w}{a}\left(1 - \frac{I}{I_{\mathrm{c}}}\right)\right] \tag{4.32a}$$

$$r \simeq 1 - 2\exp\left(-\frac{\pi w}{a}\right) \tag{4.32b}$$

$$b \simeq a\left(\frac{2\ln 2}{\pi}\right) + w\left(1 - \frac{I}{I_{\mathrm{c}}}\right), \tag{4.32c}$$

so that the succinct form

$$U_{\mathrm{ac}} \simeq U_0 \left(\frac{2a}{\pi w}\right)^2 f\left(\frac{\pi w}{2a}\frac{I}{I_c}\right) \quad \text{with } f(i) = \frac{1}{6}i^3 - \frac{\ln^2 2}{2}i \tag{4.33}$$

obtains. The cubic term herein, which dominates in the small-distance/high-current regime, describes the hysteretic AC loss as for an isolated superconductor slab (Campbell and Evetts 1972). This analogue receives support from the limit $a \to 0$ of eqn. (4.32c) too, with its linear dependence already established in section 2.2.2.1.

Figure 4.4(a) shows variation in the normalized half-width of the flux-free region b/w with the normalized transport current I/I_{c} for a range of values of the normalized distance between the plane of the strip and the longitudinally oriented flat paramagnetic shields a/w. Thus, although the overall behaviour of the curves is seen to display a tendency very much like that for the unshielded strip, their ordering reveals that penetration of magnetic flux, and hence the transition from the partially flux-filled critical state to the full critical state, is *advanced* as a/w wanes. Figure 4.4(b) portrays variation in the corresponding normalized hysteretic AC loss U_{ac}/U_0 with the normalized transport current I/I_{c} for the said range of values of the normalized distance between the plane of the strip and the longitudinally oriented flat paramagnetic shields a/w. This reveals that, whereas at large

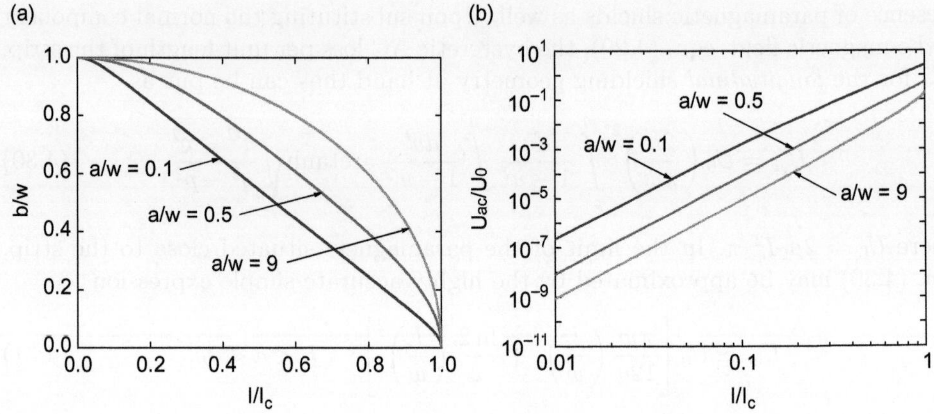

Fig. 4.4 Dependence of (a) the normalized half-width of the flux-free region b/w and (b) the normalized hysteretic AC loss U_{ac}/U_0 on I/I_c, the normalized total transport current imposed on the magnetically shielded superconductor strip in the partially flux-filled critical state, for three different values of the normalized distance between the plane of the strip and the paramagnets a/w identified on the curves, referring to the *longitudinal* paramagnet configuration of Fig. 3.6. Reproduced with permission from © IOP Publishing. All rights reserved. Genenko and Rauh 2010, Supercond. Sci. Technol. 23, 075007, doi: 10.1088/0953-2048/23/7/075007

a/w, the hysteretic AC loss is virtually like that for an unshielded strip, it *increases* substantially throughout the transport current regime, as a/w abates, the predictions of eqn. (4.33) differing indiscernibly from the exact results within the specified limitation.

4.2.2 Transverse shielding geometry

Let us now inspect the case where the surfaces of the paramagnets run perpendicular to the strip, at the distance c from the centre of the strip, as depicted in Fig. 3.8. The symmetry of this configuration with respect to the midplane $x = 0$ of the strip entails that the characteristics, eqns. (4.1) and (4.2), still apply.

4.2.2.1 Current and field distributions

In order to calculate the distributions of the sheet current and the magnetic field across the width of the superconductor strip in the partially flux-filled critical state, we again proceed by relying on eqn. (3.75) typifying the flux-free Meissner state,

$$\int_{-w}^{w} dx' J(x') \cot\left(\pi(x - x')/2c\right) = 0 \quad \text{in} \quad -w < x < w. \tag{4.34}$$

Since, like in section 4.1, the sheet current $J(x)$ within the flux-free region of the strip, $0 \leq |x| \leq b$, can be envisaged as deriving from the effect of current filaments at $x = \pm x'$ for $b \leq x' < w$ within the flux-filled margins of the strip, $b \leq |x| < w$, the ansatz

$$J(x) \to \begin{cases} \chi(x, x') & \text{for } 0 \leq |x| \leq b \\ \delta(x - x') + \delta(x + x') & \text{for } b \leq |x| < w \end{cases} \tag{4.35}$$

introducing a linear response function, $\chi(x, x')$, in eqn. (4.34) suggests itself. The change of the variables x and x' to $v = \tan(\pi x/2c)$ and $v' = \tan(\pi x'/2c)$, respectively, together with the substitutions $r = \tan(\pi b/2c)$ and $s = \tan(\pi w/2c)$ as well as the request for an even solution in x thus transform eqn. (4.34) into the Cauchy-like integral equation

$$\int_{-r}^{r} dv'' \zeta(v'', v') \frac{1}{v - v''} = -\gamma(v, v') \quad \text{in } 0 \leq |v| \leq r \tag{4.36}$$

for the auxiliary function $\zeta(v'', v') = \chi(x'', x')$, with the singular two-part inhomogeneity

$$\gamma(v, v') = \frac{\pi}{2c}\left(1 + v'^2\right)\left(\frac{1}{v - v'} + \frac{1}{v + v'}\right). \tag{4.37}$$

Owing to the formal analogy of eqns. (4.36), (4.37) with eqns. (4.5), (4.6) derived for the isolated superconductor strip, the required solution can be directly transcribed from eqn. (4.8) to read

$$\zeta(v, v') = \frac{1 + v'^2}{2c}\sqrt{\frac{v'^2 - r^2}{r^2 - v^2}}\left(\frac{1}{v - v'} - \frac{1}{v + v'}\right) + \frac{1 + v'^2}{2c}\frac{\alpha + 2}{\sqrt{r^2 - v^2}} \quad \text{for } 0 \leq |v| \leq r \tag{4.38}$$

with a free constant α. This admits the representation of the distribution of the sheet current $J(x)$ within the flux-free region of the strip,

$$J(x) = J_c\left(\frac{2c}{\pi}\right)\int_{r}^{s} dv' \zeta(v(x), v')\frac{1}{1 + v'^2} \quad \text{for } 0 \leq |x| \leq b. \tag{4.39}$$

Thus, using eqn. (4.38) in eqn. (4.39) and ensuring regularity as well as continuity of $J(x)$ at $x = \pm b$ through the choice

$$\alpha + 2 = 2\sqrt{s + r}/\sqrt{s - r} \tag{4.40}$$

furnishes the distribution of the sheet current across the entire width of the magnetically shielded strip,

$$J(x) = \begin{cases} \dfrac{2J_c}{\pi}\arctan\sqrt{\dfrac{s^2 - r^2}{r^2 - v(x)^2}} & \text{for } 0 \leq |x| \leq b \\ J_c & \text{for } b \leq |x| < w. \end{cases} \tag{4.41}$$

Substituting eqn. (4.41) into eqn. (4.12) and solving for b yields the half-width of the flux-free region in terms of I,

$$b = \frac{2c}{\pi}\arctan r \quad \text{with } r = \sqrt{\cos^2\left(\frac{\pi w}{2c}\frac{I}{I_c}\right)\sec^2\left(\frac{\pi w}{2c}\right) - 1}. \tag{4.42}$$

Likewise, eqn. (4.14) can be used to establish the corresponding distribution of the normal component of the magnetic field across the width of the magnetically shielded strip,

$$H_y(x) = \begin{cases} 0 & \text{for } 0 \le |x| \le b, y = 0 \\ H_c\,\mathrm{sgn}(x)\mathrm{arctanh}\sqrt{\dfrac{v(x)^2 - r^2}{s^2 - r^2}} & \text{for } b \le |x| < w, y = 0. \end{cases} \tag{4.43}$$

where $H_c = J_c/\pi$.

Figure 4.5(a) displays the variation of the normalized sheet current J/J_c over the width of the strip for a range of values of the normalized distance between the centre of the strip and the transversely oriented flat paramagnetic shields c/w, adopting a fixed value of the normalized total transport current I/I_c. A general trait due to sheet current redistribution is the *flattening* and *widening* of the symmetric current profile within the central flux-free region of the strip between the maxima of critical sheet current adopted in the flux-filled margins of the strip, a phenomenon which gets more pronounced as c/w abates. Figure 4.5(b) depicts the variation in the corresponding normalized magnetic field H_y/H_c over the width of the strip for

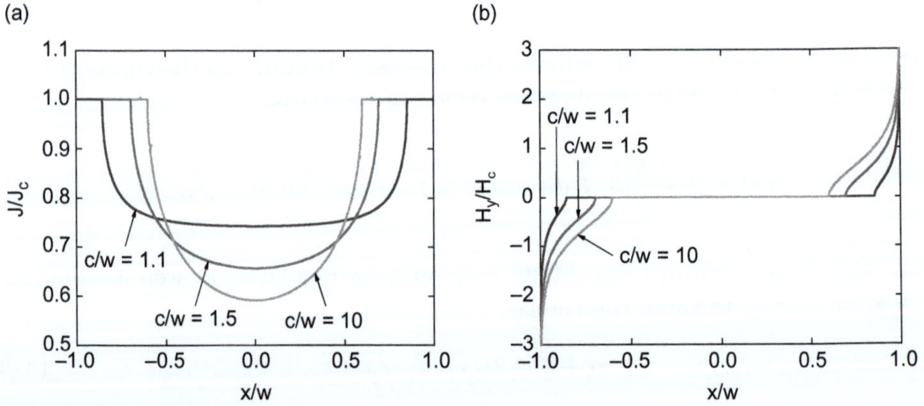

Fig. 4.5 Distribution of (a) the normalized sheet current J/J_c and (b) the normalized magnetic field H_y/H_c as functions of the normalized coordinate x/w over the width of the magnetically shielded superconductor strip in the partially flux-filled critical state for three different values of the normalized distance between the centre of the strip and the paramagnets c/w identified on the curves, when the normalized total transport current $I/I_c = 0.8$, referring to the *transverse* paramagnet configuration of Fig. 3.8.

the said range of values of the normalized distance between the centre of the strip and the transversely oriented flat paramagnetic shields c/w and the same value of the normalized total transport current I/I_c. Evidently, whereas at large c/w the field profile virtually reproduces that of an isolated strip, it *decreases* in strength while leaving an *enlarged* flux-free zone, as c/w wanes.

We note that the distributions of both the sheet current and the magnetic field are reminiscent of those of an infinite regular array of flat unshielded superconducting films, effectively assembled by the strip together with its images, as remarked on previously (cf. section 3.4.2.2).

4.2.2.2 Hysteretic AC loss

The dynamics of magnetic flux in the partially flux-filled critical state of the magnetically shielded strip, with an oscillating transport current imposed, is basically the same as that for an unshielded strip. Hence, eqn. (4.15) determines the electromagnetic energy dissipated during a steady-state cycle of the transport current in the presence of paramagnetic shields as well. Upon substituting the normal component of the magnetic field, eqn. (4.43), the hysteretic AC loss per unit length of the strip, U_{ac}, for the *transverse* shielding geometry at hand thus can be put as

$$U_{\mathrm{ac}} = U_0 \left(\frac{2c}{\pi w}\right)^2 \int_r^s \frac{dv}{1+v^2} \int_r^v \frac{dv'}{1+v'^2} \operatorname{arctanh}\sqrt{\frac{v'^2-r^2}{s^2-r^2}}, \tag{4.44}$$

where $U_0 = 2\mu_0 I_c^2/\pi$. In the limit of the magnets situated close to the strip, eqn. (4.44) may be approximated with high accuracy by the expression

$$U_{\mathrm{ac}} \simeq U_0 \left(\frac{a}{w}\right)^2 \left[\sqrt{\frac{l}{2a}\left(\frac{l}{2a}+1\right)} \operatorname{arcsec}\left(\frac{l}{a}+1\right) - \ln\left(\frac{l}{a}+1\right)\right]; \quad h < a \ll w, \tag{4.45}$$

which introduces the distance between the centre of the strip and the paramagnets $a = c - w$, apart from the flux penetration depth $l = w - b$, with b related to r via eqn. (4.42). An essentially equivalent representation uncovering the dependence on the current I, albeit confined to the range $0 \leq I/I_c \leq 1 - a/w$, arises from eqn. (4.45) when making use of the approximations

$$r \simeq s \cos\left(\frac{\pi w}{2c}\frac{I}{I_c}\right) \tag{4.46a}$$

$$s \simeq \frac{2c}{\pi a}\left[1 - \frac{\pi^2}{12}\left(\frac{a}{w}\right)^2\right] \tag{4.46b}$$

$$b \simeq c - a \sec\left(\frac{\pi w}{2c}\frac{I}{I_c}\right), \tag{4.46c}$$

so that the succinct form

$$U_{\mathrm{ac}} \simeq U_0 \left(\frac{a}{w}\right)^2 f\left(\frac{\pi w}{2c}\frac{I}{I_{\mathrm{c}}}\right) \quad \text{with } f(i) = \frac{1}{2}\, i \tan i + \ln \cos i \tag{4.47}$$

obtains. The quadratic scaling factor herein describes the hysteretic AC loss as for a superconductor strip between two longitudinal superconducting shields (Genenko 2004).

Figure 4.6(a) shows the variation of the normalized half-width of the flux-free region b/w with the normalized transport current I/I_{c} for a range of values of the normalized distance between the centre of the strip and the transversely oriented flat paramagnetic shields c/w. Thus, although the overall behaviour of the curves is seen to display a tendency very much like that for the unshielded strip, their ordering reveals that the penetration of magnetic flux, and hence the transition from the partially flux-filled critical state to the full critical state, is *retarded* as c/w wanes. Figure 4.6(b) portrays the variation in corresponding normalized hysteretic AC loss U_{ac}/U_0 with the normalized transport current I/I_{c} for the said range of values of the normalized distance between the centre of the strip and the transversely oriented flat paramagnetic shields c/w. This reveals that, whereas at large c/w the hysteretic AC loss is virtually like that for an unshielded strip, it *decreases* substantially throughout the transport current regime, as c/w abates, the predictions of eqn. (4.47) differing indiscernibly from the exact results within the specified limitation. The strong decrease in hysteretic AC loss obviously arises from both the

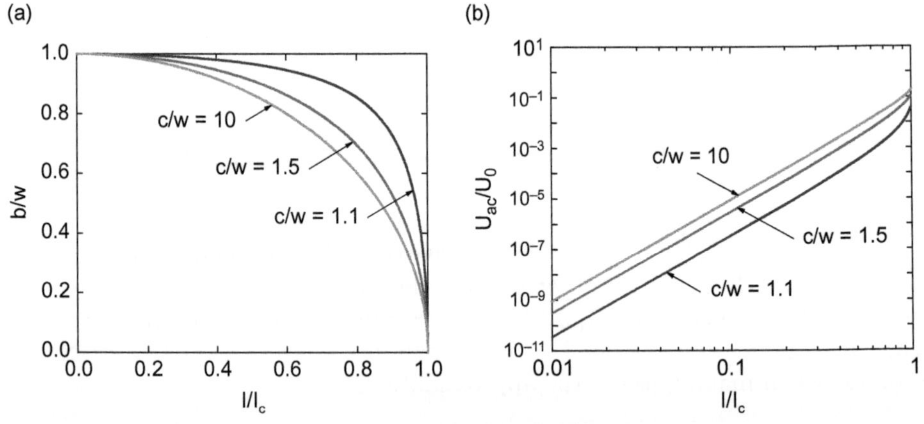

Fig. 4.6 Dependence of (a) the normalized half-width of the flux-free region b/w and (b) the normalized hysteretic AC loss U_{ac}/U_0 on I/I_{c}, the normalized total transport current imposed on the magnetically shielded superconductor strip in the partially flux-filled critical state, for three different values of the normalized distance between the centre of the strip and the paramagnets c/w identified on the curves, referring to the *transverse* paramagnet configuration of Fig. 3.8. Reproduced with permission from © IOP Publishing. All rights reserved. Genenko and Rauh 2010, Supercond. Sci. Technol. 23, 075007, doi: 10.1088/0953-2048/23/7/075007

reduction in strength of the magnetic field near the edges of the strip and the fact that a major part of the strip remains virtually flux-free even for total currents close to the critical current of the strip, as magnetic shielding takes effect.

4.3 Superconductor strip inside a cylindrical paramagnetic cavity

We go on now to further develop the considerations of section 4.1, looking at an infinitely extended, thin current-carrying type-II superconductor strip in the partially flux-filled critical state, now placed inside a cylindrical paramagnetic cavity with radius a and infinite permeability, $\mu \to \infty$, as illustrated in Fig. 3.10. The symmetry of this configuration with respect to the midplane $x = 0$ of the strip entails that the characteristics, eqns. (4.1) and (4.2), still apply. Let us present the analysis, thereby augmenting the works of Genenko and Rauh (2006, 2007).

4.3.1 Current and field distributions

The distributions of the sheet current and the magnetic field across the width of the superconductor strip in the partially flux-filled critical state are suitably obtained by reverting to eqn. (3.91) typifying the flux-free Meissner state,

$$\int_{-w}^{w} dx' J(x') Q(x, x') = 0 \quad \text{in} \quad -w < x < w, \tag{4.48}$$

with the singular two-part kernel

$$Q(x, x') = \frac{1}{x - x'} + \frac{1}{x - a^2/x'}. \tag{4.49}$$

Since, like in section 4.1, the sheet current $J(x)$ within the flux-free region of the strip, $0 \leq |x| \leq b$, can be envisaged as deriving from the effect of current filaments at $x = \pm x'$ for $b \leq x' < w$ within the flux-filled margins of the strip, $b \leq |x| < w$, the ansatz

$$J(x) \to \begin{cases} \chi(x, x') & \text{for } 0 \leq |x| \leq b \\ \delta(x - x') + \delta(x + x') & \text{for } b \leq |x| < w \end{cases} \tag{4.50}$$

introducing a linear response function, $\chi(x, x')$, in eqn. (4.48) suggests itself. The change of the variables x to $v = x(a^2 + b^2)/(a^2 + x^2)$ and x' to $v' = x'(a^2 + b^2)/(a^2 + x'^2)$ transforms eqn. (4.48) into the Cauchy-like integral equation

$$\int_{-b}^{b} dv'' \zeta(v'', v') \frac{1}{v - v''} = -\gamma(v, v') \quad \text{in } 0 \leq |v| \leq b \tag{4.51}$$

for the auxiliary function $\zeta(v'', v') = \chi(x'', x')/(dv''/dx'')$ subject to the normalization

$$\int_{-b}^{b} dv''\zeta(v'', v') = \tau, \tag{4.52}$$

involving a (positive) constant τ in the singular four-part inhomogeneity

$$\gamma(v, v') = \frac{1}{v - v'} + \frac{1}{v + v'} + \frac{\tau + 2}{v}\left(\frac{1}{\sqrt{1 - nv^2/b^2}} - 1\right) \tag{4.53}$$

with the parameter $n = \left(2ab/(a^2 + b^2)\right)^2$. We note formal correspondences between the first two terms on the right-hand side of eqn. (4.53) and those of eqn. (4.6) for the isolated superconductor strip in the partially flux-filled critical state as well as between the last two terms on the right-hand side of eqn. (4.53) and those of eqn. (3.95) for the magnetically shielded superconductor strip in the flux-free Meissner state. The general solution of eqn. (4.51) can be directly transcribed from eqn. (4.7) to yield

$$\zeta(v, v') = \frac{1}{\pi^2}\int_{-b}^{b} dv''\sqrt{\frac{b^2 - v''^2}{b^2 - v^2}}\,\gamma(v'', v')\frac{1}{v - v''} + \frac{\sigma}{\pi\sqrt{b^2 - v^2}} \quad \text{for } 0 \le |v| \le b, \tag{4.54}$$

the term labelled with a free constant σ on the right-hand side of eqn. (4.54) satisfying the homogeneous equation to eqn. (4.51). Evaluation of the integral in eqn. (4.54) with eqn. (4.53) gives

$$\begin{aligned}\zeta(v, v') = {} & \frac{1}{\pi}\sqrt{\frac{v'^2 - b^2}{b^2 - v^2}}\left(\frac{1}{v - v'} - \frac{1}{v + v'}\right) + \frac{\sigma + 1}{\pi\sqrt{b^2 - v^2}} \\ & + \frac{2\tau + 4}{\pi^2\sqrt{b^2 - v^2}}\left[K(n) + \left(\frac{b^2}{v^2} - 1\right)\Pi\left(\frac{b^2}{v^2}\middle| n\right)\right] \quad \text{for } 0 \le |v| \le b,\end{aligned} \tag{4.55}$$

where K and Π denote complete elliptic integrals of the first and third kinds, respectively (Abramowitz and Stegun 1972). From eqn. (4.55) together with the requirement of normalization, eqn. (4.52), the assignment $\sigma = -1$ ensues. Furthermore, since in the limit of the radius of the paramagnetic cavity tending to infinity, $a \to \infty$, the linear response function χ under consideration, and hence the auxiliary function ζ defined by eqn. (4.55), must duplicate the solution appropriate for the isolated superconductor strip, eqn. (4.8), the choice $\tau = \alpha$ in keeping with eqn. (4.10) is evident.

Calling upon eqn. (4.9), the above approach permits a representation of the distribution of the sheet current $J(x)$ within the flux-free region of the strip in the form

$$J(x) = J_c\,(dv/dx) \int_b^s dv'\,\zeta(v(x), v')/\,(dv'/dx') \quad \text{for } 0 \le |x| \le b, \tag{4.56}$$

involving the quantity $s = w(a^2 + b^2)/(a^2 + w^2)$. Evaluation of the integral in eqn. (4.56) with recourse to eqn. (4.55) furnishes the distribution of the sheet current across the entire width of the strip,

$$J(x) = \begin{cases} \dfrac{J_c}{\pi}\left(\dfrac{a^2}{x^2} - 1\right) F(v(x)) & \text{for } 0 \le |x| \le b \\ J_c & \text{for } b \le |x| < w, \end{cases} \tag{4.57}$$

where

$$\begin{aligned} F(v) = {} & \frac{r}{\sqrt{r^2 - v^2}} \arctan \sqrt{\left(\frac{r^2 - v^2}{r^2 - s^2}\right)\left(\frac{s^2 - b^2}{b^2 - v^2}\right)} - \arctan \sqrt{\frac{s^2 - b^2}{b^2 - v^2}} \\ & + \frac{2\sqrt{b^2 - v^2}}{b}\left[\Pi\left(\frac{b^2}{v^2}\middle| n\right) \Big/ K(n) - 1\right] \arctan \sqrt{\frac{b^2 w^2 - b^4}{a^4 - b^2 w^2}} \end{aligned} \tag{4.58}$$

with the parameter $r = (a^2 + b^2)/2a$. This result, upon substituting into eqn. (4.12), also yields the implicit relation for the half-width of the flux-free region b in terms of the total transport current I,

$$2bwK(n)(I/I_c) = \pi\left(a^2 + b^2\right) \arctan \sqrt{\frac{b^2 w^2 - b^4}{a^4 - b^2 w^2}}. \tag{4.59}$$

The corresponding distribution of the normal component of the magnetic field across the entire width of the strip, which can be calculated from eqn. (3.88b) with eqns. (4.1) and (4.57), finally comes out as

$$H_y(x) = \begin{cases} 0 & \text{for } 0 \le |x| \le b, y = 0 \\ H_c G(x) + \dfrac{1}{2\pi} \displaystyle\int_{-b}^{b} dx' J(x') Q(x, x') & \text{for } b \le |x| < w, y = 0, \end{cases} \tag{4.60}$$

where $H_c = J_c/\pi$ and

$$G(x) = \frac{1}{2}\left[\ln\left|\left(\frac{x - b}{x + b}\right)\left(\frac{x + w}{x - w}\right)\right| + \frac{a^2}{x^2} \ln\left|\left(\frac{bx + a^2}{bx - a^2}\right)\left(\frac{wx - a^2}{wx + a^2}\right)\right| + \frac{2(w - b)}{x}\right], \tag{4.61}$$

the first term together with the integral on the right-hand side of eqn. (4.60) ensuring that $H_n(x)$ vanishes at $x = \pm b$.

Figure 4.7(a) displays the variation of the normalized sheet current J/J_c over the width of the strip for a range of values of the normalized radius of the cavity a/w,

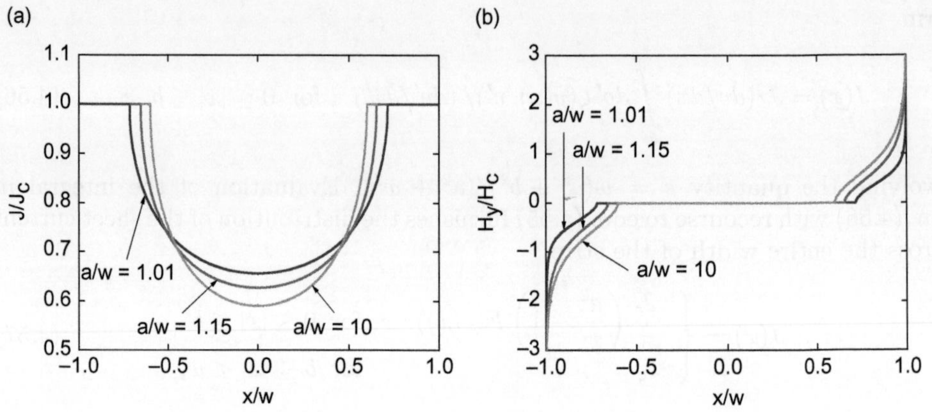

Figure 4.7 Distribution of (a) the normalized sheet current J/J_c and (b) the normalized magnetic field H_y/H_c as functions of the normalized coordinate x/w over the width of the magnetically shielded superconductor strip in the partially flux-filled critical state for three different values of the normalized radius of the cavity a/w identified on the curves, when the normalized total transport current $I/I_c = 0.8$, referring to the cylindrical paramagnet configuration of Fig. 3.10.

adopting a fixed value of the normalized total transport current I/I_c. A general trait due to sheet current redistribution is the *flattening* and *widening* of the symmetric current profile within the central flux-free region of the strip between the maxima of critical sheet current adopted in the flux-filled margins of the strip, a phenomenon which gets more distinct as a/w abates. Figure 4.7(b) depicts the variation in the corresponding normalized magnetic field H_y/H_c over the width of the strip for the said range of values of the normalized radius of the cavity a/w and the same value of the normalized total transport current I/I_c. Evidently, whereas at large a/w the field profile virtually duplicates that of an isolated strip, it *decreases* in strength while leaving an *enlarged* flux-free zone, as a/w wanes.

The above findings qualitatively resemble those for the flat semi-infinite paramagnets in the transverse shielding configuration, Figs. 4.5(a) and 4.5(b). However, due to weaker shielding by the cylindrical paramagnetic cavity, its quantitative effect on the results is less pronounced (cf. section 3.5.2).

4.3.2 Hysteretic AC loss

The dynamics of the magnetic flux in the partially flux-filled critical state of the magnetically shielded strip, with an oscillating transport current imposed, is basically the same as that for an unshielded strip. Hence, eqn. (4.15) determines the electromagnetic energy dissipated during a steady-state cycle of the transport current in the presence of paramagnetic shields as well. Upon substituting the normal component of the magnetic field, eqn. (4.60), changing the variables x and x' to

$u = x/w$ and $u' = x'/w$, respectively, as well as introducing substitution $l = b/w$, the hysteretic AC loss per unit length of the strip, U_{ac}, for the cylindrical shielding geometry at hand thus can be cast into the form

$$U_{\mathrm{ac}} = U_0 \int_l^1 du \int_l^u du' \, H_y(u')/H_c, \tag{4.62}$$

where $U_0 = 2\mu_0 I_c^2/\pi$, by due reference to eqn. (4.59).

Figure 4.8(a) shows the variation of the normalized half-width of the flux-free region b/w with the normalized transport current I/I_c for a range of values of the normalized radius of the cavity a/w. Thus, although the overall behaviour of the curves is seen to display a tendency very much like that for the unshielded strip, their ordering unfolds that penetration of magnetic flux, and hence the transition from the partially flux-filled critical state to the full critical state, is *retarded* as a/w wanes. Figure 4.8(b) portrays the variation in corresponding normalized hysteretic AC loss U_{ac}/U_0 with the normalized total transport current I/I_c for the said range of values of the normalized radius of the cavity a/w. This reveals that, whereas at large a/w the hysteretic AC loss is virtually like that for an unshielded strip, it *decreases* substantially throughout the transport current regime, as a/w abates.

In a similar vein the above findings qualitatively resemble those for the flat semi-infinite paramagnets in the transverse shielding configuration, Figs. 4.6(a) and 4.6(b). However, due to weaker shielding by the cylindrical paramagnetic cavity, quantitative effects again are less pronounced (cf. section 3.5.2).

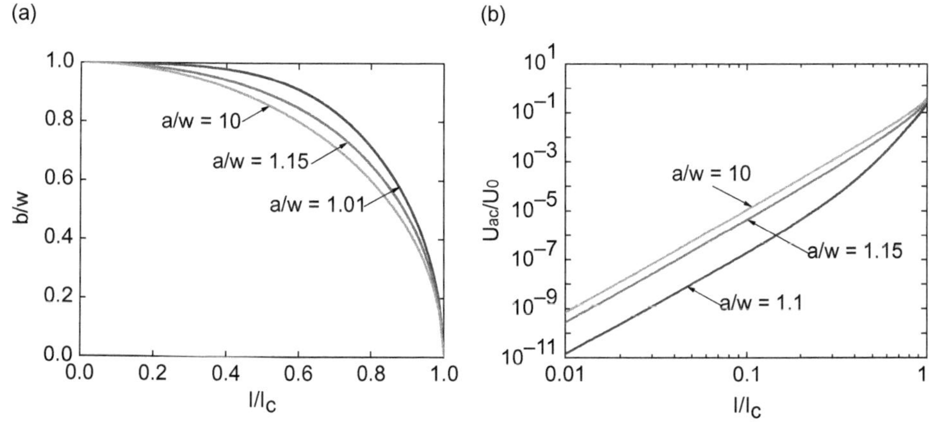

Fig. 4.8 Dependence of (a) the normalized half-width of the flux-free region b/w and (b) the normalized hysteretic AC loss U_{ac}/U_0 on I/I_c, the normalized total transport current imposed on the magnetically shielded superconductor strip in the partially flux-filled critical state, for three different values of the normalized radius of the cavity a/w identified on the curves, referring to the cylindrical paramagnet configuration of Fig. 3.10.

4.4 Superconductor strip between two curved semi-infinite paramagnets: overcritical states

We once more develop the considerations of section 4.1, addressing an infinitely extended, thin current-carrying type-II superconductor strip in the partially flux-filled critical state, now placed inside a cusp-shaped paramagnetic cavity with infinite permeability, $\mu \to \infty$, as depicted in Fig. 3.13. This is motivated by the observation that for a superconductor strip in the flux-free Meissner state, the latter type of magnetic environment facilitates redistribution of the sheet current towards the centre of the strip, bringing about potentially considerable enhancements of the maximum total transport current that can be imposed on the magnetically shielded strip (cf. section 3.7.2). The presentation that follows takes its cue from the work of Genenko, Snezhko, and Freyhardt (2000).

4.4.1 Conformal transformation

The distributions of the sheet current and the magnetic field across the width of the superconductor strip in the partially flux-filled critical state are obtained by starting from the derivations of section 4.2.2 for the flat paramagnetic environment of Fig. 3.8, generalizing them to the curved paramagnetic environment of Fig. 3.13 with the help of the conformal mapping technique. This procedure calls for some explanation since, although the condition on the distribution of the magnetic field, eqn. (4.1), is observed thereby, satisfaction of the requirement on the distribution of the sheet current, eqn. (4.2), is not by itself guaranteed. Unlike in the case of the Meissner state, the sheet current in the critical state appropriate to the curved paramagnetic environment thus cannot directly be inferred from eqn. (3.105). Nevertheless, the sheet current distribution under consideration can well be constructed from the linear response function due to the flat paramagnetic environment, if the transformation involved preserves the flux-free region of the strip.

Adhering to section 3.6.1.2, we restate that a conformal transformation effects a change of coordinates from the complex variable $s = x + iy$ to the complex variable $\sigma = \xi + i\eta$ by means of some analytic function $\sigma = f_b(s)$ with a nonzero derivative df_b/ds. This maps areas of the s-plane onto areas of the σ-plane, implying $x = x(\xi, \eta)$ and $y = y(\xi, \eta)$. Reverting to the problem at hand, let us map the flat half-domains $|x| \geq c$ of the s-plane onto some cusp-shaped half-domains of the σ-plane, with opening half-angle of their vertices $\beta = p\pi/2$, where $0 < p \leq 1$, by employing the function

$$f_b(s) = c_b\left[(1 + s/c)^p - (1 - s/c)^p\right] \quad \text{with } c_b = b/\left[(1 + b/c)^p - (1 - b/c)^p\right], \quad (4.63)$$

which introduces the parameter $c > w$. This is analytic in the whole s-plane, apart from two cuts on the real axis, $x < -c, y = 0$ and $x > c, y = 0$. Application of eqn. (4.63) transforms the points $(x = \pm c, y = 0)$ of the s-plane into the points $(\xi = \pm f_b(c), \eta = 0)$ of the σ-plane, while keeping the width and the position of the flux-free region of the strip fixed. The points $(x = \pm w, y = 0)$ of the s-plane, on the other hand, turn into the points $(\xi = \pm f_b(w), \eta = 0)$ of the σ-plane, keeping the position of the strip fixed, but adjusting its width, for $f_b(w) = w_b > w$, except in the limits $b \to w$, $c \to \infty$ or $p \to 1$, where w_b adopts the minimum value w. The initial gap between the edges of the strip and the surfaces of the magnetic shields, $a = c - w$, thus changes to the distance between the transformed edges of the strip and the vertices of the magnetic shields, $a_b = f_b(c) - w_b$. We note the link $f_b(s) = (c_b/c_w)f_w(s)$, recalling the analytic function $f_w(s)$ defined by eqn. (3.108) in the case of the Meissner state, where the flux-free region covers the entire width of the strip. Proper reinterpretation $a_w \to a_b$ and $w \to w_b$ understood, the variation in the shape of the magnetic cavity depicted in Fig. 3.13 therefore is representative here too. The symmetry of this configuration with respect to the midplane $\xi = 0$ of the strip entails that a characteristic similar to that in eqn. (4.1) applies, i.e. vanishing of the normal component of the transformed magnetic field $\hat{H}_\eta$ in the flux-free region of the strip,

$$\hat{H}_\eta(\xi) = 0 \quad \text{for } 0 \le |\xi| \le b,\ \eta = 0. \tag{4.64}$$

A characteristic similar to that in eqn. (4.2), viz. constancy of the transformed sheet current $\hat{J}$ in the flux-filled margins of the strip,

$$\hat{J}(\xi) = J_\mathrm{c} \quad \text{for } b \le |\xi| < w_b, \tag{4.65}$$

is provided as shown in section 4.4.2.

4.4.2 Current and field distributions

Since following along the lines of section 4.1, the sheet current $\hat{J}(\xi)$ within the flux-free region of the strip, $0 \le |\xi| \le b$, can be envisaged as arising from the effect of current filaments at $\xi = \pm\xi'$ for $b \le \xi' < w_b$ within the flux-filled margins of the strip, $b \le |\xi| < w_b$, the introduction of a transformed linear response function, $\hat{\chi}(\xi, \xi')$, suggests itself. This, by analogy to eqn. (4.9) upon observing the demand, eqn. (4.65), admits the representation

$$\hat{J}(\xi) = J_\mathrm{c} \int_b^{w_b} d\xi'\, \hat{\chi}(\xi, \xi') \quad \text{for } 0 \le |\xi| \le b. \tag{4.66}$$

Due extension of eqn. (3.105) shows that the relation

$$\hat{\chi}(\xi, \xi') = \chi\left[x(\xi, \eta), x(\xi', \eta)\right] \left.\frac{\partial y}{\partial \eta}\right|_{\eta=0} \tag{4.67}$$

holds, where $\chi(x, x') = \zeta(v, v')$ denotes the linear response function appropriate to the flat magnetic environment, with the auxiliary function $\zeta(v, v')$ derived in section 4.2.2.1, making use of the change of variables $x = x(\xi, \eta)$ and $x' = x(\xi', \eta)$ to $v = \tan(\pi x/2c)$ and $v' = \tan(\pi x'/2c)$, respectively, as well as of the particulars defining the map.

The rectangular cusp-shaped magnetic cavity depicted in Fig. 3.13(b) may serve as an example for further scrutiny. Thus, eqn. (4.63) with $p = 1/2$ affords an implicit transformation relationship from which the dependence $x(\xi, \eta)$ and the derivative $\partial y/\partial \eta$ at $\eta = 0$ can be deduced. Substituting these quantities into the right-hand side of eqn. (4.67) for the linear response function in the partially flux-filled critical state due to the flat magnetic environment yields the sheet current distribution across the entire width of the magnetically shielded strip in that state appropriate to the rectangular cusp-shaped magnetic cavity,

$$\hat{J}(\xi) = \begin{cases} J_c \,(d\vartheta_b/d\xi) \int_b^{w_b} d\xi' \,\zeta(v_b(\xi), v_b(\xi')) & \text{for } 0 \leq |\xi| \leq b \\ J_c & \text{for } b \leq |\xi| < w_b. \end{cases} \tag{4.68}$$

Herein,

$$\vartheta_b(\xi) = \xi \left(1 - \xi^2/4c_b^2\right)^{1/2}, \tag{4.69}$$

where from eqn. (4.38),

$$\zeta(v_b, v_b') = \frac{1 + v_b'^2}{2c} \sqrt{\frac{v_b'^2 - r^2}{r^2 - v_b^2}} \left(\frac{1}{v_b - v_b'} - \frac{1}{v_b + v_b'} \right) + \frac{1 + v_b'^2}{2c} \frac{\alpha + 2}{\sqrt{r^2 - v_b^2}} \tag{4.70}$$

with

$$v_b(\xi) = \tan\left(\pi \vartheta_b(\xi)/2c_b\right) \quad \text{and} \quad r = \tan\left(\pi b/2c\right). \tag{4.71}$$

Regularity and continuity of $\hat{J}(\xi)$ at $\xi = \pm b$ is ensured through the choice

$$\alpha + 2 = 2m/n \tag{4.72}$$

with

$$m = \int_b^{w_b} d\xi' v_b(\xi') \left(1 + v_b(\xi')^2\right) / \sqrt{v_b(\xi')^2 - r^2} \quad \text{and} \quad n = \int_b^{w_b} d\xi' \left(1 + v_b(\xi')^2\right), \tag{4.73}$$

duly extending eqn. (4.40). We note that the sheet current distribution, eqn. (4.68), can be cast into the equivalent form

$$\hat{J}(\xi) = \begin{cases} \dfrac{J_c}{\pi} \dfrac{d\vartheta_b}{d\xi} \displaystyle\int_0^{\varphi(\xi)} d\varphi' \left(1/\psi_+(\xi, \varphi')^{1/2} + 1/\psi_-(\xi, \varphi')^{1/2} \right) & \text{for } 0 \leq |\xi| \leq b \\ J_c & \text{for } b \leq |\xi| < w_b \end{cases} \tag{4.74}$$

deemed more convenient for further analysis, where

$$\psi_\pm(\xi, \varphi') = 1 \pm \frac{2}{\pi} \arctan \sqrt{r^2 + (r^2 - v_b(\xi)^2) \tan^2 \varphi'} \tag{4.75}$$

and

$$\varphi(\xi) = \arctan \sqrt{\frac{s^2 - r^2}{r^2 - v_b(\xi)^2}} \quad \text{with } s = \tan\left(\pi w/2c\right). \tag{4.76}$$

The superconducting states described by eqn. (4.74) fall into two categories: (i) sub critical states, which are always physically stable, and (ii) overcritical states, which can be either physically stable or unstable. There exists a threshold value $b = b_{\mathrm{t}}$ separating the stable and unstable regimes of the latter type of states, defined through the requirement that $d\hat{J}/d\xi = 0$ as $\xi \to \pm b$ from the flux-free region of the strip. To elucidate this fact, let us take a closer look at the situation where the distance between the transformed edges of the strip and the vertices of the magnetic shields is small, $a_b \ll w_b$, and where the flux-filled marginal parts of the strip are narrow, $w_b - b \ll w_b$, a case in point of foremost interest here. Under such conditions, the estimates $s \simeq w_b^2/\pi a_b^2 \gg 1$ and $r \simeq 2c/\pi(c-b) \gg 1$ hold, so that the derivative of the sheet current distribution, eqn. (4.74), adopts the simple approximate form

$$\frac{d\hat{J}}{d\xi} \sim \frac{J_{\mathrm{c}}}{\sqrt{r^2 - v_b(\xi)^2}} \left(1 + 2\sqrt{\frac{s}{r}} - \frac{s^2}{r^2}\right) \quad \text{as } \xi \to \pm b, \tag{4.77}$$

leaving a nonsingular modulating factor aside. Two important insights follow from eqn. (4.77) straight away: (i) since $v_b(\xi) \to \pm r$ as $\xi \to \pm b$, the sheet current distribution has a singular slope at the fronts of penetrated magnetic flux, typical of the two-dimensional geometry of thin superconductor strips (Norris 1970; Brandt and Indenbom 1993; Zeldov, Clem, McElfresh, and Darwin 1994); and (ii) the slope of the sheet current distribution changes sign when ξ crosses the points $\xi = \pm b_t$. This distribution thus reveals physical stability if $\sqrt{s/r} < \gamma_{\mathrm{t}}$, but instability if $\sqrt{s/r} > \gamma_{\mathrm{t}}$, the limit of stability being established by $\sqrt{s/r} = \gamma_{\mathrm{t}}$ with $\gamma_{\mathrm{t}} \simeq 7/5$ denoting the only positive zero of the bracketed term in eqn. (4.77). The latter condition, by means of the definitions, eqns. (4.71) and (4.76), takes the highly accurate shape

$$b_{\mathrm{t}} \simeq w_b - (\gamma_{\mathrm{t}} - 1)a_b; \qquad h < a_b \ll w_b. \tag{4.78}$$

Hence, if $b_{\mathrm{t}} < b < w_b$, overcritical states revealing stability against a collapse into the full critical state by virtue of transitional zones of subcritical sheet current in marginal parts of the strip exist; if $b < b_{\mathrm{t}} < w_b$, overcritical states lacking stability because of the absence of such zones persist.

Referring to eqn. (4.12) by analogy, eqn. (4.74) permits a direct evaluation of the total transport current

$$I = 2J_{\mathrm{c}}\,(w_b - b) + \int_{-b}^{b} d\xi'\,\hat{J}(\xi') \tag{4.79}$$

in terms of b, indicating the critical current $I_{\mathrm{c}} = 2w_b J_{\mathrm{c}}$. Since, given the premises above, the integral in eqn. (4.74) only weakly depends on ξ, eqn. (4.79) with eqn. (4.78) makes it possible to estimate the maximum total transport current at the limit of stability of overcritical states,

$$I_{\max} \simeq (w_b/2\pi a_b)\,I_{\mathrm{c}} \quad \text{for } b = b_{\mathrm{t}}; \quad h < a_b \ll w_b. \tag{4.80}$$

The distribution of the normal component of the magnetic field across the entire width of the magnetically shielded strip appropriate to the rectangular cusp-shaped magnetic cavity, which can be calculated from eqns. (3.103) to (3.105) with eqn. (3.73b) under consideration of eqns. (4.64) and (4.68) or (4.74), finally comes out as

$$\hat{H}_\eta(\xi) = \begin{cases} 0 & \text{for } 0 \leq |\xi| \leq b, \eta = 0 \\ J_c \hat{G}(\xi) + \dfrac{1}{4c_b} \displaystyle\int_{-b}^{b} d\xi' \hat{J}(\xi') \hat{Q}_-(\xi, \xi') & \text{for } b \leq |\xi| < w_b, \eta = 0, \end{cases} \tag{4.81}$$

where

$$\hat{G}(\xi) = \frac{1}{4c_b} \int_b^{w_b} d\xi' \left(\hat{Q}_+(\xi, \xi') + \hat{Q}_-(\xi, \xi') \right) \tag{4.82}$$

and

$$\hat{Q}_\pm(\xi, \xi') = (d\vartheta_b/d\xi) \cot\left[\pi \left(\vartheta_b(\xi) \pm \vartheta_b(\xi')\right)/2c_b\right], \tag{4.83}$$

the first term together with the integral on the right-hand side of eqn. (4.81) ensuring that $\hat{H}_\eta(\xi)$ vanishes at $\xi = \pm b$.

Figure 4.9(a) displays the variation of the normalized sheet current $\hat{J}/J_c$ over the width of the strip for a range of values of the normalized distance between the edges of the strip and the vertices of the magnetic shields a_b/w_b, adopting a fixed value of the normalized half-width of the flux-free region b/w_b. A general trait due to sheet current redistribution is the warping of the symmetric current profile around the central part of the strip and the increase of the total current itself, a phenomenon which gets more pronounced as a_b/w_b wanes, reaching saturation with a *nonuniform* maximum in the limiting sheet current profile of eqn. (4.74). Thus, for the two largest values of a_b/w_b, the downward slopes of the sheet current profiles at the fronts of penetrated magnetic flux give rise to *subcritical* sheet current with $\hat{J}/J_c < 1$ in the immediate vicinity of the flux-filled marginal parts of the strip, and hence to stability of the respective sheet current distributions overall. By contrast, for the two smallest values of a_b/w_b, the upward slopes of the sheet current profiles at the fronts of penetrated magnetic flux generate *overcritical* sheet current with $\hat{J}/J_c > 1$ in the entire flux-free region adjacent to the flux-filled marginal parts of the strip, and therefore instability of the respective sheet current distributions overall. Magnetic flux then moves towards the centre of the strip, and the partially flux-filled overcritical states collapse into the full critical state with $\hat{J}/J_c = 1$ prevailing all over the strip. A sheet current distribution of marginal stability, for an intermediate value of a_b/w_b, separates the two regimes; it is characterized by zero slopes of the sheet current profile at the fronts of penetrated magnetic flux. Figure 4.9(b) presents the variation of the corresponding normalized maximum total transport current $I_{\max}/I_c$ with the normalized distance between the edges of the strip and the vertices of the magnetic shields a_b/w_b. Starting from a distinct peak at the smallest value of a_b/w_b, there is a monotonic drop followed by an approach towards the respective value

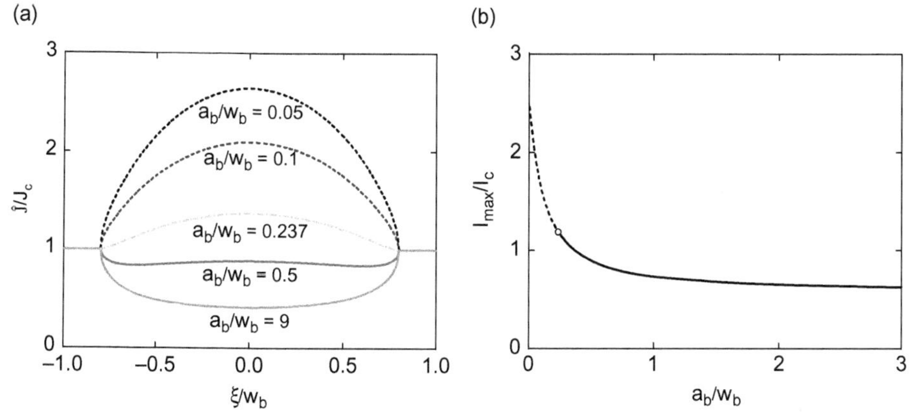

Fig. 4.9 (a) Distribution of the normalized sheet current $\hat{J}/J_c$ as a function of the normalized coordinate ξ/w_b over the width of the magnetically shielded superconductor strip in the partially flux-filled critical state for five different values of the normalized distance between the edges of the strip and the vertices of the magnetic shields a_b/w_b identified on the curves. Reproduced with permission from Genenko, Snezhko, and Freyhardt 2000 *Phys. Rev. B* **62**, pg 3453. Copyrighted by the American Physical Society, doi: https://doi.org/10.1103/PhysRevB.62.3453. (b) Variation of the corresponding normalized maximum total transport current I_{max}/I_c with the normalized distance between the edges of the strip and the vertices of the magnetic shields a_b/w_b, when the normalized half-width of the flux-free region $b/w_b = 0.8$, referring to the rectangular cusp-shaped paramagnet configuration of Fig. 3.13(b). The solid/dashed curves delineate physically stable and unstable regimes, respectively, of the partially flux-filled critical state, separated by the marginally stable critical state (dash-dotted curve or empty circle) in between.

prevailing for the isolated strip, as a_b/w_b continues to grow. Remarkably enough, although most of the peak current turns out to lack stability, some scope for stable *overcritical* total current with $I_{max}/I_c > 1$ still exists.

Figure 4.10(a) displays the variation of the normalized sheet current $\hat{J}/J_c$ over the width of the strip for a range of values of the normalized half-width of the flux-free region b/w_b, adopting a fixed value of the normalized distance between the edges of the strip and the vertices of the magnetic shields a_b/w_b. A general trait due to sheet current redistribution is the warping of the symmetric current profile around the central part of the strip up to a maximum and the initial increase of the total current itself, with a subsequent depression of the current profile and concomitant decrease of the total current too, as b/w_b augments, a phenomenon which reaches its final stage in the limiting *uniform* current-free profile of eqn. (4.74). Thus, for the two smallest values of b/w_b, the upward slopes of the sheet current profiles at the fronts of penetrated magnetic flux generate *overcritical* sheet current with $\hat{J}/J_c > 1$ in the entire flux-free region adjacent to the flux-filled marginal parts of the strip, and therefore instability of the respective sheet current distributions overall. Magnetic flux then moves towards the centre of the strip, and the partially flux-filled

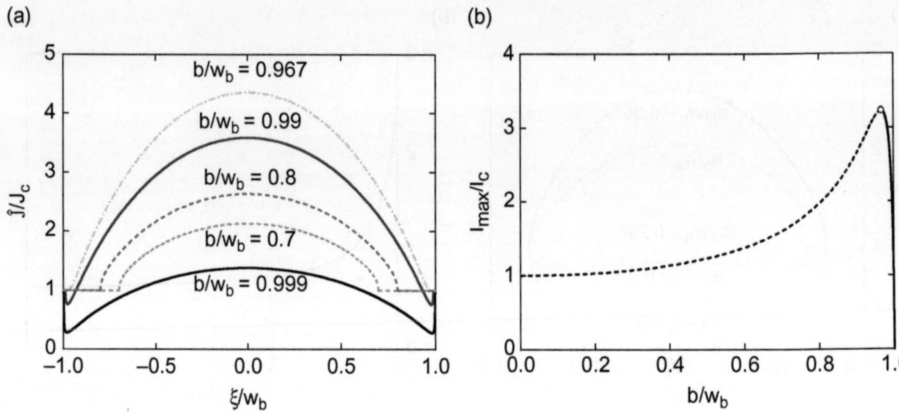

Fig. 4.10 (a) Distribution of the normalized sheet current $\hat{J}/J_c$ as a function of the normalized coordinate ξ/w_b over the width of the magnetically shielded superconductor strip in the partially flux-filled critical state for five different values of the normalized half-width of the flux-free region b/w_b identified on the curves. Reproduced with permission from Genenko, Snezhko, and Freyhardt 2000, *Phys. Rev. B* **62**, pg 3453, doi: https://doi.org/10.1103/PhysRevB.62.3453. Copyrighted by the American Physical Society. (b) Variation of the corresponding normalized maximum total transport current $I_{\max}/I_c$ with the normalized half-width of the flux-free region b/w_b, when the normalized distance between the edges of the strip and the vertices of the magnetic shields $a_b/w_b = 0.05$, referring to the rectangular cusp-shaped paramagnet configuration of Fig. 3.13(b). The solid/dashed curves delineate physically stable and unstable regimes, respectively, of the partially flux-filled critical state, separated by the marginally stable critical state (dash-dotted curve or empty circle) in between.

overcritical states collapse into the full critical state with $\hat{J}/J_c = 1$ prevailing all over the strip. By contrast, for the two largest values of b/w_b, the downward slopes of the sheet current profiles at the fronts of penetrated magnetic flux give rise to *subcritical* sheet current with $\hat{J}/J_c < 1$ in the immediate vicinity of the flux-filled marginal parts of the strip, and hence to stability of the respective sheet current distributions overall. A sheet current distribution of marginal stability, for an intermediate value of b/w_b, separates the two regimes; it is characterized by zero slopes of the sheet current profile at the fronts of penetrated magnetic flux. Figure 4.10(b) presents the variation of the corresponding normalized maximum total transport current $I_{\max}/I_c$ with the normalized half-width of the flux-free region b/w_b. Starting from unity, the value prevailing in the full critical state, at zero b/w_b, there is a monotonic rise towards a distinct maximum followed by a sharp drop to 0, as b/w_b continues to grow. Remarkably enough, although the current predicted for the greatest part of b/w_b turns out to lack stability, ample scope for stable *overcritical* total current with $I_{\max}/I_c > 1$ in a narrow region of b/w_b exists.

We reiterate, however, that high total currents associated with strong tangential components of the magnetic field around the centre of the strip might facilitate penetration of magnetic flux from the *surfaces* of the strip, and hence could severely restrict the loss-free current transport in type-II superconductors possessing small lower critical fields. An ultimate limitation, at any rate, derives from the quench of superconductivity through the depairing mechanism, as underlined in section 3.7.2.

4.4.3 Hysteretic AC loss

The dynamics of the magnetic flux in the partially flux-filled critical state of the magnetically shielded strip, with an oscillating transport current imposed, is basically the same as for an unshielded strip. Hence, eqn. (4.15) determines the electromagnetic energy dissipated during a steady-state cycle of the transport current in the presence of paramagnetic shields as well. Upon substituting the normal component of the magnetic field, eqn. (4.81), changing the variables ξ and ξ' to $\lambda = \xi/w_b$ and $\lambda' = \xi'/w_b$, respectively, as well as introducing the substitution $l_b = b/w_b$, the hysteretic AC loss per unit length of the strip, U_{ac}, for the cusp-shaped shielding geometry at hand thus can be cast into the form

$$U_{\mathrm{ac}} = U_0 \int_{l_b}^{1} d\lambda \int_{l_b}^{\lambda} d\lambda' \, \hat{H}_\eta(\lambda')/H_c, \tag{4.84}$$

where $U_0 = 2\mu_0 I_c^2/\pi$, by due reference to eqn. (4.79) for physically stable states.

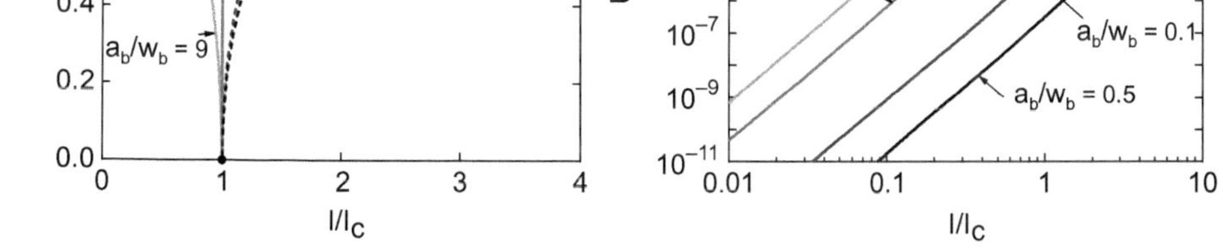

Fig. 4.11 Dependence of (a) the normalized half-width of the flux-free region b/w_b and (b) the normalized hysteretic AC loss U_{ac}/U_0 on I/I_c, the normalized total transport current imposed on the magnetically shielded superconductor strip in the partially flux-filled critical state, for four different values of the normalized distance between the edges of the strip and the vertices of the magnetic shields a_b/w_b identified on the curves, referring to the rectangular cusp-shaped paramagnet configuration of Fig. 3.13(b). The empty circles terminating some of the curves separate physically stable and unstable regimes of the partially flux-filled critical state, whereas the solid circles terminating the residual curves denote the onset of the full critical state.

Figure 4.11(a) shows the variation of the normalized half-width of the flux-free region b/w_b with the normalized transport current I/I_c for a range of values of the normalized distance between the edges of the strip and the vertices of the magnetic shields a_b/w_b. Thus, although the ordering of the curves unfolds that penetration of magnetic flux, and hence the transition from the partially flux-filled critical state to the full critical state, is *retarded*, as a_b/w_b wanes, very much like in the case of transversely oriented flat magnetic shields, important qualitative differences persist. For the two largest values of a_b/w_b typifying stable subcritical states, the curves exhibit monotonic behaviours, and hence single-valued dependences of b/w_b on I/I_c, within the critical current domain. By contrast, for the two smallest values of a_b/w_b identifying overcritical states, the curves disclose nonmonotonic behaviours, and therefore multi-valued dependences of b/w_b on I/I_c, beyond the critical current domain, the respective branches separating stable from unstable regimes. Instability here arises because of direct contact of the central flux-free overcritical region with the flux-filled margins of the strip, as Fig. 4.10(a) depicts. This generates a Lorentz-like force exerted upon magnetic vortices which dominates the pinning force, dragging the vortices towards the centre of the strip and, hence, facilitating a transition to the critical state. Figure 4.11(b) portrays the variation in corresponding normalized hysteretic AC loss U_{ac}/U_0 with the normalized transport current I/I_c for the said range of values of the normalized distance between the edges of the strip and the vertices of the magnetic shields a_b/w_b. This reveals that, whereas at large a_b/w_b the hysteretic AC loss is virtually like that for an unshielded strip, it decreases substantially throughout the transport current regime, as a_b/w_b abates, overcritical states outperforming subcritical states and extending their range by far.

5

Meissner state and critical state of a superconductor strip in ideal diamagnetic environments

The use of superconducting constituents as diamagnetic shields in *passive* large-scale power and microelectronic device applications has been a proven technology for a long time (Newhouse 1969). Magnetic shielding there serves to increase the performance of, e.g., microwave filters, high-speed digital receivers, or electrical transmission lines, with thin type-II superconductor strips representing the current-carrying elements, by suppressing the peaks of the sheet current at the edges of the strips which otherwise could promt local penetration of magnetic flux associated with a nonlinear electromagnetic response (Monaco et al. 2000; Lancaster 2006). Typical arrangements for reduction of the inhomogeneity of the sheet current distribution set out from a thin superconductor film deposited on an insulated superconducting ground plane which acts like a flat support (Muchowski and Schmid 1972; Chang 1979). Small gaps between the film and the support result in a depression of the aforementioned peaks, providing an almost uniform distribution of the sheet current across the width of the film. A similar effect finds routine application for obtaining homogenized current density profiles in magnetically shielded long Josephson junctions too (Barone and Paternò 1982). Superconducting shields, not least, help widely to master the electromagnetic compatibility task of protecting field-sensitive devices and personnel from unwanted interference with strong magnetic fields (Ott 2009).

In contrast to this, arrangements of a superconductor strip placed between ideal transversely oriented paramagnetic shields or surrounded by ideal paramagnetic cavities demonstrate that magnetic shielding is able to play an *active* role in redistributing the sheet current, significantly enhancing the total loss-free current that can be carried by the strip and considerably reducing the hysteretic AC loss (cf. Chapters 3 and 4). We here emphasize the opposite extreme of zero permeability of the magnetic constituents, $\mu = 0$, illustrated by the example of Fig. 2.7(b) for a perfect spherical diamagnet with complete expulsion of magnetic flux which a superconductor in the fully developed Meissner state reflects, its flux-filled surface

Electromagnetics of Superconductor/Paramagnet Heterostructures. Yuri Genenko and Hermann Rauh, Oxford University Press.

DOI: 10.1093/9780191782855.003.0005

sheath exempted. It turns out that a reciprocity in performance of a superconductor strip with respect to longitudinal orientation of flat diamagnetic shields, on the one hand, and transverse orientation of flat paramagnetic shields on the other, exists, covering both the flux-free Meissner state and the partially flux-filled critical state. Similar findings also carry over to convex diamagnetic and concave paramagnetic shields; further magnetic environments, more elaborate than those commonly employed, are under investigation. These correspondences could be exploited, e.g., to greatly enhance the output signal of microstrip devices for which, from a practical point of view, superconducting constituents only are favoured in many applications (Koshelets and Shitov 2000; Lancaster 2006), yet bypassing the inherent limitations which superconducting shields naturally underlie.

5.1 Superconductor strip between two flat semi-infinite diamagnets: Meissner state

We start by addressing an infinitely extended, thin current-carrying type-II superconductor strip in the Meissner state, now placed between two longitudinally oriented flat semi-infinite diamagnetic shields with permeability μ, in which case the surfaces of the diamagnets run parallel to the strip, at the distance a from the plane of the strip, as illustrated in Fig. 5.1. The account that follows augments the work of Genenko and Rauh (2003).

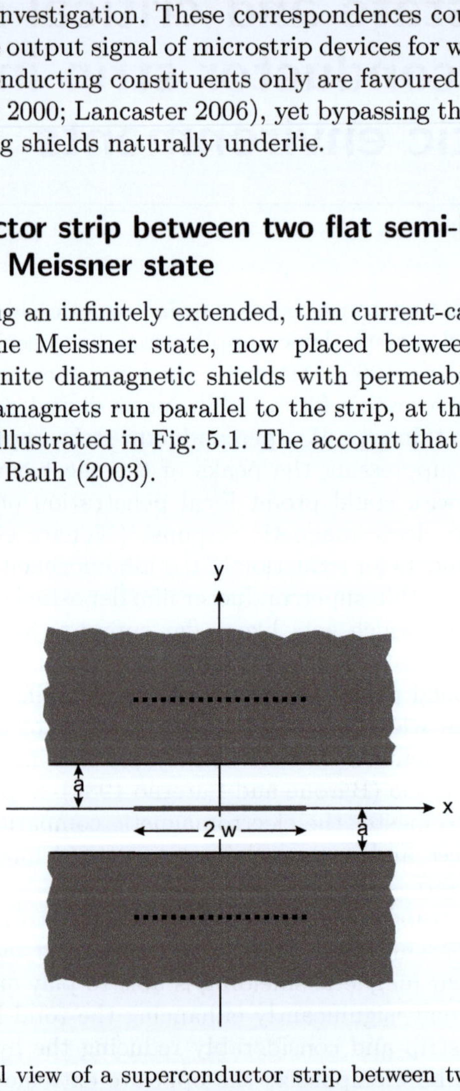

Fig. 5.1 Cross-sectional view of a superconductor strip between two flat semi-infinite diamagnets with longitudinal orientation extending infinitely in the z-direction of a Cartesian coordinate system x, y, z. The definition of the distance between the plane of the strip and the diamagnets a as well as two of the images of the strip (dotted lines) are marked.

5.1.1 Field distribution

Clearly, eqns. (2.87) and (2.88) here simplify to the Poisson equation for the single nontrivial component of the magnetic vector potential A_z,

$$\Delta A_z = \begin{cases} -\mu_0 J(x)\delta(y) & \text{in } -w \leq x \leq w \\ 0 & \text{otherwise,} \end{cases} \tag{5.1}$$

supplemented with the boundary conditions

$$\lim_{|y|\to a-} A_z(x,y) = \lim_{|y|\to a+} A_z(x,y) \quad \text{and} \quad \left.\frac{1}{\mu_0}\frac{\partial A_z}{\partial n}\right|_{|y|\to a-} = \left.\frac{1}{\mu}\frac{\partial A_z}{\partial n}\right|_{|y|\to a+}, \tag{5.2}$$

according to the explanations of section 3.4.1.1. Obviously, the latter demands rule out direct contact between the magnetic shields and the surfaces of the strip to avoid complications arising from the proximity effect (Schmidt 1997). The desired solution thus can be constructed by the method of images like before, calling in their current strength $\kappa = (\mu - \mu_0)/(\mu + \mu_0)$, along with the Green's function technique. Identifying the contributions $dI = dx' J(x')$ of a distributed current element of unit weight at $(x = x', y = 0)$ as well as of image current elements of weight $\kappa^{|n|}$ at $(x = x', y = 2na)$ for $n = \pm1, \pm2, \pm3, ...$ and integrating over x' yields the convenient representation in the vacuum space, $0 \leq |y| < a$ inclusive of the surface of the strip ∂D,

$$A_z(x,y) = \frac{\mu_0}{2\pi}\sum_{n=-\infty}^{\infty}\kappa^{|n|}\int_{-w}^{w} dx' J(x') \ln\frac{a}{\sqrt{(x-x')^2 + (y-2na)^2}}. \tag{5.3}$$

Let us establish a closed form of eqn. (5.3) in the limit of zero permeability of ideal diamagnetic shields, $\mu = 0$, where the strip and its images form an infinite current-carrying bifilar stack (Clem 2008). Thus, summing up the partial derivatives with respect to x and y of A_z, eqn. (5.3), by means of customary table sets (Prudnikov, Brychkov, and Marychev 1986) and integrating the total differential of A_z yields the exact representation in the vacuum space, $0 \leq |y| < a$ inclusive of ∂D,

$$\begin{aligned} A_z(x,y) = &\frac{\mu_0}{2\pi}\int_{-w}^{w} dx' J(x') \ln\frac{a}{\sqrt{(x-x')^2 + y^2}} \\ &+ \frac{\mu_0}{2\pi}\int_{-w}^{w} dx' J(x') \ln\left|\vartheta_+(x-x',y)\vartheta_-(x-x',y)\right|, \end{aligned} \tag{5.4}$$

where

$$\vartheta_\pm(x-x',y) = \Gamma\left[(i(x-x') \pm y)/4a + 1/2\right]/\Gamma\left[(i(x-x') \pm y)/4a\right] \tag{5.5}$$

with Euler's gamma function Γ (Abramowitz and Stegun 1972). The first term in eqn. (5.4) emulates the contribution of an isolated superconductor strip, recalling eqn. (3.4), whereas the second term in eqn. (5.4) reflects the influence of the diamagnetic shields. Application of eqn. (2.3) to eqn. (5.4) thus gives the Cartesian components of the magnetic field **H**,

$$H_x(x,y) = -\frac{1}{2\pi}\int_{-w}^{w} dx' J(x')\frac{y}{(x-x')^2+y^2}$$

$$+\frac{1}{8\pi a}\int_{-w}^{+w} dx' J(x')\mathrm{Re}\left[\phi_+(x-x',y)-\phi_-(x-x',y)\right] \qquad (5.6a)$$

$$H_y(x,y) = \frac{1}{2\pi}\int_{-w}^{w} dx' J(x')\frac{x-x'}{(x-x')^2+y^2}$$

$$+\frac{1}{8\pi a}\int_{-w}^{+w} dx' J(x')\mathrm{Im}\left[\phi_+(x-x',y)+\phi_-(x-x',y)\right], \qquad (5.6b)$$

where

$$\phi_\pm(x-x',y) = \psi\left[\left(i(x-x')\pm y\right)/4a+1/2\right]-\psi\left[\left(i(x-x')\pm y\right)/4a\right], \qquad (5.7)$$

introducing Euler's psi function ψ (Abramowitz and Stegun 1972). Since in the diamagnet space, $|y| \geq a$, the magnetic induction vanishes identically, $\mathbf{B} = 0$, eqn. (2.3) shows that A_z is constant throughout here, adopting a value consistent with eqn. (5.2). The magnetic field, for the sheet current, eqn. (5.11), is visualized in Fig. 2.13. Obviously, a congruence exists between the magnetic field **H** and the velocity field **v** of steady-state rotational flow of an incompressible perfect fluid about a strip-like vortex core placed in a longitudinally oriented flat interspace. The allied problem of steady-state rotational flow of an incompressible perfect fluid permeating a regular cascade has been considered elsewhere (Meister 1983).

5.1.2 Current distribution

As noted in section 3.4.1.2 the Meissner state of the superconductor strip requires the normal component of the magnetic field H_y to vanish across the width of the strip,

$$H_y(x) = 0 \quad \text{for} \quad -w < x < w, y = 0. \qquad (5.8)$$

Recalling eqn. (5.6b) hence provides the condition for this state

$$\int_{-w}^{w} dx' J(x')\mathrm{csch}\left(\pi(x-x')/2a\right) = 0 \quad \text{in} \quad -w < x < w. \qquad (5.9)$$

The change of the variables x and x' to $u = \tanh(\pi x/2a)$ and $u' = \tanh(\pi x'/2a)$, respectively, together with the introduction of the substitution $q = \tanh(\pi w/2a)$ and the request for an even solution in x transform eqn. (5.9) into the singular integral equation of the Cauchy type

$$\int_{-q}^{q} du' S(u') \frac{1}{u - u'} = 0 \quad \text{in} \quad -q < u < q \tag{5.10}$$

for the auxiliary function $S(u') = J(x')$. The functional form of its solution hence follows from eqn. (3.8); ensuring normalization, eqn. (3.2), finally yields the sheet current as

$$J(x) = \frac{\pi I}{4aK(m)} \frac{\cosh(\pi w/2a)}{\sqrt{[\cosh(\pi w/a) - \cosh(\pi x/a)]/2}} \quad \text{for} \quad -w < x < w, \tag{5.11}$$

bringing in the complete elliptic integral of the first kind K with the parameter $m = \tanh^2(\pi w/2a)$ (Abramowitz and Stegun 1972). Although overstretching the model requirements, it is instructive to formally test eqn. (5.11) against the apparent situation of direct strip–magnet contact, when $a \to 0$. In this case, the sheet current and its images make up an effective current-carrying 'bifilar superconductor slab' which occupies the space between the planes $x = \pm w$. Current flow permeates the whole slab, consistent with a uniform sheet current distribution in the strip, $J(x) = I/2w$ for $-w < x < w$, as demanded by the Meissner state.

Reasoning in a similar vein applicable to the isolated superconductor strip in the barrier-protected Meissner state of section 3.1 limits the validity of eqn. (5.11) to the range $-w + h \leq x \leq w - h$ and simultaneously recommends imposing the maximum value $J_{\max} = J_b$ in $|x \pm w| \leq h$. Substituting $I_{\max} = 2w\bar{J}_{\max}$ for I in eqn. (5.11) with eqn. (3.14) therefore yields the maximum average sheet current $\bar{J}_{\max}$, which may be approximated by the highly accurate simple form

$$\bar{J}_{\max} \simeq 2K(m)H_{c1}\sqrt{\frac{2h}{w}}\sqrt{\frac{\tanh(\pi w/2a)}{\pi w/2a}}; \quad h \ll w < a. \tag{5.12}$$

Comparing eqn. (5.12) with eqn. (3.15) shows that the maximum average sheet current $\bar{J}_{\max}$, and hence the maximum loss-free total current $I_{\max}$, in the presence of the longitudinally oriented flat diamagnetic shields is *larger* than the respective quantities for the unshielded strip, the strongest current enhancement $I_{\max}/I_b \simeq (1/\pi)\sqrt{2w/h}$ certainly being achieved if $a \to 0$.

Figure 5.2(a) displays the variation of the normalized sheet current J/J_b over the width of the strip for a range of values of the normalized distance between the plane of the strip and the longitudinally oriented flat diamagnetic shields a/w. A general trait due to sheet current redistribution in the strip is the steady evening out of the symmetric current profile around the central part of the strip and the increase of the total current itself, a phenomenon which gets more pronounced as a/w wanes, reaching saturation with a *uniform* maximum in the limiting sheet current profile of eqn. (5.11). Figure 5.2(b) presents the variation in the corresponding normalized maximum total transport current $I_{\max}/I_b$ with the normalized distance between the plane of the strip and the longitudinally oriented flat diamagnetic shields a/w. Starting from a distinct peak at the smallest value of a/w, there is a monotonic drop followed by an approach towards unity, the value prevailing for the isolated strip, as a/w continues to grow.

Attention should be drawn to the fact, at last, that a reciprocity in performance between the *longitudinally* oriented flat *diamagnetic* shields, on the one hand, and *transversely* oriented flat *paramagnetic* shields, on the other, exists, as a juxtaposition of Fig. 5.2 and Fig. 3.9 shows: both types of environments yield qualitatively and quantitatively similar results.

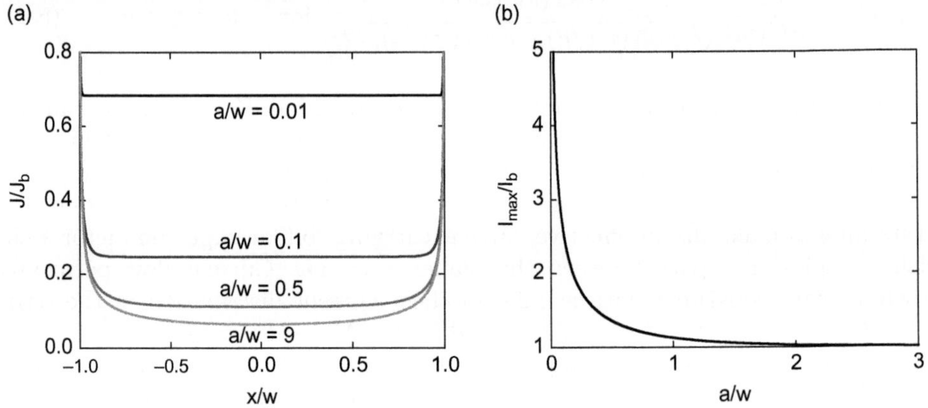

Fig 5.2 (a) Distribution of the normalized sheet current J/J_b as a function of the normalized coordinate x/w over the width of the magnetically shielded superconductor strip in the flux-free Meissner state for four different values of the normalized distance between the plane of the strip and the diamagnets a/w identified on the curves; (b) variation of the corresponding normalized maximum total transport current $I_{\max}/I_b$ with the normalized distance between the plane of the strip and the diamagnets a/w, when $h/2w = 0.001$, referring to the longitudinal diamagnet configuration of Fig. 5.1. Reproduced with permission from Genenko and Rauh 2003, *Appl. Phys. Lett.* **82**, pg 2115, Copyright 2003 AIP Publishing LLC. doi: https://doi.org/10.1063/1.1560866.

5.2 Superconductor strip between two curved semi-infinite diamagnets: Meissner state

We here revisit the infinitely extended, thin current-carrying type-II superconductor strip in the Meissner state placed between two longitudinally oriented flat semi-infinite diamagnetic shields with zero permeability, $\mu = 0$, the configuration studied in section 5.1, generalizing it to curved semi-infinite diamagnetic shields by means of the conformal mapping technique. As Fig. 5.1 suggests, we thereby allow the strip to be detached from the magnetic environment, assuming $a > 0$. The account that follows proceeds from the work of Genenko and Rauh (2003).

5.2.1 Conformal transformation

Referring to section 3.6.1.2, we note that a conformal transformation effects a change of coordinates from the complex variable $s = x + iy$ to the complex variable $\sigma = \xi + i\eta$ by means of some analytic function $\sigma = f(s)$ with a nonzero derivative df/ds. This maps areas of the s-plane onto areas of the σ-plane, implying $x = x(\xi, \eta)$ and $y = y(\xi, \eta)$. Turning to the problem at hand, let us map the flat half-domains $|y| \geq a$ of the s-plane onto some blunt wedge-shaped half-domains of the σ-plane, with asymptotic opening half-angle of their surfaces $\beta = p\pi/2$, where $0 < p \leq 1$, by again employing eqn. (3.108),

$$f_w(s) = c_w\left[(1 + s/c)^p - (1 - s/c)^p\right] \quad \text{with } c_w = w/\left[(1 + w/c)^p - (1 - w/c)^p\right], \tag{5.13}$$

which introduces the parameter $c > w$. The function $f_w(s)$ is analytic in the whole s-plane, except for two cuts on the real axis, $x < -c, y = 0$ and $x > c, y = 0$. Application of eqn. (5.13) transforms the points $(x = 0, y = \pm a)$ of the s-plane into the points $(\xi = 0, \eta = \pm f_w(ia)/i)$ of the σ-plane, while keeping the width and the position of the strip fixed. The initial gap between the plane of the strip and the surfaces of the diamagnetic shields, a, thus changes to the distance between the plane of the strip and the tips of the diamagnetic shields, $a_w = f_w(ia)/i$, that is,

$$a_w = 2c_w|\sin(p\tau)|\sec^p\tau, \tag{5.14}$$

whereas the radius of curvature of the tips themselves becomes

$$r_w = 2c_w|p\cos^2((1 - p)\tau)/(1 - p)\sin((2 - p)\tau)|\sec^p\tau, \tag{5.15}$$

implying $\tau = \arctan\gamma$ with $\gamma = a/c$. Evidently, the latter distance a_w—or, alternatively, the radius r_w—and the asymptotic opening half-angle β measured via p represent the geometrical characteristics of the blunt wedge-shaped magnetic shields; they can be installed so as to control the distribution of the sheet current in the strip by a judicious choice of the parameters w, a, c, and p.

The variation in the shape of the magnetic shields with a_w and p is revealed in Fig. 5.3. Accordingly, these shields consist of two separate parts which are elongated in the direction normal to the plane of the strip, forming a pair of averted wedges and opening asymptotically far away from the strip. For p held fixed, the tips of the wedges round off if a_w augments (Fig. 5.3(a)). The same is true if p, and hence β, shrinks, for a_w held fixed (Fig. 5.3(b)). The choice $c - w \ll w$ underlines both results. Since conformal mapping preserves the angle between two intersecting curves, the lines of the transformed magnetic field $\hat{\mathbf{H}}$ still run along the surfaces of the diamagnetic shields and the surfaces of the strip.

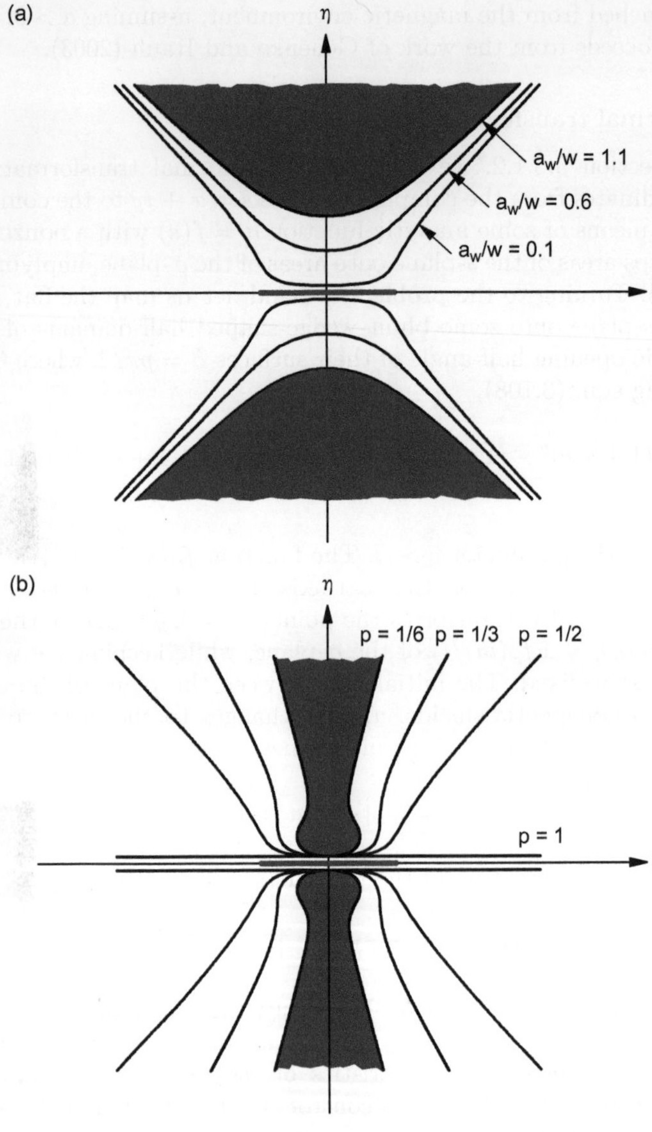

Fig 5.3 Cross-sectional view of the boundaries of the semi-spaces in the σ-plane, occupied by blunt wedge-shaped diamagnetic shields with a superconductor strip placed inside, for (a) the parameter of the conformal map $p = 1/2$ as well as three different values of the normalized distance between the plane of the strip and the tips of the magnetic shields a_w/w identified on the curves, and for (b) the normalized distance between the plane of the strip and the tips of the magnetic shields $a_w/w = 0.1$ as well as four different values of the parameter of the conformal map p identified on the curves, when $c/w = 1.001$. The width and the position of the strip are fixed.

5.2.2 Current distribution

The distribution of the sheet current across the width of the superconductor strip in the Meissner state appropriate to curved diamagnetic shields follows from

$$\hat{J}(\xi) = J[x(\xi,\eta)] \left. \frac{\partial y}{\partial \eta} \right|_{\eta=0}, \tag{5.16}$$

as stated by eqn. (3.105), making use of the distribution for the flat magnetic environment, eqn. (5.11), as well as of the particulars defining the map. We thus get the form

$$\hat{J}(\xi) = \frac{\pi I}{4\gamma c_w K(m)} \frac{d\vartheta_w}{d\xi} \frac{\cosh\left(\pi\vartheta_w(w)/2\gamma c_w\right)}{\sqrt{\left[\cosh\left(\pi\vartheta_w(w)/\gamma c_w\right) - \cosh\left(\pi\vartheta_w(\xi)/\gamma c_w\right)\right]/2}}$$
$$\text{for} \quad -w < \xi < w, \tag{5.17}$$

introducing the complete elliptic integral of the first kind K with the parameter $m = \tanh^2(\pi w/2a)$ (Abramowitz and Stegun 1972). The function, eqn. (5.13), yields the implicit relationship

$$(1 + \vartheta_w(\xi)/c_w)^p - (1 - \vartheta_w(\xi)/c_w)^p = \xi/c_w, \tag{5.18}$$

whose solution also furnishes the derivative

$$d\xi/d\vartheta_w = p/(1 + \vartheta_w(\xi)/c_w)^{1-p} + p/(1 - \vartheta_w(\xi)/c_w)^{1-p} \tag{5.19}$$

for substitution of its inverse in eqn. (5.8). Once more, though overstretching the model requirements as on the occasion of flat diamagnetic shields, eqn. (5.8) may be formally checked against the situation of direct strip–magnet contact, when $a \to 0$. Here, the magnetic field concentrates around the centre, and falls off towards the edges of the strip, entailing a cap-like sheet current distribution over the width of the strip, viz. $\hat{J}(\xi) = (I/2c_w)(c/w)(d\vartheta_w/d\xi)$ for $-w < \xi < w$. This elucidates a modulation of the Meissner state given by the ratio of the auxiliary functions, eqn. (5.19), which ultimately governs the current-carrying capability of the strip, a genuine asset of curved diamagnetic shields reminiscent of that for curved paramagnetic shields (cf. section 3.7.2). Again reasoning applicable to the magnetically shielded superconductor strip in the barrier-protected Meissner state of section 5.1 limits the validity of eqn. (5.8) to the range $-w+h \le \xi \le w-h$ and simultaneously recommends imposing the maximum value $J_{\max} = J_b$ in $|\xi \pm w| \le h$.

Figure 5.4(a) displays the variation of the normalized sheet current $\hat{J}/J_b$ over the width of the strip for a range of values of the normalized distance between the plane of the strip and the tips of the magnetic shields a_w/w, adopting a fixed value of the parameter of the conformal map p. A general trait due to sheet current redistribution in the strip is the warping of the symmetric current profile around the central part of the strip and the increase of the total current itself, a phenomenon which gets more pronounced as a_w/w wanes, reaching saturation with a *nonuniform* maximum in the limiting sheet current profile of eqn. (5.8), when the magnetic shields touch the strip. Additionally, a comparison with the situation of

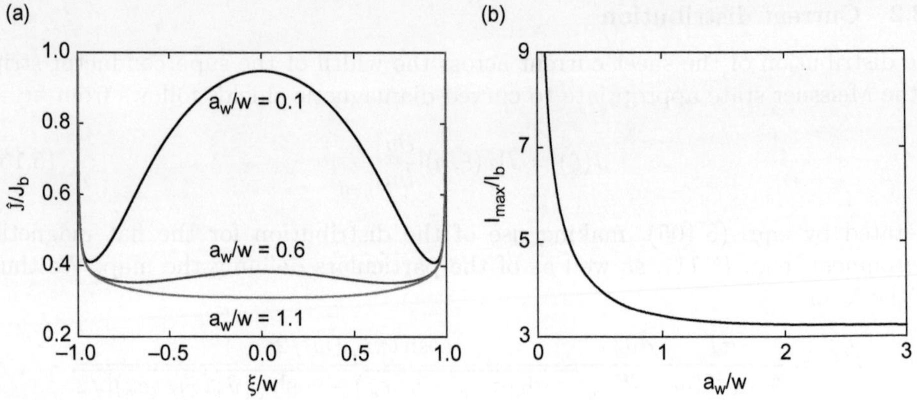

Fig 5.4 (a) Distribution of the normalized sheet current $\hat{J}/J_b$ as a function of the normalized coordinate ξ/w over the width of the magnetically shielded superconductor strip in the flux-free Meissner state for three different values of the normalized distance between the plane of the strip and the tips of the magnetic shields a_w/w identified on the curves; (b) variation in the corresponding normalized maximum total transport current $I_{\max}/I_b$ with the normalized distance between the plane of the strip and the tips of the magnetic shields a_w/w, when the parameter of the conformal map $p = 1/2$, apart from $c/w = 1.001$, referring to the blunt wedge-shaped diamagnet configuration of Fig. 5.3(a).

flat diamagnetic shields, Fig. 5.2(a), reveals that, for a_w/w and p kept fixed, the sheet current throughout the width of the strip is consistently *higher* in the case of curved diamagnetic shields. Figure 5.4(b) presents the variation in the corresponding normalized maximum total transport current $I_{\max}/I_b$ with the normalized distance between the plane of the strip and the tips of the magnetic shields a_w/w for the same value of p. Starting from a pronounced maximum at the smallest value of a_w/w, there is a sharp monotonic drop followed by a gradual settlement *above* unity, the value prevailing for the isolated strip, as a_w/w continues to grow. This result, which is intimately connected with the opening of the curved diamagnetic shields, comes down to unity in the limit of flat diamagnetic shields. A comparison with the latter situation hence shows that the maximum total transport current for fixed a_w/w is always *larger* in the case of curved diamagnetic shields. Moreover, it appears from Fig. 5.4, on the one hand, and from Fig. 3.14, on the other, that the distribution of the sheet current and the maximum total transport current too can be controlled by convex *diamagnetic* shields or concave *paramagnetic* shields with similar efficiency.

Figure 5.5(a) displays the variation of the normalized sheet current $\hat{J}/J_b$ over the width of the strip for a range of values of the parameter of the conformal map p, adopting a fixed value of the normalized distance between the plane of the strip and the tips of the magnetic shields a_w/w. A general trait due to sheet current redistribution in the strip is the formation of a hump of the symmetric current profile around the central part of the strip and the increase of the total current itself; a phenomenon which gets more pronounced as p wanes, reaching saturation in the limiting sheet current profile of eqn. (5.8), when the magnetic shields contract. Additionally, a comparison with the situation of flat diamagnetic shields, Fig. 5.2(a), again reveals that, for p and a_w/w kept fixed, the sheet current throughout the width of the strip is consistently *higher* in the case of curved diamagnetic shields. The appearance of a central maximum in particular allows for sheet currents exceeding the barrier value at the edges of the strip, $\hat{J}/J_b > 1$, separated by transitional zones of sheet current with $\hat{J}/J_b < 1$ in the marginal parts of the strip. Figure 5.5(b) presents the variation of the corresponding normalized maximum total transport current $I_{\max}/I_b$ with the parameter of the conformal map p for the same value of the normalized distance between the plane of the strip and the tips of the magnetic shields a_w/w. Starting from a distinct maximum at the smallest value of p, there is a monotonic fall towards the value appropriate to flat diamagnetic shields, as p continues to grow. A comparison with the situation of flat diamagnetic shields, Fig. 5.2(b), thus shows that the maximum total transport current for fixed a_w/w is always *larger* in the case of curved diamagnetic shields.

We note that the limitations regarding the predicted enhancement of the total transport current in the flux-free Meissner state of a type-II superconductor strip, as discussed at the end of section 3.7.2, naturally apply here too. Curved diamagnetic shields realized by superconductors in this state entail further concerns, since large sheet currents around the centre of the strip—and consequently large sheet

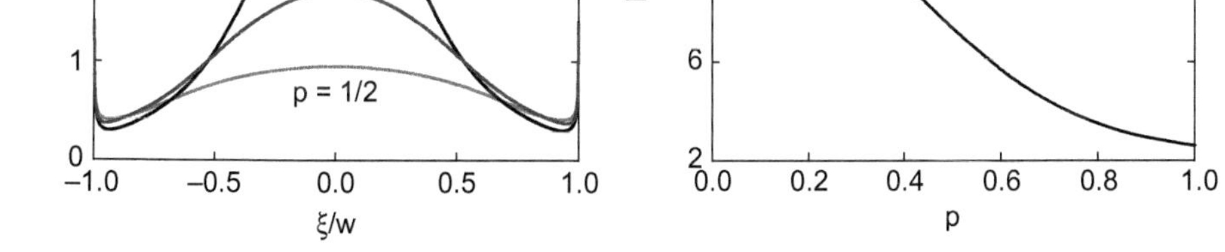

Fig 5.5 (a) Distribution of the normalized sheet current $\hat{J}/J_b$ as a function of the normalized coordinate ξ/w over the width of the magnetically shielded superconductor strip in the flux-free Meissner state for three different values of the parameter of the conformal map p identified on the curves; (b) variation of the corresponding normalized maximum total transport current $I_{\max}/I_b$ with the parameter of the conformal map p, when the normalized distance between the plane of the strip and the tips of the magnetic shields $a_w/w = 0.1$, apart from $c/w = 1.001$, referring to the blunt wedge-shaped diamagnet configuration of Fig. 5.3(b).

currents induced in the tips of the shields—may cause local penetration of magnetic flux associated with a nonlinear electromagnetic response (Monaco et al. 2000; Lancaster 2006). However, according to the criterion for stability of the Meissner state, magnetic flux is not expected to penetrate unless the local sheet current $\hat{J}$ exceeds the threshold value $\hat{J}_t = H_{c1}(\lambda_L/\xi_{GL})/\ln(\lambda_L/\xi_{GL})$, high-quality superconductor surfaces understood (Orlando and Delin 1991; Genenko et al. 1998). Recalling the barrier value $J_b = H_{c1}/2$, eqn. (3.14), we therefore conclude that superconductors with $\xi_{GL} \ll \lambda_L$ should still offer ample scope for preserving the Meissner state.

5.3 Superconductor strip between two flat semi-infinite diamagnets: critical state

We go on now to further develop the considerations of section 4.1, looking at an infinitely extended, thin current-carrying type-II superconductor strip in the partially flux-filled critical state, placed between two longitudinally oriented flat semi-infinite diamagnetic shields with zero permeability, $\mu = 0$, as illustrated in Fig. 5.1. The symmetry of this configuration with respect to the midplane $x = 0$ of the strip entails that the characteristics, eqns. (4.1) and (4.2), still apply. The presentation that follows adds to the work of Genenko (2002b).

5.3.1 Current and field distributions

In order to obtain the distributions of the sheet current and the magnetic field across the width of the superconductor strip in the partially flux-filled critical state, we proceed by resorting to eqn. (5.9) to characterize the flux-free Meissner state,

$$\int_{-w}^{w} dx' J(x') \mathrm{csch}\,(\pi(x - x')/2a) = 0 \quad \text{in} \quad -w < x < w. \tag{5.20}$$

Since, like in section 4.2.1, the sheet current $J(x)$ within the flux-free region of the strip, $0 \le |x| \le b$, can be envisaged as originating from the effect of current filaments at $x = \pm x'$ for $b \le x' < w$ within the flux-filled margins of the strip, $b \le |x| \le w$, the ansatz

$$J(x) \to \begin{cases} \chi(x, x') & \text{for} \quad 0 \le |x| \le b \\ \delta(x - x') + \delta(x + x') & \text{for} \quad b \le |x| < w \end{cases} \tag{5.21}$$

introducing a linear response function, $\chi(x, x')$, in eqn. (5.20) suggests itself. The change of the variables x and x' to $u = \tanh(\pi x/2a)$ and $u' = \tanh(\pi x'/2a)$, respectively, together with the substitutions $r = \tanh(\pi b/2a)$ and $s = \tanh(\pi w/2a)$ transforms eqn. (5.20) into the Cauchy-like integral equation

$$\int_{-r}^{r} du''\, \zeta(u'', u') \frac{1}{u - u''} = -\gamma(u, u') \quad \text{in} \quad 0 \le |u| \le r \tag{5.22}$$

for the auxiliary function $\zeta(u'', u') = \chi(x'', x')/\sqrt{1 - u''^2}$, with the singular two-part inhomogeneity

$$\gamma(u, u') = \frac{\pi}{2a}\sqrt{1 - u'^2}\left(\frac{1}{u - u'} + \frac{1}{u + u'}\right). \tag{5.23}$$

Owing to the formal analogy of eqns. (5.22), (5.23) with eqns. (4.5), (4.6) derived for the isolated superconductor strip, the general solution of eqn. (5.22) can be directly transcribed from eqn. (4.7) to yield

$$\zeta(u, u') = \frac{1}{\pi^2}\int_{-r}^{r} du''\sqrt{\frac{r^2 - u''^2}{r^2 - u^2}}\,\gamma(u'', u')\frac{1}{u - u''} + \frac{1 - u'^2}{\pi}\frac{\alpha}{\sqrt{r^2 - u^2}} \quad \text{for } 0 \le |u| \le r, \tag{5.24}$$

the term labelled with a free constant α on the right-hand side of eqn. (5.24) satisfying the homogeneous equation to eqn. (5.22). Evaluation of the integral in eqn. (5.24) with eqn. (5.23) gives

$$\zeta(u, u') = \frac{\sqrt{1 - u'^2}}{2a}\sqrt{\frac{u'^2 - r^2}{r^2 - u^2}}\left(\frac{1}{u - u'} - \frac{1}{u + u'}\right) + \frac{1 - u'^2}{2a}\frac{\alpha + 2}{\sqrt{r^2 - u^2}}$$
$$\text{for } 0 \le |u| \le r. \tag{5.25}$$

This admits the representation of the distribution of the sheet current $J(x)$ within the flux-free region of the strip,

$$J(x) = J_c\left(\frac{2a}{\pi}\right)\int_{r}^{s} du'\zeta(u(x), u')\frac{\sqrt{1 - u(x)^2}}{1 - u'^2} \quad \text{for } 0 \le |x| \le b. \tag{5.26}$$

Thus, using eqn. (5.25) in eqn. (5.26) and ensuring regularity as well as continuity of $J(x)$ at $x = \pm b$ through the choice

$$\alpha + 2 = \frac{2}{s - r}\arctan\sqrt{\frac{s^2 - r^2}{1 - s^2}} \tag{5.27}$$

furnishes the distribution of the sheet current across the entire width of the magnetically shielded strip,

$$J(x) = \begin{cases} \dfrac{2J_c}{\pi}\arctan\sqrt{\dfrac{q^2 - p^2}{p^2 - v(x)^2}} & \text{for } 0 \le |x| \le b \\ J_c & \text{for } b \le |x| < w, \end{cases} \tag{5.28}$$

upon passing to the new variable $v = \sinh(\pi x/2a)$ as well as to the substitutions $p = \sinh(\pi b/2a)$ and $q = \sinh(\pi w/2a)$. Substituting eqn. (5.28) into eqn. (4.12) and making recourse to b yields the implicit relationship for the half-width of the flux-free region in terms of I,

$$b = w\left(1 - \frac{I}{I_c}\right) + \frac{2a}{\pi^2}\int_{-p}^{p} \frac{dv'}{\sqrt{1+v'^2}} \arctan\sqrt{\frac{q^2-p^2}{p^2-v'^2}}. \tag{5.29}$$

Likewise, eqn. (4.14) can be used to establish the corresponding distribution of the normal component of the magnetic field across the width of the magnetically shielded strip,

$$H_y(x) = \begin{cases} 0 & \text{for } 0 \le |x| \le b, y = 0 \\ H_c \,\mathrm{sgn}(x) \mathrm{arctanh}\sqrt{\dfrac{v(x)^2 - p^2}{q^2 - p^2}} & \text{for } b \le |x| < w, y = 0, \end{cases} \tag{5.30}$$

where $H_c = J_c/\pi$.

Figure 5.6(a) displays the variation of the normalized sheet current J/J_c over the width of the strip for a range of values of the normalized distance between the plane of the strip and the longitudinally oriented flat diamagnetic shields a/w, adopting a fixed value of the normalized total transport current I/I_c. A general trait due to sheet current redistribution is the *flattening* and *widening* of the symmetric current profile within the central flux-free region of the strip between the maxima of critical sheet current adopted in the flux-filled margins of the strip, a phenomenon which gets more pronounced as a/w abates. Figure 5.6(b) depicts the variation of the corresponding normalized magnetic field H_y/H_c over the width

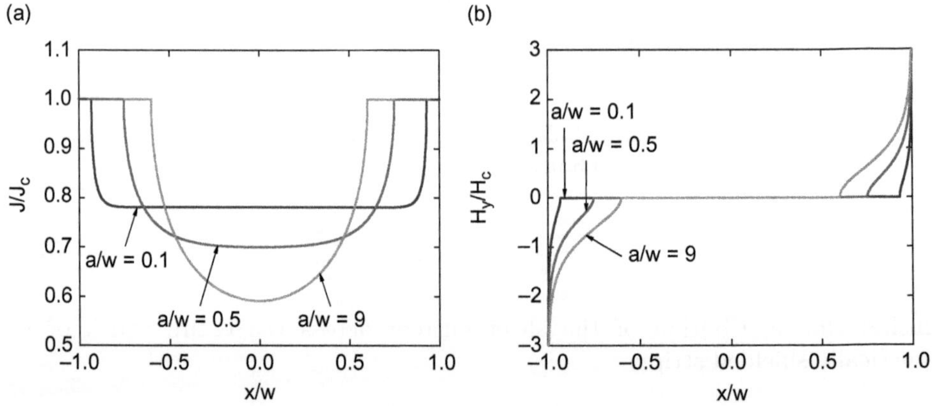

Fig 5.6 Distributions of (a) the normalized sheet current J/J_c and (b) the normalized magnetic field H_y/H_c as functions of the normalized coordinate x/w over the width of the magnetically shielded superconductor strip in the partially flux-filled critical state for three different values of the normalized distance between the plane of the strip and the diamagnets a/w identified on the curves, when the normalized total transport current $I/I_c = 0.8$, referring to the longitudinal diamagnet configuration of Fig. 5.1.

of the strip for the said range of values of the normalized distance between the plane of the strip and the longitudinally oriented flat diamagnetic shields a/w and the same value of the normalized total transport current I/I_c. Evidently, whereas at large a/w the field profile virtually reproduces that of an isolated strip, it *decreases* in strength while leaving an *enlarged* flux-free zone, as a/w wanes.

We note that the reciprocity in performance between the *longitudinally* oriented flat *diamagnetic* shields and *transversely* oriented flat *paramagnetic* shields, found in section 5.1.2 for the Meissner state, evidently carries over to the critical state: the distributions of both the sheet current and the magnetic field demonstrate similar behaviours, as a juxtaposition of Fig. 5.6 and Fig. 4.5 shows.

5.3.2 Hysteretic AC loss

The dynamics of the magnetic flux in the partially flux-filled critical state of the magnetically shielded strip, with an oscillating transport current imposed, is basically the same as for an unshielded strip. Hence, eqn. (4.15) determines the electromagnetic energy dissipated during a steady-state cycle of the transport current in the presence of diamagnetic shields as well. Upon substituting the normal component of the magnetic field, eqn. (5.30), the hysteretic AC loss per unit length of the strip, U_{ac}, for the longitudinal shielding geometry at hand thus can be put as

$$U_{\mathrm{ac}} = U_0 \left(\frac{2a}{\pi w}\right)^2 \int_p^q \frac{dv}{\sqrt{1+v^2}} \int_p^v \frac{dv'}{\sqrt{1+v'^2}} \operatorname{arctanh}\sqrt{\frac{v'^2-p^2}{q^2-p^2}}, \qquad (5.31)$$

where $U_0 = 2\mu_0 I_c^2/\pi$. In the limit of the diamagnets situated close to the strip, eqn. (5.31) may be approximated by the highly accurate expression

$$U_{\mathrm{ac}} \simeq U_0 \left(\frac{2a}{\pi w}\right)^2 \left[1 - \vartheta(l)\phi(l)\left(1+\frac{\pi l}{2a}\right) + L(\vartheta(l))\phi(l)\right]; \quad h < a \ll w. \qquad (5.32)$$

Herein,

$$\vartheta(l) = \arccos\left(\exp\left(-\frac{\pi l}{2a}\right)\right) \quad \text{and} \quad \phi(l) = \exp\left(-\frac{\pi l}{4a}\right)\sqrt{\frac{1}{2}\operatorname{csch}\left(\frac{\pi l}{2a}\right)}, \qquad (5.33)$$

apart from Lobachevsky's function L (Gradshteyn and Ryzhik 1994), introducing the flux penetration depth $l = w - b$, with b related to p via eqn. (5.29). An essentially equivalent representation unfolding the dependence on the current I, albeit limited to the range $0 \leq I/I_c < 1$, emerges from eqn. (5.32) when making use of the approximations

$$p \simeq q \cos\left(\frac{\pi}{2}\frac{I}{I_c}\right) \tag{5.34a}$$

$$q \simeq \frac{1}{2}\exp\left(\frac{\pi w}{2a}\right) \tag{5.34b}$$

$$b \simeq w + \frac{2a}{\pi}\ln\cos\left(\frac{\pi}{2}\frac{I}{I_c}\right), \tag{5.34c}$$

so that the succinct form

$$U_{\rm ac} \simeq U_0 \left(\frac{2a}{\pi w}\right)^2 f\left(\frac{\pi}{2}\frac{I}{I_{\rm c}}\right) \quad \text{with} \quad f(i) = 1 - i\cot i\,(1 - \ln\cos i) + L(i)\cot i \tag{5.35}$$

obtains. The quadratic scaling factor in eqn. (5.35) agrees with numerical computations (Dasgupta, Derov, and Sridhar 2001) and analytical estimates (Genenko 2004; Clem 2008); it also describes the hysteretic AC loss as for a superconductor strip between two transverse paramagnetic shields (cf. section 4.2.2.2).

Figure 5.7(a) shows the variation of the normalized half-width of the flux-free region b/w with the normalized transport current $I/I_{\rm c}$ for a range of values of the normalized distance between the plane of the strip and the longitudinally oriented flat diamagnetic shields a/w. Thus, although the overall behaviour of the curves is

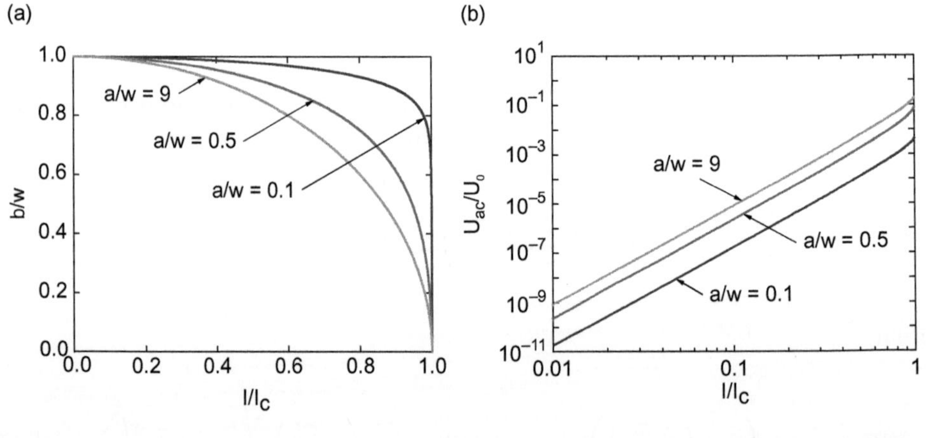

Fig 5.7 Dependence of (a) the normalized half-width of the flux-free region b/w and (b) the normalized hysteretic AC loss $U_{\rm ac}/U_0$ on I/I_c, the normalized total transport current imposed on the magnetically shielded superconductor strip in the partially flux-filled critical state, for three different values of the normalized distance between the plane of the strip and the diamagnets a/w identified on the curves, referring to the longitudinal diamagnet configuration of Fig. 5.1.

seen to display a tendency very much like that for the unshielded strip, their ordering reveals that penetration of magnetic flux, and hence the transition from the partially flux-filled critical state to the full critical state, is *retarded* as a/w wanes. Figure 5.7(b) portrays the variation of the corresponding normalized hysteretic AC loss U_{ac}/U_0 with the normalized transport current I/I_{c} for the said range of values of the normalized distance between the plane of the strip and the longitudinally oriented flat diamagnetic shields a/w. This reveals that, whereas at large a/w the hysteretic AC loss is virtually like that for an unshielded strip, it *decreases* substantially throughout the transport current regime, as a/w abates, the predictions of eqn. (5.35) differing indiscernibly from the exact results within the specified limitation. Again the strong decrease in hysteretic AC loss arises from both the reduction in strength of the magnetic field near the edges of the strip and the fact that a major part of the strip remains virtually flux-free even for total currents close to the critical current of the strip, as magnetic shielding takes effect. This analysis receives qualitative support from experiments on finite bifilar stacks of superconductor strips, with their monitored reduction of the hysteretic AC loss (Ogawa et al. 2008).

5.4 Superconductor strip between two curved semi-infinite diamagnets: overcritical states

We once more develop the considerations of section 4.1, addressing an infinitely extended, thin current-carrying type-II superconductor strip in the partially flux-filled critical state, now placed inside blunt wedge-shaped diamagnetic shields with zero permeability, $\mu = 0$, as depicted in Fig. 5.3. This is motivated by the observation that for a superconductor strip in the flux-free Meissner state, the latter type of magnetic environment facilitates redistribution of the sheet current towards the centre of the strip, bringing about potentially considerable enhancements of the maximum total transport current that can be imposed on the magnetically shielded strip (cf. section 5.2.2). The account here proceeds from the work of Genenko (2002b).

5.4.1 Conformal transformation

The distributions of the sheet current and the magnetic field across the width of the superconductor strip in the partially flux-filled critical state are obtained by starting from the derivations of section 5.3.1 for the flat diamagnetic environment of Fig. 5.1, generalizing them to the curved diamagnetic environment of Fig. 5.3 with the help of the conformal mapping technique. This procedure calls for some explanation since, although the condition on the distribution of the magnetic field, eqn. (4.1), is observed thereby, satisfaction of the requirement on the distribution of the sheet current, eqn. (4.2), is not by itself guaranteed. Unlike in the case of the Meissner state, the sheet current in the critical state appropriate to the curved diamagnetic environment thus cannot directly be inferred from eqn. (3.105). Nevertheless, the sheet current distribution under consideration can well be constructed from the linear response function due to the flat diamagnetic environment, if the transformation involved preserves the flux-free region of the strip.

Adhering to section 3.6.1.2, we restate that a conformal transformation effects a change of coordinates from the complex variable $s = x + iy$ to the complex variable $\sigma = \xi + i\eta$ by means of some analytic function $\sigma = f_b(s)$ with a nonzero derivative df_b/ds. This maps areas of the s-plane onto areas of the σ-plane, implying that $x = x(\xi, \eta)$ and $y = y(\xi, \eta)$. Reverting to the problem at hand, let us map the flat half-domains $y \geq a$ of the s-plane onto some blunt wedge-shaped half-domains of the σ-plane, with asymptotic opening half-angle of their surfaces $\beta = p\pi/2$, where $0 < p \leq 1$, by again employing eqn. (4.63),

$$f_b(s) = c_b\left[(1 + s/c)^p - (1 - s/c)^p\right] \;\text{ with }\; c_b = b/\left[(1 + b/c)^p - (1 - b/c)^p\right], \tag{5.36}$$

which introduces the parameter $c > w$. The function $f_b(s)$ is analytic in the whole s-plane, apart from two cuts on the real axis, $x < -c, y = 0$ and $x > c, y = 0$. Application of eqn. (5.36) transforms the points $(x = 0, y = \pm a)$ of the s-plane into the points $(\xi = 0, \eta = \pm f_b(ia)/i)$ of the σ-plane, while keeping the width and the position of the flux-free region of the strip fixed. The points $(x = \pm w, y = 0)$ of the s-plane, on the other hand, turn into the points $(\xi = \pm f_b(w), \eta = 0)$ of the σ-plane, keeping the position of the strip fixed, but adjusting its width, for $f_b(w) \equiv w_b > w$, except in the limits $b \to w$, $c \to \infty$ or $p \to 1$, where w_b adopts the minimum value w. The initial gap between the plane of the strip and the surfaces of the diamagnetic shields, a, thus changes to the distance between the plane of the strip and the tips of the diamagnetic shields, $a_b = f_b(ia)/i$, that is,

$$a_b = 2c_b|\sin(p\tau)|\sec^p\tau, \tag{5.37}$$

whereas the radius of curvature of the tips themselves becomes

$$r_b = 2c_b|p\cos^2((1 - p)\tau)/(1 - p)\sin((2 - p)\tau)|\sec^p\tau, \tag{5.38}$$

implying that $\tau = \arctan\gamma$ with $\gamma = a/c$. Evidently, the latter distance a_b—or, alternatively, the radius r_b—and the asymptotic opening half-angle β measured via p represent the geometrical characteristics of the blunt wedge-shaped magnetic shields; they can be installed so as to control the distribution of the sheet current in the strip by a judicious choice of the parameters w, a, c, and p. We note the link $f_b(s) = (c_b/c_w)f_w(s)$ recalling the analytic function $f_w(s)$ defined by eqn. (5.13) in the case of the Meissner state, where the flux-free region covers the entire width of the strip. Proper reinterpretation $a_w \to a_b$ and $w \to w_b$ understood, the variation of the shape of the magnetic shields depicted in Fig. 5.3 therefore is representative here too. The symmetry of this configuration with respect to the midplane $\xi = 0$ of the strip entails that a characteristic similar to that in eqn. (4.1) applies, i.e. the vanishing of the normal component of the transformed magnetic field $\hat{\text{H}}_\eta$ in the flux-free region of the strip,

$$\hat{H}_\eta(\xi) = 0 \quad \text{for } 0 \leq |\xi| \leq b,\ \eta = 0. \tag{5.39}$$

A characteristic similar to that in eqn. (4.2), viz. constancy of the transformed sheet current $\hat{J}$ in the flux-filled margins of the strip,

$$\hat{J}(\xi) = J_c \quad \text{for } b \leq |\xi| < w_b, \tag{5.40}$$

is provided as shown in the following.

5.4.2 Current and field distributions

Since, following along the lines of section 4.1, the sheet current $\hat{J}(\xi)$ within the flux-free region of the strip, $0 \leq |\xi| \leq b$, can be envisaged as arising from the effect of current filaments at $\xi = \pm\xi'$ for $b \leq \xi' < w_b$ within the flux-filled margins of the strip, $b \leq |\xi| < w_b$, the introduction of a transformed linear response function, $\hat{\chi}(\xi, \xi')$, suggests itself. This, by analogy to eqn. (4.9) upon observing the demand, eqn. (5.40), admits the representation

$$\hat{J}(\xi) = J_c \int_b^{w_b} d\xi' \, \hat{\chi}(\xi, \xi') \quad \text{for } 0 \leq |\xi| \leq b. \tag{5.41}$$

Due extension of eqn. (5.16) shows that the relation

$$\hat{\chi}(\xi, \xi') = \chi \left[x(\xi, \eta), x(\xi', \eta) \right] \left. \frac{\partial y}{\partial \eta} \right|_{\eta=0} \tag{5.42}$$

holds, where $\chi(x, x') = \zeta(u, u')\sqrt{1-u^2}$ denotes the linear response function appropriate to the flat magnetic environment, with the auxiliary function $\zeta(u, u')$ derived in section 5.3.1, making use of the change of variables $x = x(\xi, \eta)$ and $x' = x(\xi', \eta)$ to $u = \tanh(\pi x/2a)$ and $u' = \tanh(\pi x'/2a)$, respectively, as well as of the particulars defining the map.

The blunt wedge-shaped magnetic shields depicted in Fig. 5.3(b) may serve as an example for further scrutiny. Thus, eqn. (5.36) with variable p affords an implicit transformation relationship from which the dependence $x(\xi, \eta)$ and the derivative $\partial y/\partial \eta$ at $\eta = 0$ can be deduced. Substituting these quantities into the right-hand side of eqn. (5.42) for the linear response function in the partially flux-filled critical state due to the flat magnetic environment yields the sheet current distribution across the entire width of the magnetically shielded strip in that state appropriate to the blunt wedge-shaped magnetic shields,

$$\hat{J}(\xi) = \begin{cases} J_c \, (d\vartheta_b/d\xi) \int_b^{w_b} d\xi' \, \zeta(u_b(\xi), u_b(\xi')) \sqrt{1 - u_b(\xi)^2} & \text{for } 0 \leq |\xi| \leq b \\ J_c & \text{for } b \leq |\xi| < w_b. \end{cases} \tag{5.43}$$

The function, eqn. (5.36), yields the implicit relationship

$$(1 + \vartheta_b(\xi)/c_b)^p - (1 - \vartheta_b(\xi)/c_b)^p = \xi/c_b, \tag{5.44}$$

whose solution also furnishes the derivative

$$d\xi/d\vartheta_b = p/(1+\vartheta_b(\xi)/c_b)^{1-p} + p/(1-\vartheta_b(\xi)/c_b)^{1-p} \tag{5.45}$$

for substitution of its inverse in eqn. (5.43), whereas from eqn. (5.25),

$$\zeta(u_b, u_b') = \frac{\sqrt{1-u_b'^2}}{2a}\sqrt{\frac{u_b'^2 - r^2}{r^2 - u_b^2}}\left(\frac{1}{u_b - u_b'} - \frac{1}{u_b + u_b'}\right) + \frac{1-u_b'^2}{2a}\frac{\alpha+2}{\sqrt{r^2 - u_b^2}} \tag{5.46}$$

with

$$u_b(\xi) = \tanh\left(\pi\vartheta_b(\xi)/2\gamma c_b\right) \quad \text{and} \quad r = \tanh\left(\pi b/2a\right). \tag{5.47}$$

Regularity and continuity of $\hat{J}(\xi)$ at $\xi = \pm b$ is ensured through the choice

$$\alpha + 2 = 2m/n \tag{5.48}$$

with

$$m = \int_b^{w_b} d\xi' u_b(\xi')\sqrt{1-u_b(\xi')^2}/\sqrt{u_b(\xi')^2 - r^2} \quad \text{and} \quad n = \int_b^{w_b} d\xi' \left(1 - u_b(\xi')^2\right), \tag{5.49}$$

duly extending eqn. (5.27). We note that the sheet current distribution, eqn. (5.43), can be cast into the equivalent form

$$\hat{J}(\xi) = \begin{cases} \dfrac{2pJ_c}{\pi}\dfrac{d\vartheta_b}{d\xi}\displaystyle\int_0^{\varphi(\xi)} d\varphi'\,\theta(\xi,\varphi')\left(\psi_+(\xi,\varphi')^{p-1} + \psi_-(\xi,\varphi')^{p-1}\right) & \text{for } 0 \le |\xi| \le b \\ J_c & \text{for } b \le |\xi| < w_b \end{cases} \tag{5.50}$$

deemed more convenient for further analysis, where

$$\theta(\xi,\varphi') = \sqrt{1-u_b(\xi)^2}/\sqrt{1 - r^2 - (r^2 - u_b(\xi)^2)\tan^2\varphi'} \tag{5.51}$$

and

$$\psi_\pm(\xi,\varphi') = 1 \pm \frac{2\gamma}{\pi}\,\text{arctanh}\sqrt{r^2 + (r^2 - u_b(\xi)^2)\tan^2\varphi'} \tag{5.52}$$

as well as

$$\varphi(\xi) = \arctan\sqrt{\frac{s^2 - r^2}{r^2 - u_b(\xi)^2}} \quad \text{with } s = \tanh\left(\pi w/2a\right). \tag{5.53}$$

The superconducting states described by eqn. (5.50) fall into two categories: (i) subcritical states, which are always physically stable, and (ii) overcritical states, which can be either physically stable or unstable. There exists a threshold value $b = b_t$ separating the stable and unstable regimes of the latter type of states, defined through the requirement that $d\hat{J}/d\xi = 0$ as $\xi \to \pm b$ from the flux-free region of the strip. To elucidate this fact, let us take a closer look at the situation where the distance between the plane of the strip and the tips of the magnetic shields is small, $a_b \ll w_b$ for $c - w \ll w$, and where the flux-filled marginal parts of the strip are narrow, $w_b - b \ll w_b$, a case in point of foremost interest here. Under such conditions, the inequalities $a \ll w$ and $\tau \simeq \gamma \ll 1$ hold, so that eqns. (5.37), (5.38) admit the estimates $a_b \simeq 2p\gamma c_b$ and $r_b \simeq 2pc_b/(1-p)(2-p)\gamma$, and the derivative of the sheet current, eqn. (5.50), adopts the simple approximate form

$$\frac{d\hat{J}}{d\xi} \sim \frac{J_c}{\sqrt{r^2 - u_b(\xi)^2}} \left(\frac{1}{\sqrt{s^2 - r^2}} - \frac{\sqrt{s^2 - r^2}}{1 - s^2} \right) \quad \text{as } \xi \to \pm b, \tag{5.54}$$

leaving a nonsingular modulating factor aside. Two important insights follow from eqn. (5.54) straight away: (i) since $u_b(\xi) \to \pm r$ as $\xi \to \pm b$, the sheet current distribution has a singular slope at the fronts of penetrated magnetic flux, typical of the two-dimensional geometry of thin superconductor strips (Norris 1970; Brandt and Indenbom 1993; Zeldov, Clem, McElfresh, and Darwin 1994); and (ii) the slope of the sheet current distribution changes sign when ξ crosses the points $\xi = \pm b_t$. This distribution thus reveals physical stability if $2s^2 - r^2 < 1$, but instability if $2s^2 - r^2 > 1$, the limit of stability being established by $2s^2 - r^2 = 1$. The latter condition, by means of the definitions, eqns. (5.47) and (5.53), takes the moderately accurate shape

$$b_t \simeq w - (\ln 2/\pi)a; \qquad h < a \ll w,\, c - w \ll w. \tag{5.55}$$

Hence, if $b_t < b < w_b$, overcritical states revealing stability against a collapse into the full critical state by virtue of transitional zones of subcritical sheet current in marginal parts of the strip exist; if $b < b_t < w_b$, overcritical states lacking stability because of the absence of such zones persist.

Referring to eqn. (4.12) by analogy, eqn. (5.50) permits a direct evaluation of the total transport current

$$I = 2J_c\,(w_b - b) + \int_{-b}^{b} d\xi' \hat{J}(\xi') \tag{5.56}$$

in terms of b, indicating the critical current $I_c = 2w_bJ_c$. Unfortunately, a simple analytic estimate of the maximum total transport current I_{max} in terms of the geometrical characteristics of the diamagnetic shields at the limit of stability of overcritical states $b = b_t$ based on eqn. (5.56) with eqn. (5.55) does not seem to exist.

The distribution of the normal component of the magnetic field across the entire width of the magnetically shielded strip appropriate to the blunt wedge-shaped magnetic shields, which can be calculated from eqns. (3.103) to (3.105) with eqn. (5.6b) under consideration of eqns. (5.39) and (5.43) or (5.50), finally comes out as

$$\hat{H}_\eta(\xi) = \begin{cases} 0 & \text{for } 0 \le |\xi| \le b, \eta = 0 \\ J_c \hat{G}(\xi) + \dfrac{1}{4\gamma c_b} \displaystyle\int_{-b}^{b} d\xi' \hat{J}(\xi') \hat{Q}_-(\xi, \xi') & \text{for } b \le |\xi| < w_b, \eta = 0, \end{cases} \tag{5.57}$$

where

$$\hat{G}(\xi) = \frac{1}{4\gamma c_b} \int_{b}^{w_b} d\xi' \left(\hat{Q}_+(\xi, \xi') + \hat{Q}_-(\xi, \xi') \right) \tag{5.58}$$

and

$$\hat{Q}_\pm(\xi, \xi') = (d\vartheta_b/d\xi) \operatorname{csch}\left[\pi \left(\vartheta_b(\xi) \pm \vartheta_b(\xi')\right)/2\gamma c_b\right], \tag{5.59}$$

the first term together with the integral on the right-hand side of eqn. (5.57) ensuring that $\hat{H}_\eta(\xi)$ vanishes at $\xi = \pm b$.

Figure 5.8(a) displays the variation of the normalized sheet current $\hat{J}/J_c$ over the width of the strip for a range of values of the parameter of the conformal map p, adopting a fixed value of the normalized distance between the plane of the strip and the tips of the magnetic shields a_b/w_b and a fixed value of the normalized half-width of the flux-free region b/w_b. A general trait due to sheet current redistribution is the

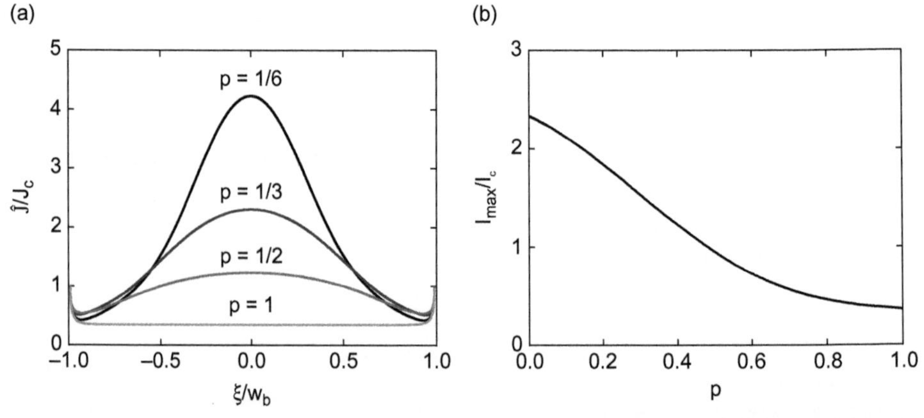

Fig 5.8 (a) Distribution of the normalized sheet current $\hat{J}/J_c$ as a function of the normalized coordinate ξ/w_b over the width of the magnetically shielded superconductor strip in the partially flux-filled critical state for the normalized distance between the plane of the strip and the tips of the magnetic shields $a_b/w_b = 0.1$ as well as four different values of the parameter of the conformal map p identified on the curves; (b) variation of the corresponding normalized maximum total transport current I_{max}/I_c with the parameter of the conformal map p, when the normalized half-width of the flux-free region $b/w_b = 0.99$, apart from $c/w = 1.001$, referring to the blunt wedge-shaped diamagnet configuration of Fig. 5.3(b). All of the curves delineate physically stable regimes of the partially flux-filled critical state.

warping of the symmetric current profile around the central part of the strip and the increase of the total current itself, a phenomenon which gets more pronounced as p wanes, reaching saturation with a *nonuniform* maximum in the limiting sheet current profile of eqn. (5.50), when the magnetic shields contract. Thus, for all values of p, the downward slopes of the sheet current profiles at the fronts of penetrated magnetic flux give rise to *subcritical* sheet current with $\hat{J}/J_c < 1$ in the immediate vicinity of the flux-filled marginal parts of the strip, and hence to stability of the respective sheet current distributions overall. Figure 5.8(b) presents the variation of the corresponding normalized maximum total transport current I_{max}/I_c with the parameter of the conformal map p. Starting from a peak at the smallest value of p, there is a quasilinear drop followed by a turn towards the value appropriate for flat diamagnetic shields, as p continues to grow. Remarkably enough, the regimes of overcritical and subcritical total current with $I_{max}/I_c > 1$ and $I_{max}/I_c < 1$, respectively, approximately show equal shares.

Figure 5.9(a) displays the variation of the normalized sheet current $\hat{J}/J_c$ over the width of the strip for a range of values of the normalized half-width of the flux-free region b/w_b, adopting a fixed value of the parameter of the conformal map p and a fixed value of the normalized distance between the plane of the strip and the tips of the magnetic shields a_b/w_b. A general trait due to sheet current distribution is

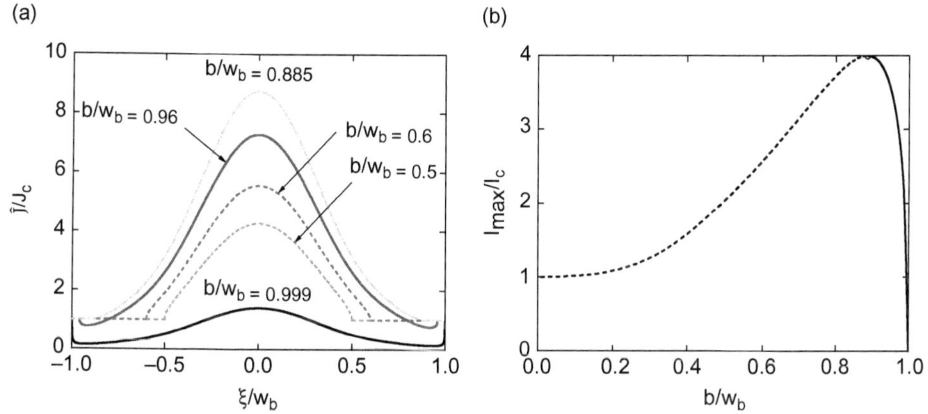

Fig 5.9 (a) Distribution of the normalized sheet current $\hat{J}/J_c$ as a function of the normalized coordinate ξ/w_b over the width of the magnetically shielded superconductor strip in the partially flux-filled critical state for the parameter of the conformal map $p = 1/6$ as well as five different values of the normalized half-width of the flux-free region b/w_b identified on the curves; (b) variation of the corresponding normalized maximum total transport current I_{max}/I_c with the normalized half-width of the flux-free region b/w_b, when the normalized distance between the plane of the strip and the vertices of the magnetic shields $a_b/w_b = 0.1$, referring to the blunt wedge-shaped diamagnet configuration of Fig. 5.3(b). The solid/dashed curves delineate physically stable and unstable regimes, respectively, of the partially flux-filled critical state, separated by the marginally stable critical state (dash-dotted curve or empty circle) in between.

the warping of the symmetric current profile around the central part of the strip up to a maximum and the initial increase in the total current itself, with a subsequent depression of the current profile and concomitant decrease in the total current too, as b/w_b augments, a phenomenon which reaches its final stage in the limiting *uniform* current-free profile of eqn. (5.50). Thus, for the two smallest values of b/w_b, the upward slopes of the sheet current profiles at the fronts of penetrated magnetic flux generate *overcritical* sheet current with $\hat{J}/J_c > 1$ in the entire flux-free region adjacent to the flux-filled marginal parts of the strip, and therefore instability of the respective sheet current distributions overall. Magnetic flux then moves towards the centre of the strip, and the partially flux-filled overcritical states collapse into the full critical state with $\hat{J}/J_c = 1$ prevailing all over the strip. By contrast, for the two largest values b/w_b, the downward slopes of the sheet current profiles at the fronts of penetrated magnetic flux give rise to *subcritical* sheet current with $\hat{J}/J_c < 1$ in the immediate vicinity of the flux-filled marginal parts of the strip, and hence to stability of the respective sheet current distributions overall. A sheet current distribution of marginal stability, for an intermediate value of b/w_b, separates the two regimes; it is characterized by zero slopes of the sheet current profile at the fronts of penetrated magnetic flux. Figure 5.9(b) presents the variation of the corresponding normalized maximum total transport current $I_{\max}/I_c$ with the normalized half-width of the flux-free region b/w_b. Starting from unity, the value prevailing in the full critical state, at zero b/w_b, there is a monotonic rise towards a broad maximum followed by a moderate drop to 0, as b/w_b continues to grow. Remarkably enough, although the current predicted for the greatest part of b/w_b turns out to lack stability, ample scope for stable *overcritical* total current with $I_{\max}/I_c > 1$ in a distinct region of b/w_b exists.

We emphasize, however, again that high total currents associated with strong tangential components of the magnetic field around the centre of the strip might facilitate penetration of magnetic flux from the *surfaces* of the strip, and hence could severely restrict the loss-free current transport in type-II superconductors possessing small lower critical fields. Curved diamagnetic shields realized by superconductors in the flux-free Meissner state entail further concerns, since large sheet currents around the centre of the strip—and consequently large sheet currents induced in the tips of the shields—may cause local penetration of magnetic flux associated with a nonlinear electromagnetic response (Monaco et al. 2000; Lancaster 2006). An ultimate limitation, at any rate, derives from the quench of superconductivity through the depairing mechanism, as underlined in section 3.7.2.

5.4.3 Hysteretic AC loss

The dynamics of the magnetic flux in the partially flux-filled critical state of the magnetically shielded strip, with an oscillating transport current imposed, is basically the same as for an unshielded strip. Hence, eqn. (4.15) determines the electromagnetic energy dissipated during a steady-state cycle of the transport current in the presence of diamagnetic shields as well. Upon substituting the normal component of the magnetic field, eqn. (5.57), changing the variables ξ and ξ' to

$\lambda = \xi/w_b$ and $\lambda' = \xi'/w_b$, respectively, as well as introducing the substitution $l_b = b/w_b$, the hysteretic AC loss per unit length of the strip, U_{ac}, for the blunt wedge-shaped shielding geometry at hand thus can be cast into the form

$$U_{ac} = U_0 \int_{l_b}^{1} d\lambda \int_{l_b}^{\lambda} d\lambda' \, \hat{H}_\eta(\lambda')/H_c, \tag{5.60}$$

where $U_0 = 2\mu_0 I_c^2/\pi$, by due reference to eqn. (5.56) for physically stable states.

Figure 5.10(a) shows the variation of the normalized half-width of the flux-free region b/w_b with the normalized transport current I/I_c for a range of values of the parameter of the conformal map p, adopting a fixed value of the normalized distance between the plane of the strip and the tips of the magnetic shields a_b/w_b. For the largest value of p typifying a subcritical state appropriate to flat magnetic shields, the curve exhibits a monotonic behaviour, and hence a single-valued dependence of b/w_b on I/I_c, within the critical current domain. By contrast, for the three smallest values of p identifying overcritical states appropriate to curved magnetic shields, the curves disclose nonmonotonic behaviours, and therefore multi-valued dependences of b/w_b on I/I_c, beyond the critical current domain, the respective branches separating stable from unstable regimes. Instability here arises because of

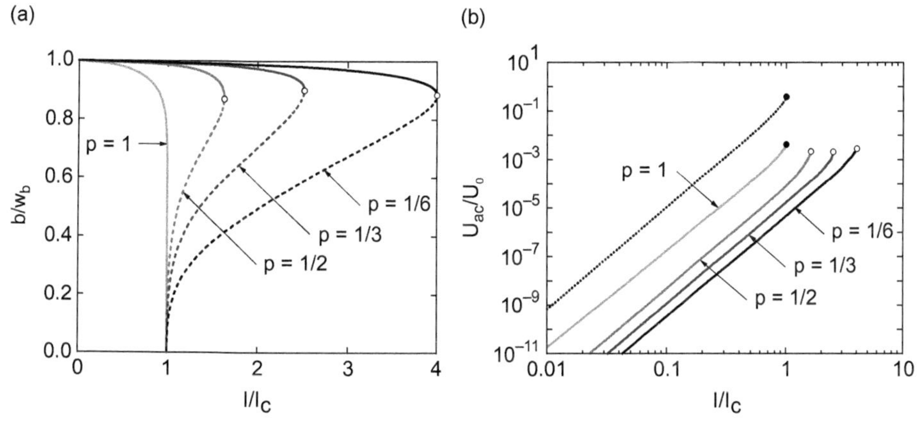

Fig 5.10 Dependence of (a) the normalized half-width of the flux-free region b/w_b and (b) the normalized hysteretic AC loss U_{ac}/U_0 on I/I_c, the normalized total transport current imposed on the magnetically shielded superconductor strip in the partially flux-filled critical state, for the normalized distance between the plane of the strip and the tips of the magnetic shields $a_b/w_b = 0.1$ as well as four different values of the parameter of the conformal map p identified on the curves, when $c/w = 1.001$, referring to the blunt wedge-shaped diamagnet configuration of Fig. 5.3(b). The empty circles terminating some of the curves separate physically stable and unstable regimes of the partially flux-filled critical state, whereas the solid circles terminating the residual curves denote the onset of the full critical state; the dotted curve relates to the isolated superconductor strip and merely serves for comparison.

direct contact of the central flux-free overcritical region with the flux-filled margins of the strip, as Fig. 5.9(a) depicts, in close analogy to paramagnetic environments (cf. section 4.4.3). This generates a Lorentz-like force exerted upon magnetic vortices which dominates the pinning force, dragging the vortices towards the centre of the strip and, hence, facilitating a transition to the critical state. Figure 5.10(b) portrays the variation in the corresponding normalized hysteretic AC loss U_{ac}/U_0 with the normalized transport current I/I_{c} for the said range of values of the parameter of the conformal map p. This reveals that, whereas at the largest p the hysteretic AC loss duplicates that for flat magnetic shields, it decreases substantially throughout the transport current regime, as p abates, overcritical states outperforming the subcritical state and extending its range by far.

6

Meissner state of thin superconductors in nonideal paramagnetic environments

The method of conditioning magnetic fields in superconductor/paramagnet heterostructures, demonstrated hitherto for the example of a thin type-II superconductor strip surrounded by various idealized open or closed paramagnetic environments (cf. Chapters 3 and 4), enjoys a wide technological application range. This includes devices that store energy in the magnetic field of superconducting solenoids (Friedman et al. 1999; Kondoh and Umeda 2002), magnetically shielded superconducting 'pancake' coils aimed at reducing the hysteretic AC loss (Zhang, Kvitkovic, Pamidi, and Coombs 2012; Ainslie, Yuan, and Flack 2013; Niu, Yong, Xia, and Zhou 2019; Song, Xia, and Ma 2019), scanning microscopes fitted with a magnetic sensor superconductor SQUID (Gudoshnikov et al. 2001), and magnetically shielded SQUIDs for magnetometer or gradiometer use (Bondarenko, Shablo, Pavlov, and Perepelkin 2002).

Another important area of practical use concerns electrical transmission lines with the major requirements for guiding magnetic fields, viz. an increase in the total transport current which the superconductor elements are able to sustain and a reduction in the concomitant hysteretic AC loss. Experimentally, these goals can be met by sophisticated magnetic environments such as horseshoe-shaped paramagnets placed around the edges of a superconductor strip (Alamgir, Gu, and Han 2005; Gu, Alamgir, Qu, and Han 2007; Krüger et al. 2013) or by propitious arrangements of magnetically shielded superconducting tapes (Nguyen et al. 2010). If deposited on a magnetic support, both the critical current of the tapes and the hysteretic AC loss show characteristic dependences on substrate magnetism as well as on the orientation of an external magnetic field (Suenaga and Li 2006; Nguyen, Sastry, and Schwartz 2007; Suenaga et al. 2008; Miyagi et al. 2008; Nguyen, Ashworth, and Willis 2009). Substantial enhancements of the critical current can result when placing magnetic barriers in the form of soft-magnetic stripes on a superconductor bridge (Vlasko-Vlasov et al. 2015).

Computational studies claiming to simulate realistic conditions must certainly take account of the finite dimensions of the magnetic shields as well as of their true

Electromagnetics of Superconductor/Paramagnet Heterostructures. Yuri Genenko and Hermann Rauh, Oxford University Press.
 DOI: 10.1093/9780191782855.003.0006

magnetic properties. While restricting the latter by assuming a constant magnetic permeability, finite geometries are examined in investigations of an iron-sheathed superconductor strip (Farinon et al. 2005; Yampolskii, Genenko, and Rauh 2007), a superconductor strip deposited on a Ni-alloy support (Miyagi, Amadutsumi, and Takahashi 2007), cylindrically assembled coated conductors with a magnetic support (Miyagi, Umabuchi, Takahashi, and Tsukamoto 2007), coated conductors with magnetic supports in various tubular configurations (Amemiya, Nakahata, Fujiwara, and Shiohara 2010; He, Xue, Yong, and Zhou 2014), and stacks of coated conductors in different transport current and applied magnetic field regimes (Wan, Huang, Yong, and Zhou 2015). Improved simulations of the electromagnetic properties of superconductor/paramagnet heterostructures ensue if the field dependence of the magnetic permeability of the substrate material is accounted for too (Claassen and Thieme 2008; Miyagi, Umabuchi, Takahashi, and Tsukamoto 2008a,b; Nguyen, Ashworth, and Willis 2009; Nguyen et al. 2010), although employing a constant, effective magnetic permeability yields sufficiently accurate predictions regarding hysteretic AC loss (Amemiya and Nakahata 2007; Ranecky et al. 2017).

Since the possibility of actively using the magnetic shielding effect to enhance the performance of superconductor/paramagnet heterostructures essentially originates from the redistribution of the current passing through the superconductor constituents, we relax our previous idealizations of the geometrical and magnetic characteristics of the shields by admitting finite permeabilities and finite spatial extents. Building on some of the model configurations for a thin superconductor strip analysed in Chapter 3 and also introducing thin superconductors of tubular and annular shapes, the emphasis here is to test the quality of results obtained when idealized conditions prevail and, at the same time, to identify regimes where new traits appear.

6.1 Superconductor strip adjacent to a flat semi-infinite paramagnet

We start by resuming the problem of an infinitely extended, thin current-carrying type-II superconductor strip in the Meissner state, placed at distance c adjacent to a flat semi-infinite paramagnetic environment with permeability μ, as depicted in Fig. 3.2. Our aim now is to study the effect of this one-sided open magnetic shield on the distribution of the sheet current across the width of the strip in the whole paramagnetic range of μ beyond the limit $\mu \to \infty$; an attempt which ultimately warrants computational means. The following account derives from the work of Genenko, Snezhko, and Freyhardt (2000).

6.1.1 Field distribution

According to the explanations of section 3.2.1, eqns. (2.90) and (2.91) here simplify to the Poisson equation for the single nontrivial component of the magnetic vector potential A_z,

$$\Delta A_z = \begin{cases} -\mu_0 J(x)\delta(y) & \text{in } c-w \le x \le c+w \\ 0 & \text{otherwise,} \end{cases} \tag{6.1}$$

supplemented with the boundary conditions

$$\lim_{x\to 0-} A_z(x,y) = \lim_{x\to 0+} A_z(x,y) \quad \text{and} \quad \left.\frac{1}{\mu}\frac{\partial A_z}{\partial n}\right|_{x=0-} = \left.\frac{1}{\mu_0}\frac{\partial A_z}{\partial n}\right|_{x=0+}. \tag{6.2}$$

Obviously, the latter demands rule out direct contact between the paramagnetic shield and the strip to avoid complications arising from the proximity effect (Schmidt 1997). The desired solution thus can be constructed by the method of images like before, calling in their current strength $\kappa = (\mu - \mu_0)/(\mu + \mu_0)$, together with the Green's function technique. Identifying the contributions $dI = dx' J(x')$ of a distributed current element of unit weight at $(x = x', y = 0)$ as well as of an image current element of weight κ at $(x = -x', y = 0)$ and integrating over x' yields the convenient representation in the vacuum space, $x > 0$ inclusive of the surface of the strip ∂D,

$$A_z(x,y) = \frac{\mu_0}{2\pi}\int_{c-w}^{c+w} dx' J(x') \left\{ \ln \frac{c}{\sqrt{(x-x')^2 + y^2}} + \kappa \ln \frac{c}{\sqrt{(x+x')^2 + y^2}} \right\}. \tag{6.3}$$

Application of eqn. (2.3) to eqn. (6.3) thus gives the Cartesian components of the magnetic field $\mathbf{H}$,

$$H_x(x,y) = -\frac{1}{2\pi}\int_{c-w}^{c+w} dx' J(x') \left[\frac{y}{(x-x')^2 + y^2} + \kappa \frac{y}{(x+x')^2 + y^2} \right] \tag{6.4a}$$

$$H_y(x,y) = \frac{1}{2\pi}\int_{c-w}^{c+w} dx' J(x') \left[\frac{x-x'}{(x-x')^2 + y^2} + \kappa \frac{x+x'}{(x+x')^2 + y^2} \right]. \tag{6.4b}$$

6.1.2 Current distribution

Naturally again the Meissner state of the superconductor strip requires the normal component of the magnetic field H_y to vanish across the width of the strip,

$$H_y(x) = 0 \quad \text{for } c-w < x < c+w, y = 0. \tag{6.5}$$

Recalling eqn. (6.4b) hence yields the condition for this state

$$\frac{1}{2\pi}\int_{c-w}^{c+w} dx' J(x') \left(\frac{1}{x-x'} + \kappa \frac{1}{x+x'} \right) = 0 \quad \text{in } c-w < x < c+w, \tag{6.6}$$

i.e. a homogeneous integral equation which duly generalizes eqn. (3.25). The first term herein represents the magnetic field created by the transport current itself and the second term delineates the magnetic field induced by the paramagnetic shield.

Since eqn. (6.6) proves cumbersome for a numerical analysis because of the singularity it contains, we invert eqn. (6.4b) in respect to the induced magnetic field H_{i} across the width of the strip to obtain

$$J(x) = \frac{I}{\pi} \frac{1}{\sqrt{(c+w-x)(x-c+w)}} + \frac{2}{\pi} \int_{c-w}^{c+w} dx' \sqrt{\frac{(c+w-x')(x'-c+w)}{(c+w-x)(x-c+w)}} \frac{H_{\mathrm{i}}(x')}{x-x'}$$

$$\text{in } c-w < x < c+w, \tag{6.7}$$

making use of the conformal mapping technique (Brandt 1992; Brandt and Indenbom 1993), upon observing eqn. (3.2). As a result, the first term in eqn. (6.7) emulates the contribution of an isolated superconductor strip, recalling *mutatis mutandis* eqn. (3.8), whereas the second term in eqn. (6.7) reflects the influence of the paramagnetic shield. Substituting H_{i} from eqn. (6.6) into eqn. (6.7) thus furnishes the sheet current as a solution of the norm-conserving regular integral equation of the Fredholm type

$$J(x) = \frac{I}{\pi} \frac{1+\kappa}{\sqrt{(c+w-x)(x-c+w)}} - \frac{\kappa}{\pi} \int_{c-w}^{c+w} dx' \sqrt{\frac{(c+w+x')(x'+c-w)}{(c+w-x)(x-c+w)}} \frac{J(x')}{x+x'}$$

$$\text{in } c-w < x < c+w, \tag{6.8}$$

which easily permits application of standard numerical discretization routines (Delves and Mohamed 1985; Press, Teukolsky, Vetterling, and Flannery 2007). One can verify that, in the limit $\mu \to \infty$, eqn. (6.8) is satisfied by the solution, eqn. (3.27). Conversely, if $\mu \to \mu_0$, eqn. (6.8) duplicates the solution for the unshielded strip, eqn. (3.8), with reference to the appropriate Cartesian coordinate system understood.

Reasoning in a similar vein to that for the isolated superconductor strip in the barrier protected Meissner state of section 3.1 limits the validity of the solution of eqn. (6.8) to the range $c-w+h \le x \le c+w-h$, with the maximum value $J_{\max}$ imposed in $c+w-h \le x \le c+w$. However, since we're focused on investigating the sensitivity of the distribution of the sheet current to shielding by the paramagnetic environment, it is assumed that $J_{\max}$ stays *below* the characteristic sheet current for entry of magnetic flux, $J_{\max} < J_b$. This implies that the total current $I = 2w\bar{J}$, expressed through the average sheet current $\bar{J}$, remains *lower* than the maximum loss-free total current, $I < I_{\max}$, too, irrespective of the edge sheet current J_{edge} which adjusts itself in $c-w \le x \le c-w+h$.

The influence of the paramagnetic environment shows up in the distribution of the sheet current J measured against its average $\bar{J}$, on the one hand, and in the edge sheet current J_{edge} calibrated against its equivalent without the magnetic shield $J_{\mathrm{edge},0}$, on the other hand, referring to the total current I. Both the geometrical parameter c and the magnetic characteristic μ play crucial roles thereby.

Figure 6.1 displays the variation of the normalized sheet current $J/\bar{J}$ over the width of the strip for a range of values of the relative magnetic permeability of the shield μ/μ_0, adopting two fixed values of the normalized distance between the centre of the strip and the shield c/w. A general trait due to sheet current redistribution in the strip is the depression of the peak of the sheet current at the left edge of the strip adjacent to the magnetic shield, a phenomenon which clearly stands out for the smaller value of c/w (Fig. 6.1(a)) and to a lesser extent for the larger value of c/w (Fig. 6.1(b)), reaching saturation with increasing μ/μ_0 in the limiting sheet current profile, eqn. (3.27).

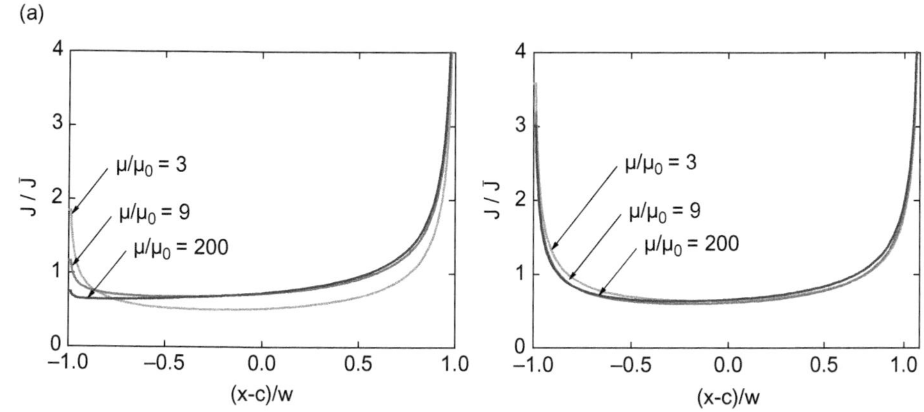

Fig. 6.1 Distribution of the normalized sheet current $J/\bar{J}$ as a function of the normalized coordinate $(x-c)/w$ over the width of the magnetically shielded superconductor strip in the flux-free Meissner state for three different values of the relative magnetic permeability of the shield μ/μ_0 identified on the curves along with the normalized distance between the centre of the strip and the paramagnet (a) $c/w = 1.01$ and (b) $c/w = 2$, when $h/2w = 0.005$, referring to the flat semi-infinite paramagnet configuration of Fig. 3.2.

Figure 6.2 illustrates the dependence of the normalized edge sheet current $J_{\text{edge}}/J_{\text{edge},0}$ on the normalized distance between the centre of the strip and the paramagnet c/w for a range of values of the relative magnetic permeability of the shield μ/μ_0 and, respectively, the dependence on the relative magnetic permeability of the shield μ/μ_0 for a range of values of the normalized distance between the centre of the strip and the paramagnet c/w. Keeping μ/μ_0 fixed and starting from a distinct minimum at the smallest value of c/w, there is a monotonic rise towards unity, the value prevailing for the isolated strip, as c/w continues to grow. Accordingly, the curves approach the step-like shape expected for the isolated strip, if μ/μ_0 wanes (Fig. 6.2(a)). Keeping c/w fixed and starting from unity at the vacuum value of μ/μ_0, there is a monotonic drop followed by a settlement at the level prevailing for an ideal paramagnetic shield, as μ/μ_0 continues to grow. This effect gets more pronounced as c/w wanes and virtually saturates from $\mu/\mu_0 \simeq 200$, proving that eqn. (3.27) adequately captures sheet current redistribution in the strip for values $\mu/\mu_0 \gtrsim 200$ (Fig. 6.2(b)).

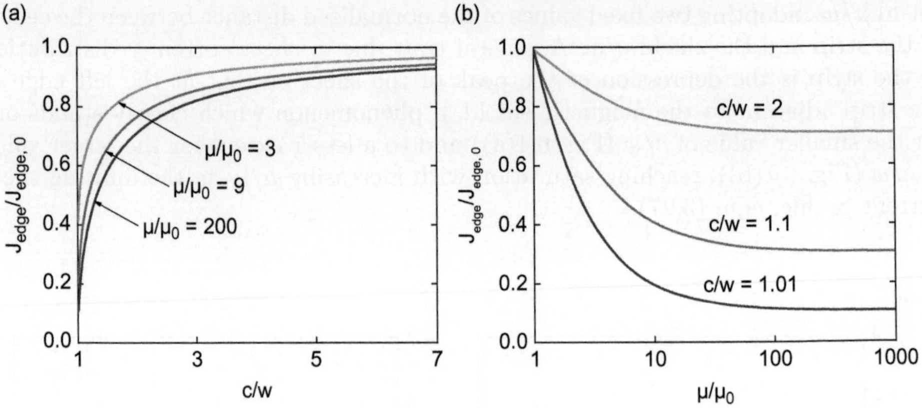

Fig. 6.2 Variation in the normalized edge sheet current $J_{\rm edge}/J_{\rm edge,0}$ of the magnetically shielded superconductor strip in the flux-free Meissner state with (a) the normalized distance between the centre of the strip and the paramagnet c/w for three different values of the relative magnetic permeability of the shield μ/μ_0 identified on the curves, and (b) the relative magnetic permeability of the shield μ/μ_0 for three different values of the normalized distance between the centre of the strip and the paramagnet c/w identified on the curves, when $h/2w = 0.0002$, referring to the flat semi-infinite paramagnet configuration of Fig. 3.2. The total current $I = 2w\bar{J}$ is fixed. Reproduced with permission from Genenko, Snezhko, and Freyhardt 2000, *Phys. Rev. B* 62, pg 3453, doi: https://doi.org/10.1103/PhysRevB.62.3453. Copyrighted by the American Physical Society.

6.2 Superconductor strip inside a cylindrical paramagnetic cavity

We continue by resuming the problem of an infinitely extended, thin current-carrying type-II superconductor strip in the Meissner state, placed inside a cylindrical paramagnetic cavity with radius a and permeability μ, as depicted in Fig. 3.10. Our aim now is to study the effect of this two-sided closed magnetic shield on the distribution of the sheet current across the width of the strip in the whole paramagnetic range of μ beyond the limit $\mu \to \infty$, an attempt which again warrants computational means. The account that follows adheres to the work of Genenko, Snezhko, and Freyhardt (2000).

6.2.1 Field distribution

According to the explanations of section 3.5.1, eqns. (2.87) and (2.88) here simplify to the Poisson equation for the single nontrivial component of the magnetic vector potential A_z,

$$\Delta A_z = \begin{cases} -\mu_0 J(x)\delta(y) & \text{in } -w \le x \le w \\ 0 & \text{otherwise,} \end{cases} \tag{6.9}$$

supplemented with the boundary conditions

$$\lim_{r\to a-} A_z(x,y) = \lim_{r\to a+} A_z(x,y) \quad \text{and} \quad \left.\frac{1}{\mu_0}\frac{\partial A_z}{\partial n}\right|_{r=a-} = \left.\frac{1}{\mu}\frac{\partial A_z}{\partial n}\right|_{r=a+}, \tag{6.10}$$

introducing cylindrical coordinates r, ϕ, z adapted to the paramagnetic cavity. Obviously, the latter demands exclude direct contact between the cavity and the edges of the strip to avoid complications arising from the proximity effect (Schmidt 1997). The desired solution thus can be constructed by the method of images, calling in their current strength $\kappa = (\mu - \mu_0)/(\mu + \mu_0)$, together with the Green's function technique. Identifying the contributions $dI = dx' J(x')$ of a distributed current element of unit weight at $(x = x', y = 0)$, the point marked M in Fig. 3.10, as well as of an image current element of weight κ at $(x = a^2/x', y = 0)$, the point marked N in Fig. 3.10, and integrating over x' yields the representation in the vacuum space, $0 \le r < a$ inclusive of the surface of the strip ∂D,

$$A_z(x,y) = \frac{\mu_0}{2\pi}\int_{-w}^{w} dx' J(x')\left\{\ln\frac{a}{\sqrt{(x-x')^2+y^2}} + \kappa \ln\frac{a}{\sqrt{(x-a^2/x')^2+y^2}}\right\}. \tag{6.11}$$

Applying eqn. (2.3) to eqn. (6.11) thus gives the Cartesian components of the magnetic field $\mathbf{H}$,

$$H_x(x,y) = -\frac{1}{2\pi}\int_{-w}^{w} dx' J(x')\left[\frac{y}{(x-x')^2+y^2} + \kappa\frac{y}{(x-a^2/x')^2+y^2}\right] \tag{6.12a}$$

$$H_y(x,y) = \frac{1}{2\pi}\int_{-w}^{w} dx' J(x')\left[\frac{x-x'}{(x-x')^2+y^2} + \kappa\frac{x-a^2/x'}{(x-a^2/x')^2+y^2}\right]. \tag{6.12b}$$

6.2.2 Current distribution

As noted earlier, the Meissner state of the superconductor strip requires the normal component of the magnetic field H_y to vanish across the width of the strip,

$$H_y(x) = 0 \quad \text{for} \quad -w < x < w, y = 0. \tag{6.13}$$

Recalling eqn. (6.12b) hence yields the condition for this state

$$\frac{1}{2\pi}\int_{-w}^{w} dx' J(x')\left(\frac{1}{x-x'} + \kappa\frac{1}{x-a^2/x'}\right) = 0 \quad \text{in} \quad -w < x < w, \tag{6.14}$$

i.e. a homogeneous integral equation which duly generalizes eqn. (3.91) with eqn. (3.92). The first term herein represents the magnetic field created by the transport current itself and the second term delineates the magnetic field induced by the paramagnetic shield. Since eqn. (6.14) proves cumbersome for a numerical analysis because of the singularity it contains, we invert eqn. (6.12b) in respect to the induced magnetic field H_i across the width of the strip to obtain

$$J(x) = \frac{I}{\pi}\frac{1}{\sqrt{(w-x)(x+w)}} + \frac{2}{\pi}\int_{-w}^{w} dx' \sqrt{\frac{(w-x')(x'+w)}{(w-x)(x+w)}}\frac{H_\mathrm{i}(x')}{x-x'}$$

$$\text{in } \; -w < x < w, \tag{6.15}$$

making use of the conformal mapping technique (Brandt 1992; Brandt and Indenbom 1993), upon observing eqn. (3.2). As a result, the first term in eqn. (6.15) emulates the contribution of an isolated superconductor strip, recalling eqn. (3.8), whereas the second term in eqn. (6.15) reflects the influence of the paramagnetic shield. Substituting H_i from eqn. (6.14) in eqn. (6.15) thus furnishes the sheet current as a solution of the norm-conserving regular integral equation of the Fredholm type

$$J(x) = \frac{I}{\pi}\frac{1+\kappa}{\sqrt{(w-x)(x+w)}} + \frac{\kappa}{\pi}\int_{-w}^{w} dx' \sqrt{\frac{(a^2-x'^2)(x'^2+a^2)}{(w-x)(x+w)}}\frac{J(x')/x'}{x-a^2/x'}$$

$$\text{in } \; -w < x < w, \tag{6.16}$$

which easily permits application of standard numerical discretization routines (Delves and Mohamed 1985; Press, Teukolsky, Vetterling, and Flannery 2007). One can verify that, in the limit $\mu \to \infty$, eqn. (6.16) is satisfied by the solution, eqn. (3.98). Conversely, if $\mu \to \mu_0$, eqn. (6.16) duplicates the solution for the unshielded strip, eqn. (3.8).

Reasoning in a similar vein to that for the isolated superconductor strip in the barrier protected Meissner state of section 3.1 limits the validity of the solution of eqn. (6.16) to the range $-w+h \leq x \leq w-h$, with the maximum value J_max imposed in $|x \pm w| \leq h$. However, since we focus on investigating the sensitivity of the distribution of the sheet current to shielding by the paramagnetic environment, it is assumed that J_max stays *below* the characteristic sheet current for entry of magnetic flux, $J_\mathrm{max} < J_b$. This implies that the total current $I = 2w\bar{J}$, expressed through the average sheet current $\bar{J}$, remains *lower* than the maximum loss-free total current, $I < I_\mathrm{max}$, too.

The influence of the paramagnetic environment shows up in the distribution of the sheet current J measured against its average $\bar{J}$, on the one hand, and in the maximum sheet current J_max calibrated against its equivalent without the magnetic shield $J_\mathrm{max,0}$, on the other, referring to the total current I. Both the geometrical parameter a and the magnetic characteristic μ play crucial roles thereby.

Figure 6.3 displays the variation in the normalized sheet current $J/\bar{J}$ over the width of the strip for a range of values of the relative magnetic permeability of the shield μ/μ_0, adopting two fixed values of the normalized radius of the cavity a/w. A general trait due to sheet current redistribution in the strip is the depression of the

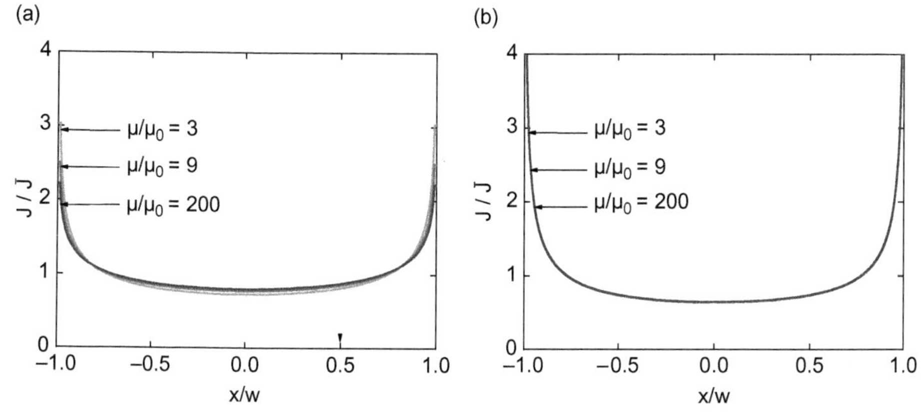

Fig. 6.3 Distribution of the normalized sheet current $J/\bar{J}$ as a function of the normalized coordinate x/w over the width of the magnetically shielded superconductor strip in the flux-free Meissner state for three different values of the relative magnetic permeability of the shield μ/μ_0 identified on the curves along with the normalized radius of the cavity (a) $a/w = 1.01$ and (b) $a/w = 2$, when $h/2w = 0.005$, referring to the cylindrical paramagnet configuration of Fig. 3.10.

peaks of the sheet current at both edges of the strip, a phenomenon which clearly stands out for the smaller value of a/w (Fig. 6.3(a)) and to a lesser extent for the larger value of a/w (Fig. 6.3(b)), reaching saturation with increasing μ/μ_0 in the limiting sheet current profile, eqn. (3.98).

Figure 6.4 illustrates the dependence of the normalized maximum sheet current $J_{\max}/J_{\max,0}$ on the normalized radius of the cavity a/w and on the relative magnetic permeability of the shield μ/μ_0 (Figs. 6.4(a) and 6.4(b), respectively). Keeping μ/μ_0 fixed and starting from a distinct minimum at the smallest value of a/w, there is a monotonic rise towards unity, the value prevailing for the isolated strip, as a/w continues to grow. Accordingly, the curves approach the step-like shape expected for the isolated strip, if μ/μ_0 wanes (Fig. 6.4(a)). Keeping a/w fixed and starting from unity at the vacuum value of μ/μ_0, there is a monotonic drop followed by a settlement at the level prevailing for an ideal paramagnetic shield, as μ/μ_0 continues to grow. This effect gets more pronounced as a/w wanes and virtually saturates from $\mu/\mu_0 \simeq 200$, proving eqn. (3.98) adequately captures sheet current redistribution in the strip for values $\mu/\mu_0 \gtrsim 200$ (Fig. 6.4(b)).

Although a comparison with the situation of a flat paramagnetic shield, Fig. 6.2, unfolds similar tendencies regarding the depression of the peaks of the sheet current for open and closed paramagnetic environments, important differences nevertheless come to the fore. Thus, the rise in the normalized maximum sheet current in the presence of the cylindrical cavity (Fig. 6.4(a)) is noticeably steeper and saturates faster with the respective geometrical parameter than the rise of the normalized

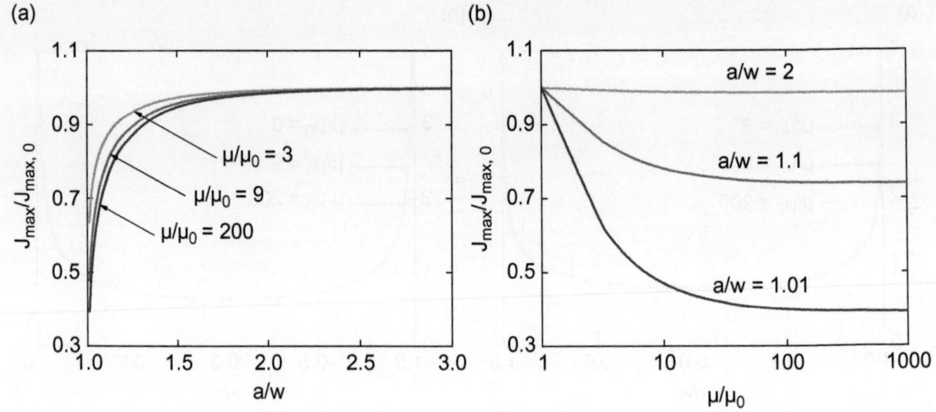

Fig. 6.4 Variation in the normalized maximum sheet current $J_{\mathrm{max}}/J_{\mathrm{max},0}$ of the magnetically shielded superconductor strip in the flux-free Meissner state with (a) the normalized radius of the cavity a/w for three different values of the relative magnetic permeability of the shield μ/μ_0 identified on the curves, and (b) the relative magnetic permeability of the shield μ/μ_0 for three different values of the normalized radius of the cavity a/w identified on the curves, when $h/2w = 0.0002$, referring to the cylindrical paramagnet configuration of Fig. 3.10. The total current $I = 2w\bar{J}$ is fixed.

edge sheet current in the presence of the flat shield (Fig. 6.2(a)). On the other hand, albeit saturating at about the same rate, the drop of the normalized edge sheet current with the magnetic characteristic in the presence of the flat shield (Fig. 6.2(b)) is much steeper than the drop in the normalized maximum sheet current in the presence of the cylindrical environment (Fig. 6.4(b)). This means that the distribution of the sheet current can be more effectively controlled by the open paramagnetic shield of Fig. 3.2, rather than by the closed paramagnetic cavity of Fig. 3.10, a result which further underpins the discussion at the end of section 3.7.2.

6.3 Superconducting tubular wire adjacent to a flat semi-infinite paramagnet

Modifying the superconductor component of the heterostructure of section 6.1, we now consider an infinitely extended, thin current-carrying type-II superconducting tubular wire with radius r_s in the Meissner state placed at distance c adjacent to a flat semi-infinite paramagnetic environment with permeability μ, as depicted in Fig. 6.5. Our aim is to study the effect of this one-sided open magnetic shield on the distribution of the sheet current along the circumference of the tubular wire in the whole paramagnetic range of μ, an attempt which ultimately warrants computational means. The following presentation derives from the works of Yampolskii, Genenko, and Rauh (2004a,b).

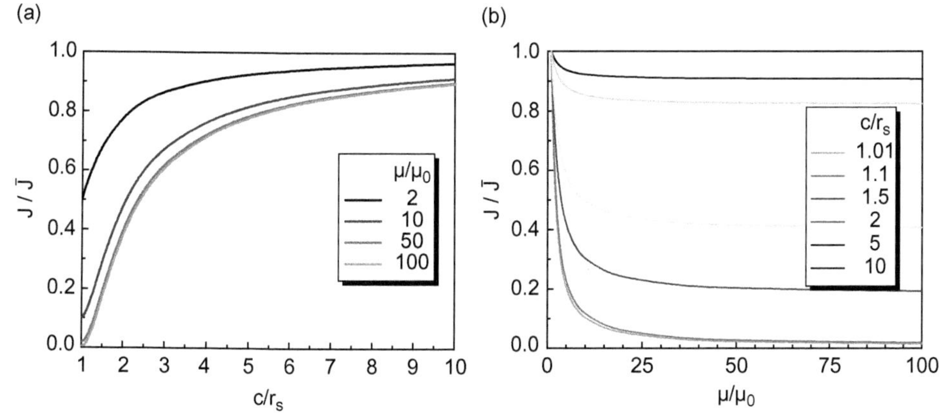

Fig. 6.7 Variation of the normalized minimum sheet current $J_{\min}/\bar{J}$ of the magnetically shielded superconducting tubular wire in the flux-free Meissner state with (a) the normalized distance between the centre of the tubular wire and the paramagnet c/r_s for different values of the relative magnetic permeability of the shield μ/μ_0, specified in the legend, and (b) the relative magnetic permeability of the shield μ/μ_0 for different values of the normalized distance between the centre of the strip and the paramagnet c/r_s, specified in the legend, referring to the flat semi-infinite paramagnet configuration of Fig. 6.5. Reproduced from S. V. Yampolskii, Y. A. Genenko, and H. Rauh, Distribution of the sheet current in a magnetically shielded superconducting filament, *Physica C: Superconductivity*, Volume 415, Issue 4, pgs 151–157, copyright (2004), with permission from Elsevier, doi.org/10.1016/j.physc.2004.05.012

either of these quantities. In the limit of zero wire–magnet distance, the minimum value adopts a finite magnitude and finite derivative, unlike in the case of shielded strips, where divergent derivatives persist; see Fig. 6.2(b). The variation of the minimum value of the sheet current at zero azimuth with the relative permeability, for different values of the distance between the wire and the magnet relative to the radius of the wire prescribed, is portrayed in Fig. 6.7(b). This shows that the minimum value of the sheet current decreases with the relative permeability due to current redistribution in the wire, but increases with the wire–magnet distance, the effect saturating again for large values of either of these quantities. As before, the asymptotic behaviour with respect to the wire–magnet distance is evident from eqn. (6.25).

We note that the phenomenon of depression of the sheet current on the side of the tubular wire adjacent to the surface of the magnet, with a minimum value at zero azimuth, can easily be understood by recalling the current distributions observed in thin superconductor strips (see sections 3.2 and 6.1). The results demonstrate a dependence of the sheet current on the wire–magnet distance very much like that for the case of magnetically shielded strips, though without current peaks typical of strips. Substantial current redistributions in the wire can already occur for low values of the relative permeability of the magnet, when the distance between the wire and the magnet is short, with evidence of saturation at moderately high values of this quantity, similar to the findings for magnetically shielded strips. We therefore expect that an asymmetric magnetic shielding configuration is indeed capable of redistributing the current in a superconducting tubular wire effectively.

6.4 Superconductor ring between two coaxial cylindrical paramagnets

Uniting assets of the magnetic environments appraised in sections 6.1 and 6.2, we proceed by considering a thin current-carrying type-II superconductor ring of width $2w$ in the Meissner state with inner radius r_1 and outer radius r_2, placed between two infinitely extended, coaxial cylindrical paramagnets with inner radius R_1, permeability μ_1 and outer radius R_2, permeability μ_2, as illustrated in Fig. 6.8.

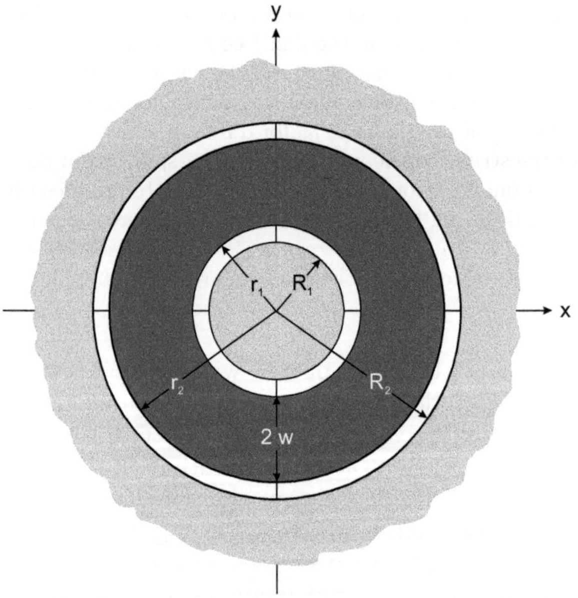

Fig. 6.8 Cross-sectional view of a flat superconductor ring (dark shading) located between two coaxial cylindrical paramagnets (light shading) which themselves extend infinitely in the z-direction of a Cartesian coordinate system x, y, z. The definition of the radii of the ring r_1 and r_2 as well as the definition of the radii of the paramagnets R_1 and R_2 are marked.

The ring is thought to occupy the cross-sectional plane $z = 0$, breaking the translational symmetry in the z-direction of the Cartesian coordinate system x, y, z. Affording circular sheet current flow about a current-free core, such a heterostructure may obviously serve as a crude model of a superconducting 'pancake' having merely one turn. Our aim is to study the effect of its two-sided closed magnetic shield on the distribution of the circular sheet current across the width of the ring in the whole paramagnetic range of μ_1, μ_2, an endeavour which again warrants computational means. The presentation that follows emanates from the works of Genenko, Rauh, and Snezhko (2001, 2002).

6.4.1 Field distribution

According to the explanations of sections 3.2.1 and 3.5.1, eqns. (2.90) and (2.91) here simplify to the Poisson equation for the single nontrivial component of the magnetic vector potential A_ϕ,

$$\Delta A_\phi - \frac{1}{r^2} A_\phi = \begin{cases} -\mu_0 J(r)\delta(z) & \text{in } r_1 \leq r \leq r_2 \\ 0 & \text{otherwise} \end{cases} \tag{6.26}$$

in cylindrical coordinates r, ϕ, z adapted to the superconductor ring, supplemented with the boundary conditions

$$\lim_{r\to R_1-} A_\phi(r,z) = \lim_{r\to R_1+} A_\phi(r,z) \quad \text{and} \quad \frac{1}{\mu_1}\frac{\partial A_\phi}{\partial r}\bigg|_{r\to R_1-} = \frac{1}{\mu_0}\frac{\partial A_\phi}{\partial r}\bigg|_{r\to R_1+} \tag{6.27a}$$

$$\lim_{r\to R_2-} A_\phi(r,z) = \lim_{r\to R_2+} A_\phi(r,z) \quad \text{and} \quad \frac{1}{\mu_0}\frac{\partial A_\phi}{\partial r}\bigg|_{r\to R_2-} = \frac{1}{\mu_2}\frac{\partial A_\phi}{\partial r}\bigg|_{r\to R_2+} . \tag{6.27b}$$

The latter demands obviously rule out direct contact between the paramagnets and the edges of the ring to avoid complications arising from the proximity effect (Schmidt 1997). Needless to say that, in addition, A_ϕ and $\partial A_\phi/\partial z$ must vanish as $z \to \pm\infty$. The desired solution thus can be constructed using the method of Fourier transforms together with the Green's function technique.

Expressing A_ϕ in terms of its one-dimensional Fourier transform $\tilde{A}_\phi^k$ gives

$$A_\phi(r,z) = \frac{1}{\pi}\int_0^\infty dk\, \tilde{A}_\phi^k(r)\cos(kz). \tag{6.28}$$

Identifying the contribution $dI = dr' J(r')$ of a distributed current element of unit weight at $(r = r', z = 0)$ and integrating over r' yields the representation in the vacuum space, $R_1 \leq r \leq R_2$ inclusive of the surface of the ring ∂D,

$$\tilde{A}_\phi^k(r) = \begin{cases} \mu_0 \int_{r_1}^{r_2} dr' J(r') f_k^<(r,r') & \text{for } R_1 \leq r < r_1 \\ \mu_0 \int_{r_1}^{r} dr' J(r') f_k^>(r,r') + \mu_0 \int_{r}^{r_2} dr' J(r') f_k^<(r,r') & \text{for } r_1 \leq r \leq r_2 \\ \mu_0 \int_{r_1}^{r_2} dr' J(r') f_k^>(r,r') & \text{for } r_2 < r \leq R_2 \end{cases} \tag{6.29}$$

with

$$f_k^<(r,r') = a_k^<(r') I_1(kr) + b_k^<(r') K_1(kr) \tag{6.30a}$$

$$f_k^>(r,r') = a_k^>(r') I_1(kr) + b_k^>(r') K_1(kr), \tag{6.30b}$$

where I_1 and K_1 denote modified Bessel functions of the first and second kinds, respectively, of order one (Abramowitz and Stegun 1972). The coefficients in eqns. (6.30a), (6.30b) read

$$a_k^<(r') = c_k^{(1)}(r')kR_1\,[I_1'(kR_1)K_1(kR_1)\mu_0/\mu_1 - I_1(kR_1)K_1'(kR_1)] \tag{6.31a}$$

$$b_k^<(r') = c_k^{(1)}(r')kR_1(1-\mu_0/\mu_1)I_1(kR_1)I_1'(kR_1) \tag{6.31b}$$

$$a_k^>(r') = c_k^{(2)}(r')kR_2(\mu_0/\mu_2 - 1)K_1(kR_2)K_1'(kR_2) \tag{6.31c}$$

$$b_k^>(r') = c_k^{(2)}(r')kR_2\,[I_1'(kR_2)K_1(kR_2) - I_1(kR_2)K_1'(kR_2)\mu_0/\mu_2] \tag{6.31d}$$

with

$$c_k^{(1)}(r') = u_k(r')r'/kR_1D_k(R_1,R_2) \tag{6.32a}$$

$$c_k^{(2)}(r') = v_k(r')r'/kR_2D_k(R_1,R_2), \tag{6.32b}$$

where

$$u_k(r') = K_1(kr')/kR_2 + (\mu_0/\mu_2 - 1)E_k(R_2,r')K_1'(kR_2) \tag{6.33a}$$

$$v_k(r') = I_1(kr')/kR_1 - (1-\mu_0/\mu_1)E_k(R_1,r')I_1'(kR_1) \tag{6.33b}$$

and

$$\begin{aligned} D_k(R_1,R_2) &= 1/k^2R_1R_2 - (1-\mu_0/\mu_1)I_1'(kR_1)K_1(kR_1)/kR_2 \\ &\quad + (1-\mu_0/\mu_1)I_1'(kR_1)E_k(R_1,R_2)(\mu_0/\mu_2 - 1)K_1'(kR_2) \\ &\quad - (\mu_0/\mu_2 - 1)I_1(kR_2)K_1'(kR_2)/kR_1 \end{aligned} \tag{6.34}$$

with

$$E_k(r,r') = K_1(kr)I_1(kr') - I_1(kr)K_1(kr'), \tag{6.35}$$

I_1' and K_1' meaning derivatives of the modified Bessel functions with respect to their arguments. Applying eqn. (2.3) to eqn. (6.28) with eqn. (6.29) thus gives the Fourier representation of the cylindrical components of the magnetic field $\mathbf{H}$,

$$H_r(r,z) = \frac{1}{\pi}\int_0^\infty dk\,\tilde{H}_r^k(r)\sin(kz) \tag{6.36a}$$

$$H_z(r,z) = \frac{1}{\pi}\int_0^\infty dk\,\tilde{H}_z^k(r)\cos(kz), \tag{6.36b}$$

where

$$\tilde{H}_r^k(r) = \begin{cases} k\int_{r_1}^{r_2} dr' J(r') f_k^<(r,r') & \text{for } R_1 \le r < r_1 \\ k\int_{r_1}^{r} dr' J(r') f_k^>(r,r') + k\int_{r}^{r_2} dr' J(r') f_k^<(r,r') & \text{for } r_1 \le r \le r_2 \\ k\int_{r_1}^{r_2} dr' J(r') f_k^>(r,r') & \text{for } r_2 < r \le R_2 \end{cases} \tag{6.37a}$$

$$\tilde{H}_z^k(r) = \begin{cases} k\int_{r_1}^{r_2} dr' J(r') g_k^<(r,r') & \text{for } R_1 \le r < r_1 \\ k\int_{r_1}^{r} dr' J(r') g_k^>(r,r') + k\int_{r}^{r_2} dr' J(r') g_k^<(r,r') & \text{for } r_1 \le r \le r_2 \\ k\int_{r_1}^{r_2} dr' J(r') g_k^>(r,r') & \text{for } r_2 < r \le R_2 \end{cases} \tag{6.37b}$$

with

$$g_k^<(r,r') = a_k^<(r') I_0(kr) - b_k^<(r') K_0(kr) \tag{6.38a}$$

$$g_k^>(r,r') = a_k^>(r') I_0(kr) - b_k^>(r') K_0(kr), \tag{6.38b}$$

I_0 and K_0 denoting modified Bessel functions of the first and second kinds, respectively, of order zero (Abramowitz and Stegun 1972).

6.4.2 Current distribution

As for a superconductor strip the Meissner state of the superconductor ring necessitates the normal component of the magnetic field H_z to vanish across the width of the ring,

$$H_z(r) = 0 \quad \text{for } r_1 < r < r_2, z = 0. \tag{6.39}$$

Recalling eqn. (6.36b) with eqn. (6.37b) and changing the order of integration hence furnishes the condition for this state

$$\int_{r_1}^{r_2} dr'\, J(r') Q(r,r') = 0 \quad \text{in } r_1 < r < r_2, \tag{6.40}$$

i.e. a homogeneous Fredholm integral equation with the singular two-part kernel

$$Q(r,r') = \int_0^\infty dk\, k \left[g_k^>(r,r')\theta(r-r') + g_k^<(r,r')\theta(r'-r) \right], \tag{6.41}$$

Heaviside's step function θ herein selecting terms according to their dependence on r or r'. Naturally, the sheet current as a solution of eqn. (6.40) must obey the normalization

$$\int_{r_1}^{r_2} dr'\, J(r') = I. \tag{6.42}$$

A numerical treatment of eqn. (6.40) obviously requires application of elaborate computational procedures, the results of which can be verified by comparison with known extreme cases.

Particularly, one can verify that, in the limit $\mu_1, \mu_2 \to \infty$, eqn. (6.41) permits the succinct representation

$$Q(r, r') = \int_0^\infty \frac{dk}{F_k(R_1, R_2)} \frac{\partial^2 S_k(R_1, R_2; r, r')}{\partial(kR_1)\partial(kR_2)} \tag{6.43}$$

with

$$F_k(R_1, R_2) = I_1'(kR_1)K_1'(kR_2) - K_1'(kR_1)I_1'(kR_2) \tag{6.44}$$

and

$$S_k(R_1, R_2; r, r') = G_k(R_2, r)r' \frac{\partial G_k(R_1, r')}{\partial r'} \theta(r - r') + G_k(R_1, r)r' \frac{\partial G_k(R_2, r')}{\partial r'} \theta(r' - r), \tag{6.45}$$

where

$$G_k(r, r') = I_1(kr)K_0(kr') + K_1(kr)I_0(kr'). \tag{6.46}$$

Conversely, if $\mu_1, \mu_2 \to \mu_0$, eqn. (6.41) reduces to the expression for the unshielded ring,

$$Q(r, r') = \int_0^\infty dk\, kr' \left[I_0(kr)K_1(kr')\theta(r' - r) - K_0(kr)I_1(kr')\theta(r - r')\right], \tag{6.47}$$

which admits the closed form

$$Q(r, r') = \frac{1}{2} \left[\frac{K(m)}{r + r'} - \frac{E(m)}{r - r'}\right], \tag{6.48}$$

duplicating a result used in previous studies of this limit elsewhere (Mikheenko und Kuzovlev 1993; Babaei Brojeny and Clem 2003). K and E herein denote complete elliptic integrals of the first and second kinds, respectively, with the parameter $m = 4rr'/(r + r')^2$ (Abramowitz and Stegun 1972).

Reasoning in a similar vein to that for the isolated superconductor strip in the barrier-protected Meissner state of section 3.1 limits the validity of the solution of eqn. (6.40) to the range $r_1 + h \le r \le r_2 - h$, with the maximum value $J_{\max}$ imposed in $r_1 \le r \le r_1 + h$. However, since we focus on investigating the sensitivity of the distribution of the sheet current to shielding by the paramagnetic environment, it is assumed that $J_{\max}$ stays *below* the characteristic sheet current for entry of magnetic flux, $J_{\max} < J_b$. This implies that the total current $I = 2w\bar{J}$, expressed through the average sheet current $\bar{J}$, remains *lower* than the maximum loss-free total current, $I < I_{\max}$, too, irrespective of the edge sheet current J_{edge} which adjusts itself in $r_2 - h \le r \le r_2$.

The influence of the paramagnetic environment shows up in the distribution of the sheet current J measured against its average $\bar{J}$, referring to the total current I. Nevertheless, it should be noted that, for topological reasons alone, sheet current flow in the superconductor ring is bound to exhibit a bias across its width, revealing a higher strength of the sheet current at the inner edge of the ring, J_{max}, and a lower strength of the sheet current at the outer edge of the ring, J_{edge}, even with magnetic shielding from both sides absent. The action of paramagnetic shields naturally modifies this trait. Given the shape of the ring, both the geometrical and the magnetic characteristics of these shields play crucial roles thereby.

Cumbersome expressions for coefficients in eqns. (6.40) and (6.41) supplemented by eqns. (6.31)-(6.35) and (6.38) make the numerical analysis in a general case difficult. To study the principal trends in the behaviour of current distributions when the geometrical and material parameters are varied, the following numerical procedure was used (Genenko, Rauh, and Snezhko 2001). The region of the superconductor ring was pervaded by $n+1$ concentric circular paths of radii $\rho_\nu = r_1 + \nu\Delta\rho$ with $\nu = 0, 1, \ldots, n$, where $\Delta\rho = (r_2 - r_1)/n$, carrying the respective currents $i_0 = J(\rho_0)\Delta\rho/2$ and $i_\nu = J(\rho_\nu)\Delta\rho$ for $\nu = 1, 2, \ldots, n-1$ as well as $i_n = J(\rho_n)\Delta\rho/2$. From these currents, the values of the field component $\bar{H}_z^{(\nu)} = H_z(\bar{\rho}_\nu, z = 0)$, created along circles between these paths, with radii $\bar{\rho}_\nu = (\rho_{\nu-1} + \rho_\nu)/2$ for $\nu = 1, 2, \ldots, n$, were determined. The functional $\mathcal{F}(\{i_\nu\}) = |\bar{H}_z^{(1)}(\{i_\nu\})| + |\bar{H}_z^{(2)}(\{i_\nu\})| + \ldots |\bar{H}_z^{(n)}(\{i_\nu\})|$ was then minimized with regard to the set of currents $\{i_\nu\}$ so as to reach the Meissner state, where $\bar{H}_z^{(\nu)} = 0$ for $\nu = 1, 2, \ldots, n$, given the constraint that the total current, $i = i_0 + i_1 + \ldots + i_n$, is fixed. In fact, as was discussed earlier, the Meissner state is protected if the magnitude of the edge current remains under some critical value which depends on material properties and the geometry of the edge (Indenbom et al. 1994). However, it proves mathematically more convenient to explore the redistribution of a fixed total current under different conditions and then to estimate its magnitude from the physical conditions prevailing at the edges of the ring, which is the position adopted here.

Figure 6.9 displays the variation in the normalized sheet current $J/\bar{J}$ over the width of the ring for a range of values of the relative permeability of the *inner* paramagnet μ_1/μ_0, with shielding by the outer paramagnet put to rest (Fig. 6.9(a)), and a range of values of the relative permeability of the *outer* paramagnet μ_2/μ_0, with shielding by the inner paramagnet put to rest (Fig. 6.9(b)), adopting fixed values of the normalized radii of the paramagnets R_1/r_1 and R_2/r_1. As a general result, sheet current redistribution in the ring entails a depression of the peak of the sheet current near the edge of the ring facing that paramagnet whose relative permeability augments, and a concomitant rise of the peak of the sheet current near the edge of the ring facing the opposite paramagnet whose shielding act is quenched. This effect prevails in the case of shielding from outside, rather than inside, and virtually saturates from $\mu_1/\mu_0 \simeq 200$ or $\mu_2/\mu_0 \simeq 200$, respectively, a choice which proves representative for values of μ_1/μ_0 or μ_2/μ_0, respectively, beyond these.

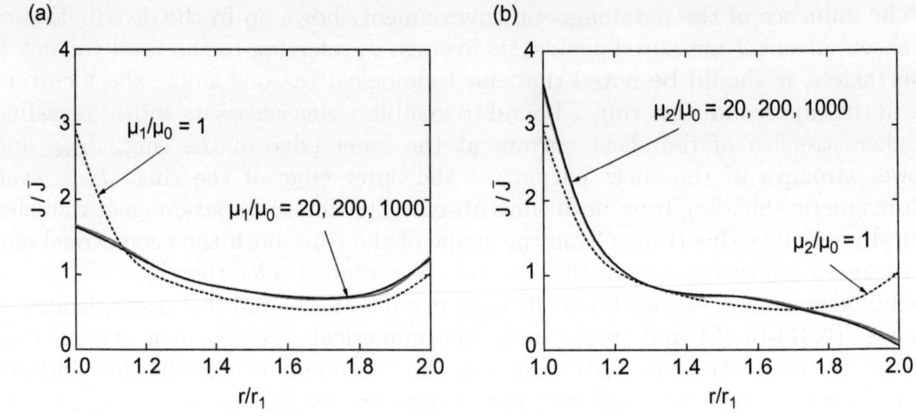

Fig. 6.9 Distribution of the normalized sheet current $J/\bar{J}$ as a function of the normalized radius r/r_1 over the width of the magnetically shielded superconductor ring in the flux-free Meissner state for (a) the relative permeability of the outer paramagnet $\mu_2/\mu_0 = 1$ as well as three different values of the relative permeability of the *inner* paramagnet μ_1/μ_0 identified on the curves, and (b) the relative permeability of the inner paramagnet $\mu_1/\mu_0 = 1$ as well as three different values of the relative permeability of the *outer* paramagnet μ_2/μ_0 identified on the curves, when the ratio of the radii of the superconductor ring $r_2/r_1 = 2$, the normalized radius of the inner paramagnet $R_1/r_1 = 0.99$, and the normalized radius of the outer paramagnet $R_2/r_1 = 2.01$ in conjunction with $h/2w = 0.01$, referring to the coaxial cylindrical paramagnet configuration of Fig. 6.8. The dotted curves relate to the isolated superconductor ring and merely serve for comparison. Reprinted from Y. A. Genenko, H. Rauh, and A. Snezhko, Novel design of a smart magnet/superconductor heterostructure, *Physica C: Superconductivity*, Volumes 372—376, Part 3, pgs 1389–1393, copyright (2002), with permission from Elsevier, doi.org/10.1016/S0921-4534(02)01036-5

Figure 6.10 shows the variation in the normalized sheet current $J/\bar{J}$ over the width of the ring for a range of values of the normalized radius of the *inner* paramagnet R_1/r_1, with the normalized radius of the outer paramagnet kept fixed (Fig. 6.10(a)), and a range of values of the normalized radius of the *outer* paramagnet R_2/r_1, with the normalized radius of the inner paramagnet kept fixed (Fig. 6.10(b)), adopting a common value of the relative permeabilities of the paramagnets μ_1/μ_0 and μ_2/μ_0. As a general result, sheet current redistribution in the ring causes a depression in the peak of the sheet current near the edge of the ring facing that paramagnet which approaches the respective edge of the ring, and a concomitant rise in the peak of the sheet current near the edge of the ring adjacent to the opposite paramagnet whose radius remains unchanged. Again, this effect prevails in the case of shielding from outside, rather than inside, and virtually saturates below $R_2/r_1 \simeq 2.01$, i.e. when the outer paramagnet almost touches the ring, yielding a monotonic decrease in the distribution of the sheet current from inside out.

We note that, because of solving eqn. (6.39) by the approximate minimization method, the peaks at the ring edges in Figs. 6.9 and 6.10 are not as pronounced

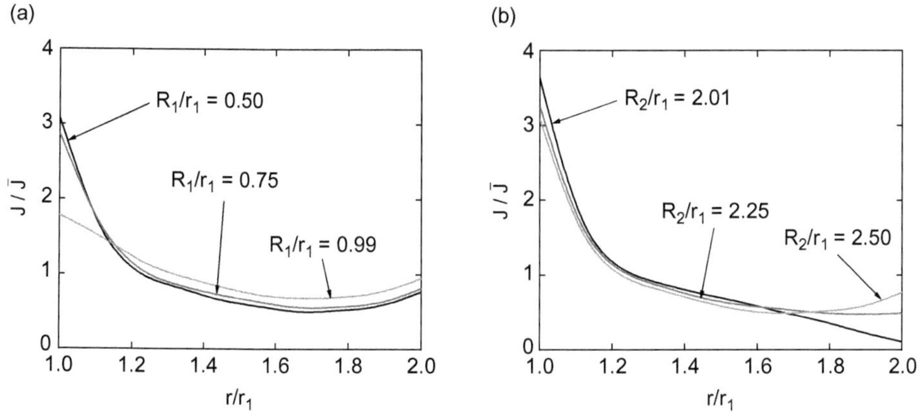

Fig. 6.10 Distribution of the normalized sheet current $J/\bar{J}$ as a function of the normalized radius r/r_1 over the width of the magnetically shielded superconductor ring in the flux-free Meissner state for (a) the normalized radius of the outer paramagnet $R_2/r_1 = 2.01$ as well as three different values of the normalized radius of the *inner* paramagnet R_1/r_1 identified on the curves, and (b) the normalized radius of the inner paramagnet $R_1/r_1 = 0.99$ as well as three different values of the normalized radius of the *outer* paramagnet R_2/r_1 identified on the curves, when the ratio of the radii of the superconductor ring $r_2/r_1 = 2$, the relative permeability of the inner paramagnet $\mu_1/\mu_0 = 200$, and the relative permeability of the outer paramagnet $\mu_2/\mu_0 = 200$ in conjunction with $h/2w = 0.01$, referring to the coaxial cylindrical paramagnet configuration of Fig. 6.8. Reprinted from Y. A. Genenko, H. Rauh, and A. Snezhko, Novel design of a smart magnet/superconductor heterostructure, *Physica C: Superconductivity*, Volumes 372–376, Part 3, pgs 1389–1393, copyright (2002), with permission from Elsevier, doi.org/10.1016/S0921-4534(02)01036-5

as at the edges of magnetically shielded strips. A comparison with the situation of a superconductor strip adjacent to a flat semi-infinite paramagnetic environment, Fig. 6.1, unfolds similar tendencies, but also important differences regarding the modification of the current profile for circular and straight current flow in the presence of paramagnetic shields. Thus, an asymmetry of the distribution of the sheet current across the width of both the ring (Fig. 6.9) and the strip (Fig. 6.1) exists. However, whereas this bias is caused by topology in the former case, and hence persists even for an unshielded ring, it is induced in the latter case, and therefore hinges upon the paramagnetic shield. Furthermore, the decrease in the sheet current in the ring shielded by the outer cylindrical paramagnet (Fig. 6.9(b)) is slower and saturates faster with the respective permeability than the decrease in the sheet current in the strip shielded by the flat paramagnetic environment (Fig. 6.1(a)). This means that the variation in the sheet current can be more effectively controlled by the outer paramagnetic shield of Fig. 6.8, rather than by the adjacent paramagnetic shield of Fig. 3.2.

We note that the change in the peaks of the edge sheet current—apparent from Figs. 6.9 and 6.10 for the shielded ring in the flux-free Meissner state—procures an overall increase in the maximum total current $I_{\max}$. Its magnitude may be comparable with that of the critical current of a shielded ring in the partially flux-filled critical state, $I_c = 2wJ_c$, if only $J_c \le J_b$. Nonetheless, even when distinct edge barriers are absent, almost the entire ring may remain devoid of magnetic flux if $I_{\max}$ is slightly less than I_c, as the findings of section 4.2 for a strip suggest. Such a condition may favour AC and high-field applications of the shielded ring, presuming that electromagnetic energy losses will be reduced, like in the case of a strip.

6.5 Superconductor strip adjacent to a flat paramagnetic wall

Modifying the paramagnet component of the heterostructure of section 6.1, we end by considering an infinitely extended, thin current-carrying type-II superconductor strip in the Meissner state, placed at distance c adjacent to a flat paramagnetic wall of thickness d_m and permeability μ, as depicted in Fig. 6.11. Our aim now is to study the effect of this one-sided open magnetic shield on the distribution of the sheet current across the width of the strip in the whole geometrical range of

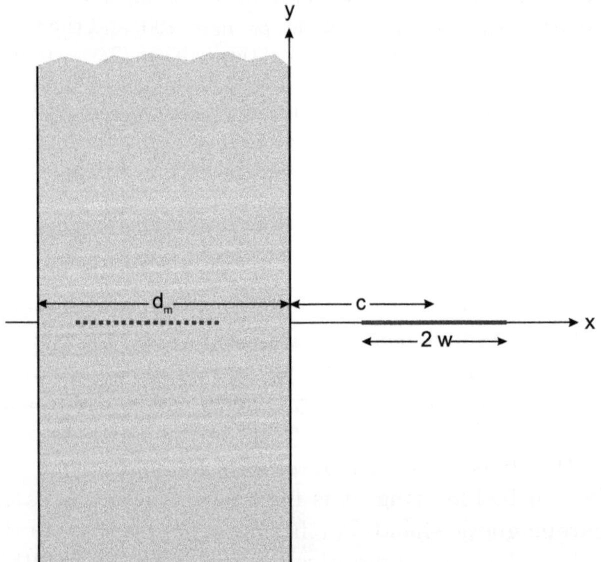

Fig 6.11 Cross-sectional view of a superconductor strip (dark shading) adjacent to a flat paramagnetic wall (light shading) extending infinitely in the z-direction of a Cartesian coordinate system x, y, z. The definition of the distance between the centre of the strip and the wall c and the definition of the thickness of the wall d_m as well as one of the images of the strip (dotted line) are marked.

d_{m} beyond the limit $d_{\mathrm{m}} \to \infty$ and in the whole paramagnetic range of μ beyond the limit $\mu \to \infty$, an attempt which ultimately warrants computational means. The following account originates from the work of Genenko and Snezhko (2002).

6.5.1 Field distribution

According to the explanations of section 6.1.1, eqns. (2.90) and (2.91) here simplify to the Poisson equation for the single nontrivial component of the magnetic vector potential A_z,

$$\Delta A_z = \begin{cases} -\mu_0 J(x)\delta(y) & \text{in } c-w \leq x \leq c+w \\ 0 & \text{otherwise,} \end{cases} \tag{6.49}$$

supplemented with the boundary conditions

$$\lim_{x\to 0-} A_z(x,y) = \lim_{x\to 0+} A_z(x,y) \quad \text{and} \quad \frac{1}{\mu}\frac{\partial A_z}{\partial n}\bigg|_{x=0-} = \frac{1}{\mu_0}\frac{\partial A_z}{\partial n}\bigg|_{x=0+} \tag{6.50a}$$

$$\lim_{x\to -d_{\mathrm{m}}-} A_z(x,y) = \lim_{x\to -d_{\mathrm{m}}+} A_z(x,y) \quad \text{and} \quad \frac{1}{\mu_0}\frac{\partial A_z}{\partial n}\bigg|_{x=-d_{\mathrm{m}}-} = \frac{1}{\mu}\frac{\partial A_z}{\partial n}\bigg|_{x=-d_{\mathrm{m}}+}. \tag{6.50b}$$

The first of the latter demands obviously rules out direct contact between the paramagnetic shield and the strip to avoid complications arising from the proximity effect (Schmidt 1997). The desired solution thus can be constructed by the method of images like before, although now expanded to two image sets, calling in their current strength $\kappa = (\mu-\mu_0)/(\mu+\mu_0)$, together with the Green's function technique. Identifying the contributions $dI = dx' J(x')$ of a distributed current element of unit weight at $(x = x', y = 0)$ as well as of image current elements of weight κ^{2n+1} at $(x = -x' - 2nd_{\mathrm{m}}, y = 0)$ and weight $-\kappa^{2n+1}$ at $(x = -x' - 2(n+1)d_{\mathrm{m}}, y = 0)$ for $n = 0, 1, 2, \ldots$ and integrating over x' yields the convenient representation in the vacuum space, $x > 0$ inclusive of the surface of the strip ∂D,

$$A_z(x,y) = \frac{\mu_0}{2\pi}\int_{c-w}^{c+w} dx' J(x') \left\{ \ln\frac{c}{\sqrt{(x-x')^2+y^2}} + \sum_{n=0}^{\infty} \kappa^{2n+1}\left[\ln\frac{c}{\sqrt{s_n(x+x')^2+y^2}} - \ln\frac{c}{\sqrt{s_{n+1}(x+x')^2+y^2}}\right]\right\} \tag{6.51}$$

with the auxiliary functions

$$s_n(x+x') = x + x' + 2nd_{\mathrm{m}}; \quad n = 0, 1, 2, \ldots. \tag{6.52}$$

Applying eqn. (2.3) to eqn. (6.51) thus gives the Cartesian components of the magnetic field **H**,

$$H_x(x,y) = -\frac{1}{2\pi}\int_{c-w}^{c+w} dx'\,J(x')\left[\frac{y}{(x-x')^2+y^2}\right.$$
$$\left.+\sum_{n=0}^{\infty}\kappa^{2n+1}\left(\frac{y}{s_n(x+x')^2+y^2}-\frac{y}{s_{n+1}(x+x')^2+y^2}\right)\right] \tag{6.53a}$$

$$H_y(x,y) = \frac{1}{2\pi}\int_{c-w}^{c+w} dx'\,J(x')\left[\frac{x-x'}{(x-x')^2+y^2}\right.$$
$$\left.+\sum_{n=0}^{\infty}\kappa^{2n+1}\left(\frac{s_n(x+x')}{s_n(x+x')^2+y^2}-\frac{s_{n+1}(x+x')}{s_{n+1}(x+x')^2+y^2}\right)\right]. \tag{6.53b}$$

6.5.2 Current distribution

Again the Meissner state of the superconductor strip requires the normal component of the magnetic field H_y to vanish across the width of the strip,

$$H_y(x) = 0 \quad \text{for } c-w<x<c+w, y=0. \tag{6.54}$$

Recalling eqn. (6.53b) hence yields the condition for this state

$$\frac{1}{2\pi}\int_{c-w}^{c+w} dx'\,J(x')\left[\frac{1}{x-x'}+\sum_{n=0}^{\infty}\kappa^{2n+1}\left(\frac{1}{s_n(x+x')}-\frac{1}{s_{n+1}(x+x')}\right)\right]=0$$
$$\text{in } c-w<x<c+w, \tag{6.55}$$

i.e. a homogeneous integral equation which duly generalizes eqn. (6.6). The first term herein represents the magnetic field created by the transport current itself and the second term delineates the magnetic field induced by the paramagnetic shield. Since eqn. (6.55) proves cumbersome for a numerical analysis because of the singularity it contains, we invert eqn. (6.53b) in respect to the induced magnetic field H_i across the width of the strip to obtain

$$J(x) = \frac{I}{\pi}\frac{1}{\sqrt{(c+w-x)(x-c+w)}}+\frac{2}{\pi}\int_{c-w}^{c+w} dx'\sqrt{\frac{(c+w-x')(x'-c+w)}{(c+w-x)(x-c+w)}}\frac{H_\text{i}(x')}{x-x'}$$
$$\text{in } c-w<x<c+w, \tag{6.56}$$

making use of the conformal mapping technique (Brandt 1992; Brandt and Indenbom 1993). As a result, the first term in eqn. (6.56) emulates the contribution of an isolated superconductor strip, recalling *mutatis mutandis* eqn. (3.8), whereas the second term in eqn. (6.56) reflects the influence of the paramagnetic shield. Substituting H_i from eqn. (6.55) in eqn. (6.56) thus furnishes the sheet current as a solution of the norm-conserving regular integral equation of the Fredholm type

$$J(x) = \frac{I}{\pi}\frac{1}{\sqrt{(c+w-x)(x-c+w)}} - \frac{1}{\pi}\frac{1}{\sqrt{(c+w-x)(x-c+w)}}\int_{c-w}^{c+w} dx'\, J(x')S(x,x') \quad \text{in } c-w<x<c+w \tag{6.57}$$

with the reduced two-part kernel

$$S(x,x') = \sum_{n=0}^{\infty} \kappa^{2n+1}\left(\sqrt{\frac{s_n(c+x')^2-w^2}{s_n(x+x')^2}} - \sqrt{\frac{s_{n+1}(c+x')^2-w^2}{s_{n+1}(x+x')^2}}\right), \tag{6.58}$$

which easily permits application of standard numerical discretization routines (Delves and Mohamed 1985; Press, Teukolsky, Vetterling, and Flannery 2007). One can verify that, in the limit $d_{\mathrm{m}} \to \infty$, eqn. (6.57) reverts to eqn. (6.8). Conversely, if $\mu \to \mu_0$, eqn. (6.57) duplicates the solution for the unshielded strip, eqn. (3.8), reference to the appropriate Cartesian coordinate system being understood.

Reasoning in a similar vein to that for the isolated superconductor strip in the barrier-protected Meissner state of section 3.1 limits the validity of the solution of eqn. (6.57) to the range $c-w+h \leq x \leq c+w-h$, with the maximum value $J_{\max}$ imposed in $c+w-h \leq x \leq c+w$. However, since we focus on investigating the sensitivity of the distribution of the sheet current to shielding by the paramagnetic environment, it is assumed that $J_{\max}$ stays *below* the characteristic sheet current for entry of magnetic flux, $J_{\max} < J_b$. This implies that the total current $I = 2w\bar{J}$, expressed through the average sheet current $\bar{J}$, remains *lower* than the maximum loss-free total current, $I < I_{\max}$, too, irrespective of the edge sheet current J_{edge} which adjusts itself in $c-w \leq x \leq c-w+h$.

The influence of the paramagnetic environment shows up in the distribution of the sheet current J measured against its average $\bar{J}$, on the one hand, and in the edge sheet current J_{edge} calibrated against its equivalent without the magnetic shield $J_{\mathrm{edge},0}$, on the other, given the total current I. The geometrical parameters c and d_{m} as well as the magnetic characteristic μ play crucial roles thereby.

In the following eqn. (6.57) is studied numerically for different values of the parameters involved. We will be interested in general trends of behaviour of the edge sheet current peaks with respect to the parameters d_{m}, c, and μ. Current distributions are shown in Fig. 6.12 for the fixed thickness of the magnet $d_{\mathrm{m}}/w = 2$ and distance to the superconductor $c/w = 1.01$, but for the different magnetic permeabilities of the media taking the values $\mu/\mu_0 = 3$, 9, 200. The current peak on the side adjacent to the magnet exhibits a clear tendency to decrease when the permeability ratio μ/μ_0 increases as in the case of a semi-infinite paramagnet, Fig. 6.1.

The question that is most vital for the utilization of the shielding effect is the influence of the finite thickness of the magnet on the peak reduction. Figure 6.13

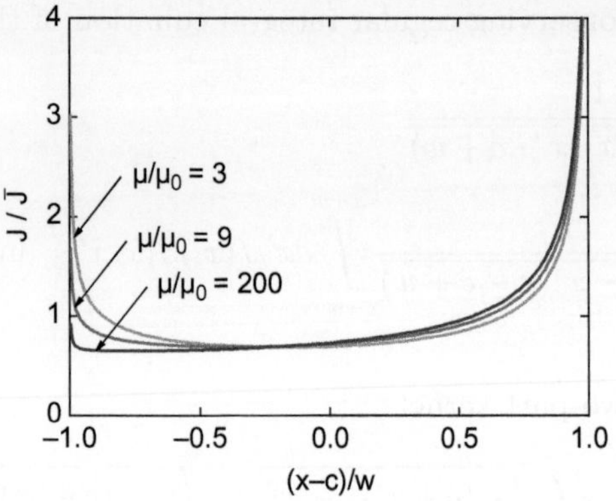

Fig 6.12 Distribution of the normalized sheet current $J/\bar{J}$ as a function of the normalized coordinate $(x-c)/w$ over the width of the magnetically shielded superconductor strip in the flux-free Meissner state for three different values of the relative magnetic permeability of the shield μ/μ_0 identified on the curves along with the normalized thickness of the wall $d_\mathrm{m}/w = 2$, when the normalized distance between the centre of the strip and the wall $c/w = 1.01$ in conjunction with $h/2w = 0.005$, referring to the flat paramagnetic wall configuration of Fig. 6.11. Reproduced with permission from Genenko and Snezhko 2002, *J. Appl. Phys.* 92, pg 357, copyright 2002, AIP Publishing LLC. doi.org/10.1063/1.1480112

demonstrates the peak value saturation as the magnet thickness increases. Figures 6.13(a) and 6.13(b) correspond to different magnet–superconductor distances c while the different curves are plotted for different magnetic permeabilities $\mu/\mu_0 = 3, 9, 200$. The dotted lines in 6.13(a) and 6.13(b) show where the peak saturates within an accuracy of 1%. The saturated peak values are smaller when the magnet–superconductor distance is smaller or when the permeability ratio μ/μ_0 is larger. We note the remarkable fact that at $\mu/\mu_0 = 200$ and $c/w = 1.01$ saturation has already occurred by a magnet thickness of $d_\mathrm{m}/w = 0.74$.

Current peaks exhibit a sharp dependence on the distance $c-w$, which is plotted in Fig. 6.14(a) for a fixed value of magnet thickness and different magnetic permeabilities, a feature established for infinite shielding configurations in sections 6.1 and 6.2. There are, however, essential differences from these cases. The shielding effect was shown to depend weakly on the image current strength κ in infinite magnet configurations (Genenko, Snezhko, and Freyhardt 2000; Genenko et al. 2000a) where it had already reached a substantial amplitude by $\mu/\mu_0 = 30$. For finite magnetic shields this is not the case. We note first the surprisingly sharp dependence of the current peak value on the magnet thickness d_m in the region of small d_m/w in Fig. 6.13. To study this practically interesting fact we plot in Fig. 6.14(b) the peak dependence on the image current strength κ for a thin shielding magnet of the thickness $d_\mathrm{m}/w = 0.001$. In contrast to the cases with infinite shields, Figs. 6.2 and 6.4, this dependence is sharp at $\kappa \to 1$. We observe that a substantial peak reduction is reached when $(d_\mathrm{m}/w)(\mu/\mu_0) \simeq 1$. This means that thin paramagnetic surroundings may also be used to control current distributions in superconductor

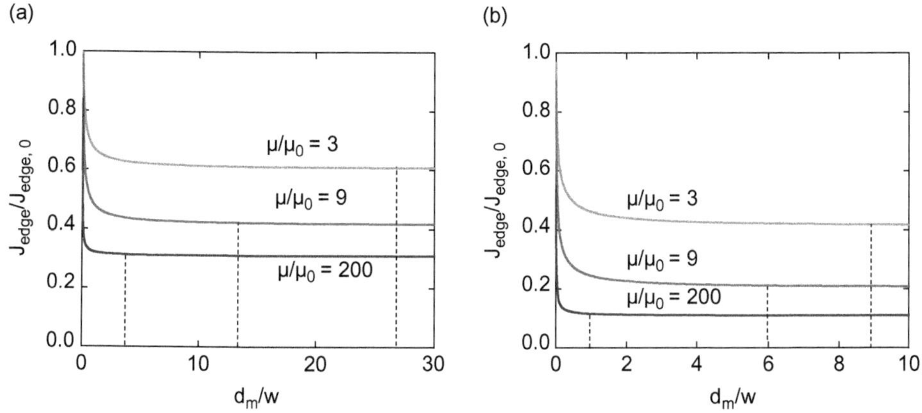

Fig 6.13 Variation in the normalized edge sheet current $J_{\text{edge}}/J_{\text{edge},0}$ of the magnetically shielded superconductor strip in the flux-free Meissner state with the normalized thickness of the wall d_{m}/w for three different values of the relative magnetic permeability of the shield μ/μ_0 identified on the curves for the normalized distance between the centre of the strip and the wall (a) $c/w = 1.1$ and (b) $c/w = 1.01$, when $h/2w = 0.005$, referring to the flat semi-infinite paramagnet configuration of Fig. 6.11. Reproduced with permission from Genenko and Snezhko 2002 *J. Appl. Phys.* 92, pg 357, copyright 2002, AIP Publishing LLC. doi: https://doi.org/10.1063/1.1480112

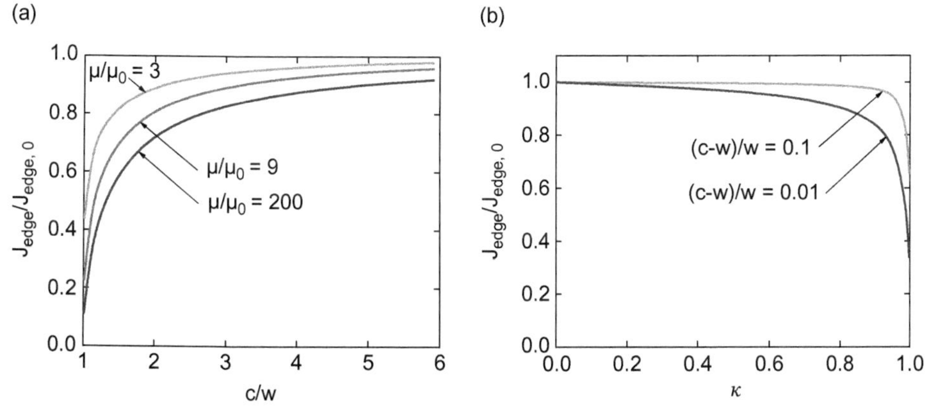

Fig 6.14 (a) Dependence of the normalized current peak magnitude at the side of the strip adjacent to the magnet versus the normalized distance c/w to the magnet of the normalized thickness $d_{\text{m}}/w = 2$ and different relative magnetic permeabilities indicated in the plot. (b) The same normalized current peak versus the image current strength κ for a thin magnet of the normalized thickness $d_{\text{m}}/w = 0.001$. Different curves correspond to different normalized distances $(c - w)/w$ shown in the plot. Reproduced with permission from Genenko and Snezhko 2002 *J. Appl. Phys.* 92, pg 357, copyright 2002, AIP Publishing LLC. doi: https://doi.org/10.1063/1.1480112

strips but the relative magnetic permeability needs to be as large as the inverse thickness of magnets in units of the superconductor strip width, $w/d_{\rm m}$.

To make the last important statement quantitative we introduce a peak reduction coefficient $\nu = J_{\rm edge,0}/J_{\rm edge}$, equal to the ratio of the current peak value in the isolated strip to the peak value in the shielded strip. This parameter characterizes the effectiveness of the shielding; in particular, it is proportional to the enhancement ratio of the total current in flux-free strips protected from flux entry by an edge barrier. Figure 6.15(a) shows that the plots of constant ν in the plane $(\mu/\mu_0, w/d_{\rm m})$ are virtually straight lines, which means that the shielding effect by $d_{\rm m} \ll w$ is controlled uniquely by the combined parameter $(\mu/\mu_0)(d_{\rm m}/w)$. Noting that the effective magnetic charge has opposite signs on the surfaces of the magnetic shield and is proportional to $(\mu/\mu_0 - 1)$ (Genenko, Snezhko, and Usoskin 2004a), it appears as if a thin shield behaves as a magnetic dipole layer of strength $\eta = (\mu/\mu_0)(d_{\rm m}/w)$, assuming $\mu/\mu_0 \gg 1$ (Jackson 1998; Lehner 2010).

The reduction coefficient $\nu = \nu_0 \simeq 5$ is a maximum which may be achieved for the chosen value of the magnet–superconductor distance $c/w = 1.01$ and cutting at the strip edge at $h/2w = 0.002$. Up to the value ν_0 the peak reduction effect may be well fitted by the expression

$$\nu = C + \ln(\eta) \tag{6.59}$$

with $C \simeq 3$. This logarithmic trend remains for another choice of the distance $c/w = 1.001$ as is shown in Fig. 6.15(b) where $C \simeq 5$ is taken.

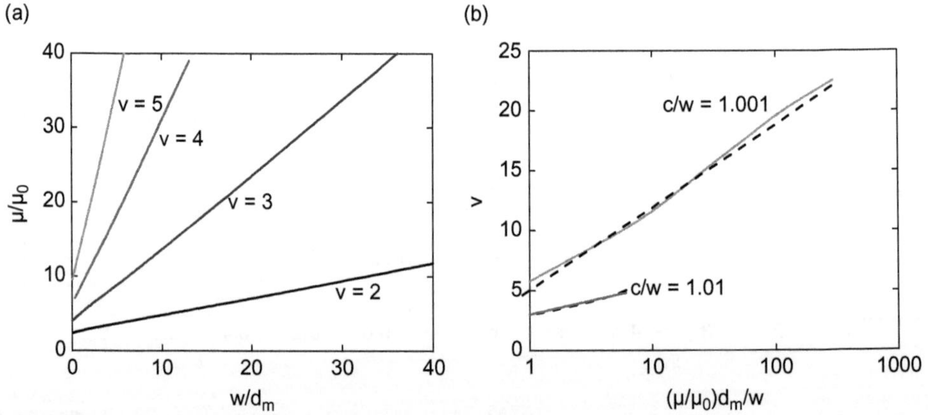

Fig 6.15 (a) The lines of equal strength of the shielding effect, characterized by the current peak reduction ν for a fixed distance $c/w = 1.01$. (b) The current peak reduction ν as a function of the magnetic dipole layer strength $\eta = (\mu/\mu_0)(d_{\rm m}/w)$ (solid lines) and corresponding fitting by formula (6.59) (dashed lines) for different values of the normalized distance c/w indicated in the plot. Reproduced with permission from Genenko, Snezhko, and Usoskin, Finite magnetic shields effect on the transport current in superconductor strips, *Physica C: Superconductivity*, 401, Issues 1–4, 2004, doi.org/10.1016/j.physc.2003.09.045.

An attempt to experimentally observe the effect of paramagnetic shielding on the transport current in a superconductor strip was made in the work (Genenko, Snezhko, and Usoskin 2004b). Here we present direct measurements of the critical transport current in a strip affected by a magnetic environment in two model configurations. Current–voltage characteristics of a thin YBCO film of thickness $d = 100\,nm$ located between two ferrite slabs parallel ($M \parallel F$) or perpendicular ($M \perp F$) to their surfaces exhibit nontrivial trends (Fig. 6.16). Without ferrites the voltage curve (F) presents a conventional behaviour. In the configuration $M \parallel F$ the voltage onset current decreases, indicating a reduction in the first flux penetration critical current. In the configuration $M \perp F$ the voltage onset occurs at higher currents which agrees in principle with the expected enhancements of the critical current. By further increasing the current the voltage curves for the perpendicular and parallel configurations change their positions with respect to the voltage of the isolated strip. The reasons for this behaviour are not understood and require more detailed investigations.

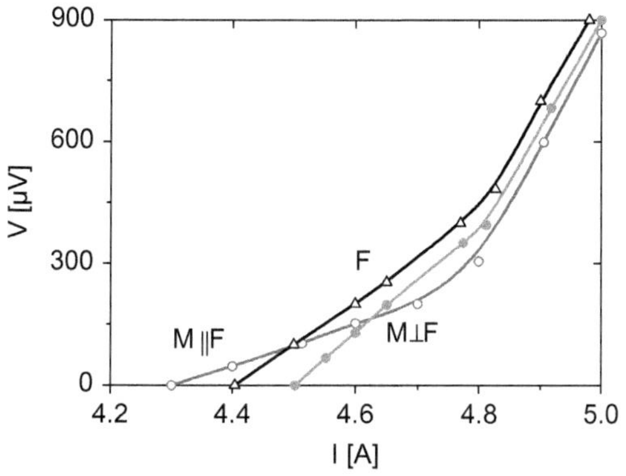

Fig 6.16 Current–voltage dependences for an isolated YBCO strip (F) and for the strip located between the two ferrite plates with parallel ($M \parallel F$) or perpendicular ($M \perp F$) orientation to the strip plane. Contact measurements are performed in the self-field at 77 K using AC transport current of frequency 100 kHz. Reproduced with permission from Genenko, Snezhko, and Usoskin, Finite magnetic shields effect on the transport current in superconductor strips, *Physica C: Superconductivity*, 401, Issues 1–4, 2004, doi.org/10.1016/j.physc.2003.09.045.

The observed effect is much weaker than is expected from the theory in section 3.4, assuming the important role of the edge barrier able to protect the Meissner state in the thin film studied. Indeed, according to formula (3.78) one can anticipate the enhancement of the dissipation-free current in the Meissner state of such a strip by a factor of 2 (Genenko, Snezhko, and Usoskin 2004b). In fact, the observed enhancement is only about 5% which may be caused by the following. First of all, the transport properties of the film have been apparently dominated by the pinning rather than by the edge barrier. In the flux-penetrated critical state, however, the total critical current is not enhanced in the considered shielding geometry if the field-independent critical current density is assumed, as was shown in Section 4.2.1. It is the field dependence of the critical current density that may result in the observed weak positive effect on the total critical current, since magnetic shielding reduces the effect of the magnetic self-field on the transport current (Majoros, Glowacki, and Campbell 2000a; Alamgir, Gu, and Han 2005; Gömöry et al. 2006) The effect on AC losses, predicted in section 4.2.1, could be much more remarkable but, unfortunately, it was not investigated in that experiment.

In conclusion, in this section, we have considered theoretically and experimentally the effect of paramagnetic shields of finite thickness on current distributions in a long superconducting strip. We have established the effect of current peak reduction at the strip edge adjacent to a paramagnetic wall which may be utilized in the superconducting micro-strip resonators. The main result of the study is that the effect of peak reduction saturates rapidly when the thickness of the magnet wall increases. Depending on the magnetic permeability and the distance to the magnet this saturation may be reached at a relatively small thickness, which makes it possible to employ thin paramagnetic films of high permeability in order to control current distributions in superconducting strips. Although the results for magnetic shields of finite thickness present substantial progress towards real experimental configurations compared with the infinite geometries investigated in previous chapters, one dimension of the considered shield still remains infinite.

7

Critical characteristics of bulk superconductors in nonideal paramagnetic environments

The investigations of Chapters 3 to 6 elucidate that current distributions in thin type-II superconductors are highly sensitive to paramagnetic, or diamagnetic, environments, a fact especially prominent in superconductor strips, as far as the flux-free Meissner state and the partially flux-filled critical state are concerned, bearing witness to augmented strengths of the magnetic field and sharp peaks of the sheet current near the edges of the strips. Suitably designed magnetic environments, however, facilitate the reduction of these peaks by attenuating the local magnetic field, giving rise to an increase in the total loss-free current that can be carried by the strips, as frequently demonstrated in this book already. Strong redistribution of the sheet current sparked off by current-filled, yet flux-free central regions in thin, large aspect ratio superconductor strips here turns out to be the key. The obvious question that thus crops up is whether such a redistribution is limited to thin superconductors of a planar shape alone, or whether a similar effect can occur in bulk superconductors of more sophisticated geometries too.

Three-dimensional arrangements of composite superconductor cables and tapes, at any rate, show a significant influence of magnetic shielding on the critical current and the hysteretic AC loss (Glowacki and Majoros 2000; Majoros, Glowacki, and Campbell 2000b; Soltanian et al. 2001; Horvat, Wang, Soltanian, and Dou 2002; Sumption et al. 2002a; Fareed, Robert, and Ruiz 2019; Kapolka and Ruiz 2022). Iron-sheathed superconductor filaments of MgB_2/Fe cables represent ideal objects for exploring the magnetic shielding effect due to the simplicity of their fabrication. They reveal, in particular, increased critical currents and reduced hysteretic AC losses over a wide range of the strength of an applied magnetic field. Cylindrical paramagnetic shells may, to a certain extent, protect a single filament from an applied magnetic field and from the magnetic field due to the presence of other filaments. However, for reasons of symmetry, they cannot protect the identified filament from a self-induced magnetic field caused by an imposed transport current like magnetically shielded superconductor strips (Genenko, Usoskin, and Freyhardt 1999; Genenko, Snezhko, and Freyhardt 2000). Indeed, a single cylindrical filament

Electromagnetics of Superconductor/Paramagnet Heterostructures. Yuri Genenko and Hermann Rauh, Oxford University Press.
 DOI: 10.1093/9780191782855.003.0007

already has the optimum shape in the sense that the magnetic field created by the transport current merely exhibits a tangential component at the surface of the filament. Hence, taking into account the remaining filaments' contribution to the magnetic field, asymmetric paramagnetic shells should also be examined with a view to optimizing the magnetic shielding effect. This primarily necessitates resolving the fundamental point as to whether nonlocal interactions of the named sort are capable of substantially altering current distributions in three-dimensional superconductors where no current peaks occur.

In any case, magnetic environments have a significant impact on the penetration of magnetic flux into superconducting constituents and, particularly, on the critical conditions of the first magnetic flux entry in superconductors. These critical conditions represented by the critical field and the critical current of the first magnetic flux penetration into a bulk superconductor are the subject of study in the current chapter.

7.1 Cylindrical superconductor filament with a paramagnetic sheath

Aiming at an estimate of the surface penetration current density first, we start by considering an infinitely extended, cylindrical type-II superconductor filament of radius r_{s} in the flux-free Meissner state, covered with a coaxial paramagnetic sheath of thickness d_{m}, of permeability μ, and subject to an applied magnetic field of strength H_{a} in the y-direction of a Cartesian coordinate system x, y, z, as Fig. 7.1 depicts. Apart from unfolding unique magnetic properties, this superconductor/paramagnet heterostructure paves the way for rigorous analyses of the critical characteristics set forth in section 7.3. The presentation that follows complements the work of Genenko, Yampolskii, and Pan (2004).

7.1.1 Field distribution

In the superconductor domain $0 \leq r < r_{\mathrm{s}}$, the magnetic induction due to the applied magnetic field, $\mathbf{B}$, obeys the classical London result, eqn. (2.12),

$$\mathbf{B} + \lambda_{\mathrm{L}}^2 \nabla \times (\nabla \times \mathbf{B}) = 0 \tag{7.1}$$

by recourse to Ampère's law, eqn. (2.19), introducing cylindrical coordinates r, ϕ, z adapted to the magnetically sheathed filament. The magnetic field $\mathbf{H}$ outside the superconductor domain $r \geq r_{\mathrm{s}}$ and the magnetic induction $\mathbf{B}$ in the entire space satisfy the homogeneous Maxwell equations

$$\nabla \times \mathbf{H} = 0 \quad \text{and} \quad \nabla \cdot \mathbf{B} = 0, \tag{7.2}$$

where $\mathbf{B} = \mu \mathbf{H}$ in the paramagnet domain $r_{\mathrm{s}} \leq r < r_{\mathrm{s}} + d_{\mathrm{m}}$ and $\mathbf{B} = \mu_0 \mathbf{H}$ otherwise. This suggests the introduction of a scalar magnetic potential ψ to yield the representation of the magnetic field,

$$\mathbf{H} = -\nabla \psi, \tag{7.3}$$

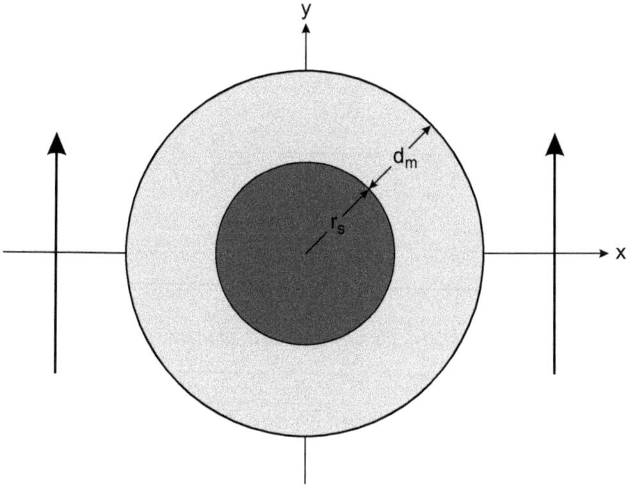

Fig. 7.1 Cross-sectional view of a cylindrical superconductor filament (dark shading) covered with a coaxial paramagnetic sheath (light shading) extending infinitely in the z-direction of a Cartesian coordinate system x, y, z. The definition of the radius of the superconductor filament r_s and the definition of the thickness of the paramagnetic sheath d_m as well as the direction of the applied magnetic field $\mathbf{H}_a$ (bold arrows) are marked.

identically satisfying the first equation (7.2), whereas the second equation (7.2) entails the Laplace equation

$$\Delta\psi = 0. \tag{7.4}$$

The assumption of an insulating nonmagnetic buffer layer of infinitesimal thickness at the superconductor/paramagnet interface, as often practically justified (Pan, Zhou, Liu, and Dou 2003), avoids complications arising from the proximity effect (Schmidt 1997). Continuity of both the normal component of the magnetic induction $\mathbf{B}$ and the tangential component of the magnetic field $\mathbf{H}$ across the superconductor/paramagnet interface and, respectively, the paramagnet/vacuum interface hence demands that the boundary conditions

$$\lim_{r\to r_s-} B_r(r,\phi) = -\mu \frac{\partial\psi}{\partial r}\bigg|_{r=r_s+} \quad \text{and} \quad \lim_{r\to r_s-} B_\phi(r,\phi) = -\mu_0 \frac{1}{r}\frac{\partial\psi}{\partial\phi}\bigg|_{r=r_s+} \tag{7.5a}$$

$$\mu\frac{\partial\psi}{\partial r}\bigg|_{r=r_s+d_m-} = \mu_0\frac{\partial\psi}{\partial r}\bigg|_{r=r_s+d_m+} \quad \text{and} \quad \frac{1}{r}\frac{\partial\psi}{\partial\phi}\bigg|_{r=r_s+d_m-} = \frac{1}{r}\frac{\partial\psi}{\partial\phi}\bigg|_{r=r_s+d_m+} \tag{7.5b}$$

hold; the asymptotic condition that $\mathbf{H}$ must tend to $\mathbf{H}_a$ far away from the paramagnet/vacuum interface becomes

$$\psi(r,\phi) \to -H_a\, r\sin\phi \quad \text{as } r\to\infty. \tag{7.6}$$

The nontrivial cylindrical components of the magnetic induction $\mathbf{B}$ in the superconductor domain $0 \leq r < r_s$, governed by eqn. (7.1) subject to the second equation (7.2), must fulfill the set of equations

$$\left.\begin{aligned} B_r - \lambda_{\mathrm{L}}^2 \left(\Delta B_r - \frac{1}{r^2} B_r - \frac{2}{r^2} \frac{\partial B_\phi}{\partial \phi} \right) = 0 \\ B_\phi - \lambda_{\mathrm{L}}^2 \left(\Delta B_\phi - \frac{1}{r^2} B_\phi + \frac{2}{r^2} \frac{\partial B_r}{\partial \phi} \right) = 0. \end{aligned}\right\} \tag{7.7}$$

Hence, substituting the ansatz

$$B_r(r, \phi) = (f(r) - g(r)) \sin \phi \tag{7.8a}$$

$$B_\phi(r, \phi) = (f(r) + g(r)) \cos \phi \tag{7.8b}$$

in eqn. (7.7) with auxiliary functions $f(r)$ and $g(r)$ admits separation of f and g in the resulting set of equations, leading to the appropriate solution for the respective components of $\mathbf{B}$,

$$B_r(r, \phi) = \mu_0 H_{\mathrm{a}} A_{\mathrm{s}} \left[I_0 \left(r/\lambda_{\mathrm{L}} \right) - I_2 \left(r/\lambda_{\mathrm{L}} \right) \right] \sin \phi \tag{7.9a}$$

$$B_\phi(r, \phi) = \mu_0 H_{\mathrm{a}} A_{\mathrm{s}} \left[I_0 \left(r/\lambda_{\mathrm{L}} \right) + I_2 \left(r/\lambda_{\mathrm{L}} \right) \right] \cos \phi, \tag{7.9b}$$

where $A_{\mathrm{s}} = (4\mu/\mu_0)/D_{\mathrm{s}}(r_{\mathrm{s}}, d_{\mathrm{m}})$ with

$$\begin{aligned} D_{\mathrm{s}}(r_{\mathrm{s}}, d_{\mathrm{m}}) = {} & [(\mu/\mu_0 + 1)^2 - (\mu/\mu_0 - 1)^2/(d_{\mathrm{m}}/r_{\mathrm{s}} + 1)^2] I_0 \left(r_{\mathrm{s}}/\lambda_{\mathrm{L}} \right) \\ & + \left(\mu^2/\mu_0^2 - 1 \right) [1 - 1/(d_{\mathrm{m}}/r_{\mathrm{s}} + 1)^2] I_2 \left(r_{\mathrm{s}}/\lambda_{\mathrm{L}} \right). \end{aligned} \tag{7.10}$$

I_0 and I_2 herein denote the modified Bessel functions of the first kind of orders zero and two, respectively (Abramowitz and Stegun 1972). Furthermore, the nontrivial cylindrical components of the magnetic field $\mathbf{H}$ in the paramagnet domain $r_{\mathrm{s}} \leq r < r_{\mathrm{s}} + d_{\mathrm{m}}$, derived from the appropriate solution of eqn. (7.4) by means of eqn. (7.3), are

$$H_r(r, \phi) = 2H_{\mathrm{a}}[A_{\mathrm{m}}^{+} - A_{\mathrm{m}}^{-} r_{\mathrm{s}}^2/r^2] \sin \phi \tag{7.11a}$$

$$H_\phi(r, \phi) = 2H_{\mathrm{a}}[A_{\mathrm{m}}^{+} + A_{\mathrm{m}}^{-} r_{\mathrm{s}}^2/r^2] \cos \phi, \tag{7.11b}$$

where

$$A_{\mathrm{m}}^{+} = [(\mu/\mu_0 + 1) I_0(r_{\mathrm{s}}/\lambda_{\mathrm{L}}) + (\mu/\mu_0 - 1) I_2(r_{\mathrm{s}}/\lambda_{\mathrm{L}})]/D_{\mathrm{s}}(r_{\mathrm{s}}, d_{\mathrm{m}}) \tag{7.12a}$$

$$A_{\mathrm{m}}^{-} = [(\mu/\mu_0 - 1) I_0(r_{\mathrm{s}}/\lambda_{\mathrm{L}}) + (\mu/\mu_0 + 1) I_2(r_{\mathrm{s}}/\lambda_{\mathrm{L}})]/D_{\mathrm{s}}(r_{\mathrm{s}}, d_{\mathrm{m}}). \tag{7.12b}$$

Likewise, the nontrivial cylindrical components of the magnetic field $\mathbf{H}$ in the vacuum domain $r \geq r_{\mathrm{s}} + d_{\mathrm{m}}$, derived from the appropriate solution of eqn. (7.4) via eqn. (7.3), read

$$H_r(r, \phi) = H_{\mathrm{a}}[1 + A_{\mathrm{v}}(r_{\mathrm{s}} + d_{\mathrm{m}})^2/r^2] \sin \phi \tag{7.13a}$$

$$H_\phi(r, \phi) = H_{\mathrm{a}}[1 - A_{\mathrm{v}}(r_{\mathrm{s}} + d_{\mathrm{m}})^2/r^2] \cos \phi, \tag{7.13b}$$

where $A_v = D_v(r_s, d_m)/D_s(r_s, d_m)$ with

$$D_v(r_s, d_m) = [(\mu/\mu_0 - 1)^2 - (\mu/\mu_0 + 1)^2/(d_m/r_s + 1)^2] I_2(r_s/\lambda_L) + (\mu^2/\mu_0^2 - 1)[1 - 1/(d_m/r_s + 1)^2] I_0(r_s/\lambda_L). \quad (7.14)$$

We note that the above representations admit various instructive limits too. Thus, if examining the domain $0 \le r < r_s$ and taking $\mu \to \mu_0$ in eqn. (7.9), the nontrivial cylindrical components of the magnetic induction inside an unshielded superconductor filament,

$$B_{UF,r}(r, \phi) = \mu_0 H_a [1 - I_2(r/\lambda_L)/I_0(r/\lambda_L)] \sin\phi \quad (7.15a)$$

$$B_{UF,\phi}(r, \phi) = \mu_0 H_a [1 + I_2(r/\lambda_L)/I_0(r/\lambda_L)] \cos\phi, \quad (7.15b)$$

obtain. Conversely, if formally letting $\lambda_L \to \infty$ in eqn. (7.9), the single nonzero Cartesian component of the magnetic induction inside a hollow paramagnetic cylinder,

$$B_{HC,y} = 4\mu H_a/[(\mu/\mu_0 + 1)^2 - (\mu/\mu_0 - 1)^2/(d_m/r_s + 1)^2], \quad (7.16)$$

prevails, as expected from theoretical considerations otherwise (Jackson 1998).

7.1.2 Magnetization

The magnetic induction in the superconductor domain and the magnetic field in the paramagnet domain fully determine the magnetic dipole moment per unit length of either of the two constituents and, hence, of the superconductor/paramagnet heterostructure as a whole.

7.1.2.1 Superconductor filament

The density of the induced Meissner current in the superconductor domain $0 \le r < r_s$, following from eqn. (7.9) together with Ampère's law, eqn. (2.19), only reveals the single nontrivial cylindrical or Cartesian component

$$j_z(r, \phi) = (2H_a/\lambda_L) A_s I_1(r/\lambda_L) \cos\phi, \quad (7.17)$$

where I_1 denotes the modified Bessel function of the first kind of order one (Abramowitz and Stegun 1972). This resolves the thermodynamic average magnetization of the superconductor filament,

$$\langle \mathbf{M}_F \rangle = \mathbf{m}_F/\pi r_s^2, \quad (7.18)$$

with the magnetic dipole moment per unit length of the filament of length l defined by

$$\mathbf{m}_F = \lim_{l \to \infty} \frac{1}{2l} \int dv\, (\mathbf{r} \times \mathbf{j}) \Big|_{0 \le r < r_s} \quad (7.19)$$

in terms of the supercurrent density $\mathbf{j}$, according to eqn. (A.10). Employing the explicit result, eqn. (A.17), thus furnishes the single nontrivial Cartesian component of the *diamagnetic* magnetization of the filament

$$\langle M_{\mathrm{F},y}\rangle = -2H_{\mathrm{a}}A_{\mathrm{s}}I_2\left(r_{\mathrm{s}}/\lambda_{\mathrm{L}}\right). \tag{7.20}$$

A practically important approximation of eqn. (7.20) deserves mentioning,

$$\langle M_{\mathrm{F},y}\rangle \simeq -4H_{\mathrm{a}}/[\mu/\mu_0 + 1 - (\mu/\mu_0 - 1)/(d_{\mathrm{m}}/r_{\mathrm{s}} + 1)^2]; \quad r_{\mathrm{s}} \gg \lambda_{\mathrm{L}}. \tag{7.21}$$

Conversely, if $\mu \to \mu_0$, eqn. (7.20) yields the limiting form for an unshielded superconductor filament,

$$\langle M_{\mathrm{UF},y}\rangle = -2H_{\mathrm{a}}I_2(r_{\mathrm{s}}/\lambda_{\mathrm{L}})/I_0(r_{\mathrm{s}}/\lambda_{\mathrm{L}}), \tag{7.22}$$

which may be used for normalization of eqn. (7.20),

$$\langle M_{\mathrm{F},y}\rangle/\langle M_{\mathrm{UF},y}\rangle = A_{\mathrm{s}}I_0(r_{\mathrm{s}}/\lambda_{\mathrm{L}}). \tag{7.23}$$

From eqn. (7.23) it appears that the inequality $\langle M_{\mathrm{F},y}\rangle/\langle M_{\mathrm{UF},y}\rangle \le 1$ testifying to the shielding effect of the paramagnetic sheath strictly holds.

Figure 7.2 illustrates the dependence of the normalized thermodynamic average magnetization of the superconductor filament $\langle M_{\mathrm{F},y}\rangle/\langle M_{\mathrm{UF},y}\rangle$ on the normalized thickness of the paramagnetic sheath $d_{\mathrm{m}}/r_{\mathrm{s}}$ for a range of values of its relative magnetic permeability μ/μ_0, adopting two extreme values of the normalized radius of the filament $r_{\mathrm{s}}/\lambda_{\mathrm{L}}$. A general trait due to magnetic shielding is the reduction in magnitude of the magnetization with either increasing $d_{\mathrm{m}}/r_{\mathrm{s}}$ or increasing μ/μ_0, the effect being weaker for the thinner filament (Fig. 7.2(a)), but stronger for the thicker filament (Fig. 7.2(b)).

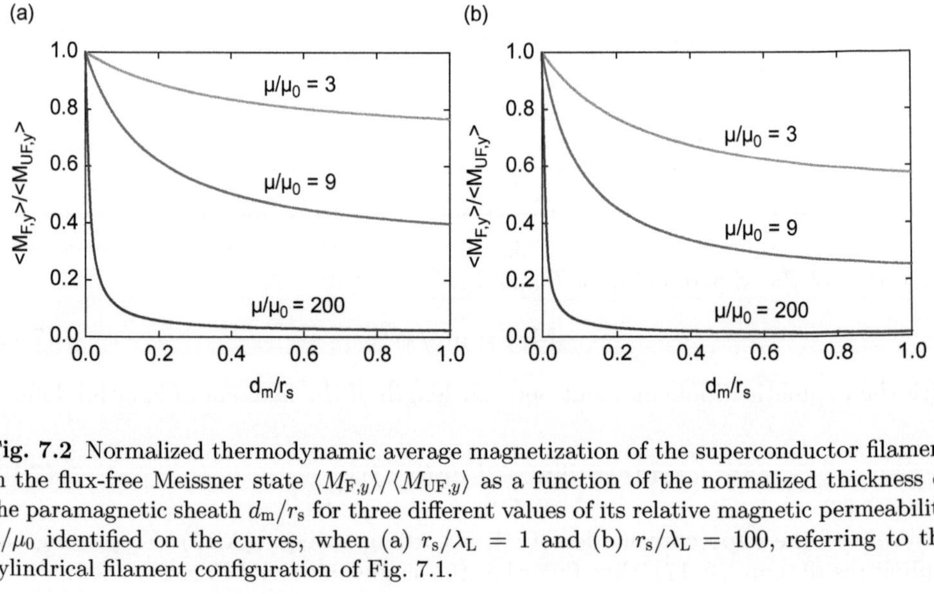

Fig. 7.2 Normalized thermodynamic average magnetization of the superconductor filament in the flux-free Meissner state $\langle M_{\mathrm{F},y}\rangle/\langle M_{\mathrm{UF},y}\rangle$ as a function of the normalized thickness of the paramagnetic sheath $d_{\mathrm{m}}/r_{\mathrm{s}}$ for three different values of its relative magnetic permeability μ/μ_0 identified on the curves, when (a) $r_{\mathrm{s}}/\lambda_{\mathrm{L}} = 1$ and (b) $r_{\mathrm{s}}/\lambda_{\mathrm{L}} = 100$, referring to the cylindrical filament configuration of Fig. 7.1.

7.1.2.2 Paramagnetic sheath

We proceed from the thermodynamic average magnetization of the paramagnetic sheath,

$$\langle \mathbf{M}_{\mathrm{S}} \rangle = \mathbf{m}_{\mathrm{S}} / \pi((r_{\mathrm{s}} + d_{\mathrm{m}})^2 - r_{\mathrm{s}}^2), \tag{7.24}$$

with the magnetic dipole moment per unit length of the sheath defined by

$$\mathbf{m}_{\mathrm{S}} = \lim_{l \to \infty} \frac{1}{l} \int dv\, \mathbf{M} \Big|_{r_{\mathrm{s}} \leq r < r_{\mathrm{s}} + d_{\mathrm{m}}} \tag{7.25}$$

in terms of the local magnetization $\mathbf{M}$. Making recourse to the constitutive relation $\mathbf{B} = \mu_0(\mathbf{H} + \mathbf{M})$ as well as to the assumption $\mathbf{B} = \mu \mathbf{H}$ and implementing the solution for $\mathbf{H}$, eqn. (7.11), thus yields the single nontrivial Cartesian component of the *paramagnetic* magnetization of the sheath,

$$\langle M_{\mathrm{S},y} \rangle = 2H_{\mathrm{a}}(\mu/\mu_0 - 1)A_{\mathrm{m}}^{+}. \tag{7.26}$$

A practically important approximation of eqn. (7.26) deserves mentioning,

$$\langle M_{\mathrm{S},y} \rangle \simeq 2H_{\mathrm{a}}(\mu/\mu_0 - 1)/[\mu/\mu_0 + 1 - (\mu/\mu_0 - 1)/(d_{\mathrm{m}}/r_{\mathrm{s}} + 1)^2]; \quad r_{\mathrm{s}} \gg \lambda_{\mathrm{L}}. \tag{7.27}$$

Conversely, if $\lambda_{\mathrm{L}} \to \infty$, eqn. (7.26) yields the limiting form for a *hollow paramagnetic cylinder*,

$$\langle M_{\mathrm{HC},y} \rangle = 2H_{\mathrm{a}}(\mu^2/\mu_0^2 - 1)/[(\mu/\mu_0 + 1)^2 - (\mu/\mu_0 - 1)^2/(d_{\mathrm{m}}/r_{\mathrm{s}} + 1)^2], \tag{7.28}$$

which may be used for normalization of eqn. (7.26),

$$\langle M_{\mathrm{S},y} \rangle / \langle M_{\mathrm{HC},y} \rangle = A_{\mathrm{m}}^{+}[(\mu/\mu_0 + 1)^2 - (\mu/\mu_0 - 1)^2/(d_{\mathrm{m}}/r_{\mathrm{s}} + 1)^2]/(\mu/\mu_0 + 1). \tag{7.29}$$

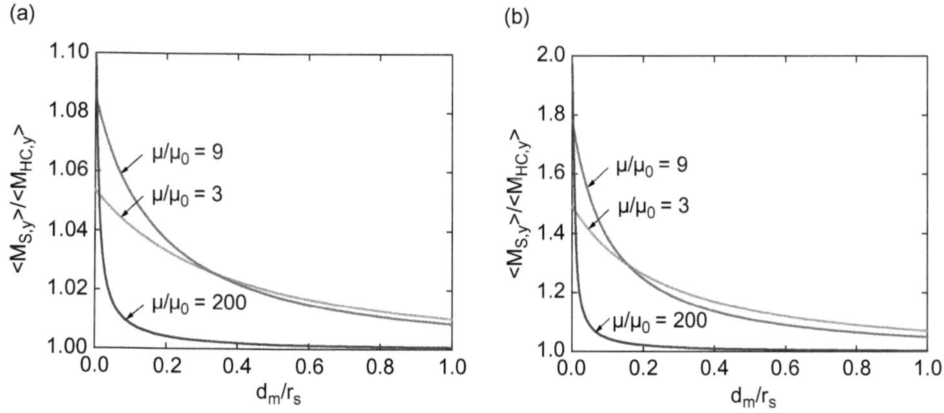

Fig. 7.3 Normalized thermodynamic average magnetization of the paramagnetic sheath $\langle M_{\mathrm{S},y} \rangle / \langle M_{\mathrm{HC},y} \rangle$ as a function of the normalized thickness of the sheath $d_{\mathrm{m}}/r_{\mathrm{s}}$ for three different values of its relative magnetic permeability μ/μ_0 identified on the curves, when (a) $r_{\mathrm{s}}/\lambda_{\mathrm{L}} = 1$ and (b) $r_{\mathrm{s}}/\lambda_{\mathrm{L}} = 100$, referring to the cylindrical filament configuration of Fig. 7.1.

From eqn. (7.29) it appears that the inequality $\langle M_{S,y}\rangle/\langle M_{HC,y}\rangle \geq 1$ testifying to the magnetic flux expelled from the superconductor filament strictly holds.

Figure 7.3 illustrates the dependence of the normalized thermodynamic average magnetization of the paramagnetic sheath $\langle M_{S,y}\rangle/\langle M_{HC,y}\rangle$ on the normalized thickness of the sheath d_m/r_s for a range of values of its relative magnetic permeability μ/μ_0, adopting two extreme values of the normalized radius of the superconductor filament r_s/λ_L. A general trait due to magnetic shielding is the rise in magnitude of the magnetization with increasing μ/μ_0 at small d_m/r_s, the effect being weaker for the thinner filament (Fig. 7.3(a)), but stronger for the thicker filament (Fig. 7.3(b)).

7.1.2.3 Superconductor/paramagnet cloak

The opposite magnetic properties revealed for the two constituents of the magnetically shielded superconductor filament suggest pondering whether mutual cancellation of their thermodynamic average magnetizations that would give rise to the cloaking of this heterostructure can occur, rendering it a magnetically neutral tool. Conditions of the named sort do exist, adding the superconductor filament with a paramagnetic sheath to the range of more sophisticated electromagnetic metamaterial structures for which the phenomenon of cloaking has been predicted first (Leonhardt 2006; Pendry, Schurig, and Smith 2006; Anlage 2011). The essential point here is the shielding of some part of space from penetration of an applied electromagnetic field by the cloaking structure such that the cloak and the shielded part leave the applied field undisturbed and therefore remain 'invisible' (Sanchez, Navau, Prat-Camps, and Chen 2011; Gömöry et al. 2012; Narayana and Sato 2012; Yampolskii and Genenko 2014a,b; Gömöry, Solovyov, and Šouc 2015; Peña-Roche, Genenko, and Badía-Majós 2016).

From eqns. (7.20) and (7.26), the sum of the thermodynamic average magnetizations $\langle M_{F,y}\rangle$ and $\langle M_{S,y}\rangle$ weighted with their areal shares, i.e. the magnetic dipole moment per unit length of the entire heterostructure made up of the individual moments $m_{F,y}$ and $m_{S,y}$, is seen to obey the relation

$$m_{F,y} + m_{S,y} = 2\pi(r_s + d_m)^2 H_a A_v, \tag{7.30}$$

with the constant A_v from eqn. (7.13) measuring the disturbance of the applied magnetic field $\mathbf{H}_a$. The prerequisite for magnetic invisibility therefore comes down to A_v vanishing, whence setting

$$\begin{aligned}&[(\mu/\mu_0 + 1)^2 - (\mu/\mu_0 - 1)^2/(d_m/r_s + 1)^2] I_2\,(r_s/\lambda_L)\\ &= (\mu^2/\mu_0^2 - 1)[(d_m/r_s + 1)^2 - 1] I_0\,(r_s/\lambda_L)\end{aligned} \tag{7.31}$$

in view of eqn. (7.14). Since, according to eqns. (7.9), (7.11), and (7.13), the heterostructure under test does not possess magnetic multipole moments of any order higher than 1, satisfaction of eqn. (7.31) proves necessary and sufficient for the magnetically shielded superconductor filament to be unobservable by purely magnetic means. This cloak of cylindrical design thus shows properties of an 'anti-magnet' discussed elsewhere (Mawatari, Navau, and Sanchez 2012; Prat-Camps, Sanchez, and Navau 2013). A practically useful approximation of eqn. (7.31) deserves mentioning at last,

$$\mu/\mu_0 \simeq [(d_{\rm m}/r_{\rm s}+1)^2+1]/[(d_{\rm m}/r_{\rm s}+1)^2-1]; \quad r_{\rm s} \gg \lambda_{\rm L}. \tag{7.32}$$

Figure 7.4 shows the variation of the relative magnetic permeability of the paramagnetic sheath μ/μ_0 with the normalized thickness of the sheath $d_{\rm m}/r_{\rm s}$, satisfying the condition of magnetic invisibility for two extreme values of the normalized radius of the superconductor filament $r_{\rm s}/\lambda_{\rm L}$. Starting from a distinct maximum at the smallest value of $d_{\rm m}/r_{\rm s}$, there is a sharp monotonic drop followed by a gradual approach towards unity, the value prevailing in the limit of an infinitely extended sheath, as $d_{\rm m}/r_{\rm s}$ continues to grow. The effect turns out to be more pronounced for the thinner filament, but less pronounced for the thicker filament.

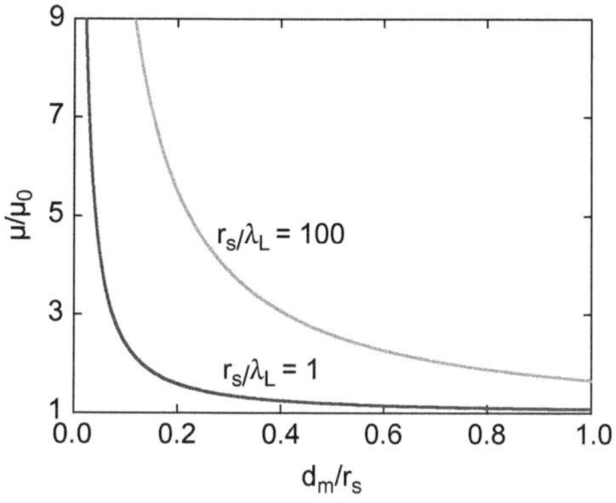

Fig. 7.4 Dependence of the relative magnetic permeability of the paramagnetic sheath μ/μ_0 on the normalized thickness of the sheath $d_{\rm m}/r_{\rm s}$ allowing for magnetic invisibility for two values of the normalized radius of the superconductor filament $r_{\rm s}/\lambda_{\rm L}$ identified on the curves, referring to the cylindrical filament configuration of Fig. 7.1.

7.1.3 Surface penetration current density

Although a rigorous analysis of the critical characteristics of the superconductor/paramagnet heterostructure must take explicit account of the nonequilibrium phenomenon of the penetration of magnetic flux into the superconductor constituent, the density of the induced Meissner current j_z together with the thermodynamic average magnetization of the superconductor filament $\langle M_{{\rm F},y}\rangle$ derived in section 7.1.2.1 can provide useful estimates, nevertheless.

7.1.3.1 Analytical approximations

Starting from eqns. (7.17) and (7.20), the succinct form of the maximum supercurrent density $j_{\max}$ attained at $r = r_s$ for $\phi = 0$ or $\phi = \pm\pi$,

$$j_{\max} \simeq \left|\langle M_{F,y}\rangle\right|/\lambda_L; \quad r_s \gg \lambda_L, \tag{7.33}$$

ensues, by reference to eqn. (7.21), the strength of the applied magnetic field within the Meissner regime $H_a < H_p$ being fixed. If H_a in turn augments, $j_{\max}$ grows towards a maximum as H_a adopts the surface penetration field H_p, at which magnetic flux first enters across the superconductor/paramagnet interface. Likewise, a succinct form of the average surface penetration current density $j_p = 2I_p/\pi r_s^2$ obtains, with the maximum loss-free current at the onset of the penetration of magnetic flux I_p given by the surface integral of eqn. (7.17) over the right-half cross section of the superconductor filament, H_a again being identified with H_p. Utilization of eqn. (7.20) then yields by reference to eqn. (7.21).

$$j_p \simeq 4\left|\langle M_{F,y}\rangle\right|/\pi r_s; \quad r_s \gg \lambda_L, \tag{7.34}$$

Since a determination of the surface penetration field H_p lies beyond the ansatz of the flux-free Meissner state, with its prediction of a linear, and hence reversible, dependence of $\langle M_{F,y}\rangle$ on H_a, this quantity may be inferred by recourse to supplementary information contained, e.g., in the initial transient of the thermodynamic average magnetization which Fig. 2.6 typically displays. Deviation from linear dependence at the onset of irreversibility due to the entry of magnetic flux thus serves to establish H_p.

7.1.3.2 Numerical estimates

Let us, for the purpose of demonstration, examine a cylindrical, iron-sheathed MgB_2 filament at the temperature $T = 20$ K, with geometrical dimensions $r_s = 500$ μm, $d_m = 250$ μm and material characteristics $\lambda_L = 140$ nm, $\mu/\mu_0 = 46$ (Canfield, Bud'ko, and Finnemore 2003). The initial transient, or 'virgin' branch, of the empirical thermodynamic average magnetization affords $H_p \simeq 2.3 \times 10^5$ A/m (Genenko, Yampolskii, and Pan 2004). Equation (7.33) with eqn. (7.21) hence allows for the maximum supercurrent density at the filament's surface $j_{\max} \simeq 2.5 \times 10^{11}$ A/m^2. Equation (7.34) with eqn. (7.21), on the other hand, yields the estimate for the surface penetration current density $j_p \simeq 8.8 \times 10^7$ A/m^2, in fair agreement with results deduced from measurements of the hysteresis width of the magnetization loop (Horvat, Wang, Soltanian, and Dou 2002; Pan, Zhou, Liu, and Dou 2003; Pan and Dou 2004).

7.2 Bean–Livingston barrier at a flat superconductor/paramagnet interface

An infinitely extended, straight magnetic vortex introduced into a flux-free, homogeneous type-II superconductor that fills *all space* proves thermodynamically stable, as soon as the strength of a uniform applied magnetic field $\mathbf{H}_a$ exceeds the lower critical field of the superconductor bulk H_{c1} (cf. section 2.1.3.1). The bounding surface of a superconductor of *restricted* size, on the other hand, inevitably creates barriers of various kinds, protecting the Meissner state and delaying the first entry of magnetic flux up to the surface penetration field H_p above H_{c1}, a fact which already shows nonequilibrium traits associated with irreversibility alone, even if the superconductor were perfect otherwise, as noted in section 2.1. While geometrical barriers underlie the considerations of thin superconductors in Chapters 3, 5, and 6, attention is drawn to the Bean–Livingston energy barrier here. This type of barrier derives from the positive Gibbs free energy of magnetic vortices subject to the inwardly pushing action of the applied magnetic field; it exists with superconductors of any shape and, hence, must be included in a proper analysis of the marginal Meissner state.

The phenomenon of the Bean–Livingston barrier has been studied extensively hitherto for superconductors facing vacuum (Bean and Livingston 1964; Galaiko 1966; Petukhov and Chechetkin 1973; Koshelev 1992; Samokhvalov 1995; Genenko et al. 1998). However, due to the induced magnetization, the presence of a paramagnetic constituent in superconductor/paramagnet composites may alter the nucleation characteristics of a magnetic vortex at the superconductor/paramagnet interface along with the surface penetration field itself, depending on the particular geometry under test (Horvat, Wang, Soltanian, and Dou 2002; Pan, Zhou, Liu, and Dou 2003). We therefore investigate this effect first for the case of fluctuation penetration of a semi-circular magnetic vortex loop with radius a at the flat interface between a semi-infinite type-II superconductor in the marginal Meissner state and a semi-infinite paramagnet with permeability μ and subject to an applied magnetic field of strength H_a in the y-direction of a Cartesian coordinate system x, y, z, as Fig. 7.5 depicts. The analysis that follows adds to the work of Genenko, Rauh, and Yampolskii (2005).

7.2.1 Field distribution

In the superconductor domain $x > 0$, the magnetic induction due to the applied magnetic field, $\mathbf{B}$, is delineated by the classical London result, eqn. (2.12), together with eqn. (2.19). The magnetic induction due to the magnetic vortex loop, $\mathbf{B}_V$, supplements this making up the total magnetic induction

$$\mathbf{B}_{\text{tot}} = \mathbf{B} + \mathbf{B}_V. \tag{7.35}$$

From eqns. (2.20) and (2.23), $\mathbf{B}_V$ is determined by the generalized London equation

$$\mathbf{B}_V + \lambda_L^2 \nabla \times (\nabla \times \mathbf{B}_V) = \mathbf{V} \tag{7.36}$$

with the vorticity

$$\mathbf{V} = \Phi_0 \int_C \mathrm{d}\mathbf{l}\, \delta(\mathbf{r} - \mathbf{l}), \tag{7.37}$$

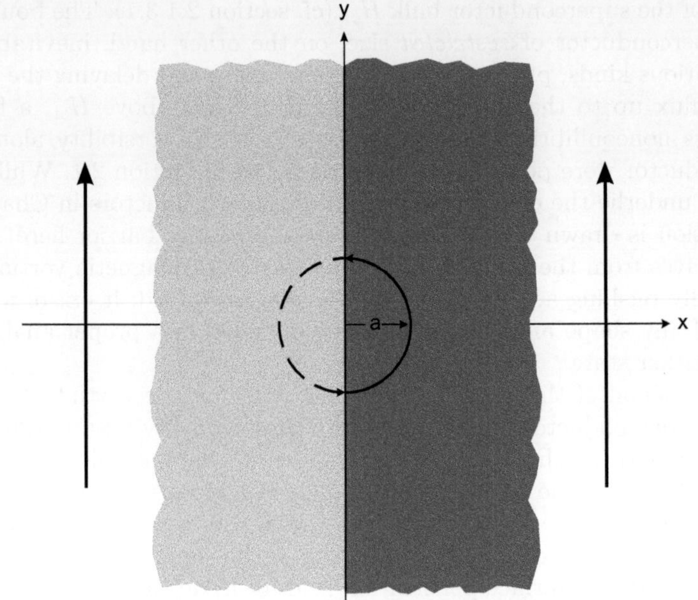

Fig. 7.5 Cross-sectional view of a semi-circular magnetic vortex loop (full line) situated in the x, y-plane at the flat interface between a semi-infinite superconductor (dark shading) and a semi-infinite paramagnet (light shading) extending infinitely in the z-direction of a Cartesian coordinate system x, y, z. The definition of the radius of the magnetic vortex loop a as well as its image (dotted line) and the direction of the applied magnetic field $\mathbf{H}_{\mathrm{a}}$ (bold arrows) are marked.

the line integral extending along the core of the vortex loop C. In the paramagnet domain $x \leq 0$, the magnetic field due to the vortex loop, $\mathbf{H}_{\mathrm{V}}$, complements the applied magnetic field $\mathbf{H}_{\mathrm{a}}$ making up the total magnetic field

$$\mathbf{H}_{\mathrm{tot}} = \mathbf{H}_{\mathrm{a}} + \mathbf{H}_{\mathrm{V}}. \tag{7.38}$$

We focus on the distributions of the vortex magnetic field $\mathbf{H}_{\mathrm{V}}$ therein and the vortex magnetic induction $\mathbf{B}_{\mathrm{V}}$ in the entire space, proceeding from the homogeneous Maxwell equations

$$\nabla \times \mathbf{H}_{\mathrm{V}} = 0 \quad \text{and} \quad \nabla \cdot \mathbf{B}_{\mathrm{V}} = 0, \tag{7.39}$$

where $\mathbf{B}_{\mathrm{V}} = \mu \mathbf{H}_{\mathrm{V}}$ in the paramagnet domain $x \leq 0$ and $\mathbf{B}_{\mathrm{V}} = \mu_0 \mathbf{H}_{\mathrm{V}}$ in the superconductor domain $x > 0$. This suggests the introduction of a scalar magnetic potential ψ_{V} to yield the representation of the magnetic field due to the vortex loop,

$$\mathbf{H}_{\mathrm{V}} = -\nabla \psi_{\mathrm{V}}, \tag{7.40}$$

identically satisfying the first equation (7.39), whereas the second equation (7.39) entails the Laplace equation

$$\Delta\psi_{\mathrm{V}} = 0. \tag{7.41}$$

Furthermore, we invoke the presence of an insulating nonmagnetic buffer layer of infinitesimal thickness at the superconductor/paramagnet interface to avoid complications arising from the proximity effect (Schmidt 1997). Continuity of both the normal component of the vortex magnetic induction $\mathbf{B}_{\mathrm{V}}$ and the tangential components of the vortex magnetic field $\mathbf{H}_{\mathrm{V}}$ across the superconductor/paramagnet interface hence demands that the boundary conditions

$$\lim_{x\to 0+} B_{\mathrm{V},x}(x,y,z) = -\mu \left.\frac{\partial\psi_{\mathrm{V}}}{\partial x}\right|_{x=0-} \tag{7.42a}$$

$$\lim_{x\to 0+} B_{\mathrm{V},y}(x,y,z) = -\mu_0 \left.\frac{\partial\psi_{\mathrm{V}}}{\partial y}\right|_{x=0-} \tag{7.42b}$$

$$\lim_{x\to 0+} B_{\mathrm{V},z}(x,y,z) = -\mu_0 \left.\frac{\partial\psi_{\mathrm{V}}}{\partial z}\right|_{x=0-} \tag{7.42c}$$

$$\lim_{x\to 0+} B_y(x) = \mu_0 H_{\mathrm{a}} \tag{7.42d}$$

hold. In addition, the magnetic induction and the scalar magnetic potential due to the vortex loop, $\mathbf{B}_{\mathrm{V}}$ and ψ_{V}, along with their derivatives must vanish asymptotically far away from the magnetic vortex loop. The required solutions thus can be constructed using the method of Fourier transforms.

The total magnetic induction $\mathbf{B}_{\mathrm{tot}}$ in the superconductor domain $x > 0$ reads

$$\mathbf{B}_{\mathrm{tot}}(x,y,z) = \mu_0 H_{\mathrm{a}} \exp\left(-x/\lambda_{\mathrm{L}}\right)\mathbf{e}_y + \mathbf{B}_{\mathrm{V}}(x,y,z), \tag{7.43}$$

where $\mathbf{B}_{\mathrm{V}}$ denotes the appropriate solution of eqn. (7.36) subject to the second equation (7.39). Expressing $\mathbf{B}_{\mathrm{V}}$ in terms of its two-dimensional Fourier transform $\tilde{\mathbf{B}}_{\mathrm{V}}^{k_y,k_z}$ yields

$$\mathbf{B}_{\mathrm{V}}(x,y,z) = \frac{1}{2\pi}\int_{-\infty}^{\infty} dk_y \,\frac{1}{\pi}\int_{0}^{\infty} dk_z \tilde{\mathbf{B}}_{\mathrm{V}}^{k_y,k_z}(x)\exp\left(ik_y y\right)\cos\left(k_z z\right), \tag{7.44}$$

with the Cartesian components of $\tilde{\mathbf{B}}_{\mathrm{V}}^{k_y,k_z}$ given by

$$\tilde{B}_{\mathrm{V},x}^{k_y,k_z}(x) = \frac{1}{\lambda_{\mathrm{L}}^2}\frac{\mu}{\mu_0 k + \mu q}\int_0^\infty dx'\, \tilde{V}_x^{k_y,k_z}(x')\exp\left(-q|x+x'|\right)$$

$$-\frac{1}{2q\lambda_{\mathrm{L}}^2}\int_0^\infty dx'\, \tilde{V}_x^{k_y,k_z}(x')\left(\exp\left(-q|x+x'|\right) - \exp\left(-q|x-x'|\right)\right) \quad (7.45a)$$

$$\tilde{B}_{\mathrm{V},y}^{k_y,k_z}(x) = \frac{i}{\lambda_{\mathrm{L}}^2}\frac{\mu_0\,(k_y/k)}{\mu_0 k + \mu q}\int_0^\infty dx'\, \tilde{V}_x^{k_y,k_z}(x')\exp\left(-q|x+x'|\right)$$

$$-\frac{1}{2q\lambda_{\mathrm{L}}^2}\int_0^\infty dx'\, \tilde{V}_y^{k_y,k_z}(x')\left(\exp\left(-q|x+x'|\right) - \exp\left(-q|x-x'|\right)\right) \quad (7.45b)$$

$$\tilde{B}_{\mathrm{V},z}^{k_y,k_z}(x) = \frac{i}{\lambda_{\mathrm{L}}^2}\frac{\mu_0\,(k_z/k)}{\mu_0 k + \mu q}\int_0^\infty dx'\, \tilde{V}_x^{k_y,k_z}(x')\exp\left(-q|x+x'|\right), \quad (7.45c)$$

where $k = \sqrt{k_y^2 + k_z^2}$ and $q = \sqrt{k^2 + 1/\lambda_{\mathrm{L}}^2}$. Likewise, expressing $\mathbf{V}$ from eqn. (7.37) in terms of its two-dimensional Fourier transform $\tilde{\mathbf{V}}^{k_y,k_z}$ yields

$$\mathbf{V}(x,y,z) = \frac{1}{2\pi}\int_{-\infty}^{\infty} dk_y\, \frac{1}{\pi}\int_0^\infty dk_z \tilde{\mathbf{V}}^{k_y,k_z}(x)\exp\left(ik_y y\right)\cos\left(k_z z\right) \quad (7.46)$$

with the explicit representation of the nontrivial Cartesian components of $\tilde{\mathbf{V}}^{k_y,k_z}$ for the semi-circular magnetic vortex loop,

$$\tilde{V}_x^{k_y,k_z}(x) = 2i\Phi_0\,\theta(a-x)\sin\left(k_y\sqrt{a^2-x^2}\right) \quad (7.47a)$$

$$\tilde{V}_y^{k_y,k_z}(x) = 2\Phi_0\,\theta(a-x)\,x\cos\left(k_y\sqrt{a^2-x^2}\right)\Big/\sqrt{a^2-x^2}. \quad (7.47b)$$

The appropriate solution of eqn. (7.41) for the scalar magnetic potential ψ_{V} in the paramagnet space $x \leq 0$, expressed through its two-dimensional Fourier transform $\tilde{\psi}_{\mathrm{V}}^{k_y,k_z}$, reads

$$\psi_{\mathrm{V}}(x,y,z) = \frac{1}{2\pi}\int_{-\infty}^{\infty} dk_y\, \frac{1}{\pi}\int_{-\infty}^{\infty} dk_z\, \tilde{\psi}_{\mathrm{V}}^{k_y,k_z}(x)\exp\left(ik_y y\right)\cos\left(k_z z\right), \quad (7.48)$$

where

$$\tilde{\psi}_{\mathrm{V}}^{k_y,k_z}(x) = \frac{1}{\lambda_{\mathrm{L}}^2}\frac{1/k}{\mu_0 k + \mu q}\int_0^\infty dx'\, \tilde{V}_x^{k_y,k_z}(x')\exp\left(kx - qx'\right). \quad (7.49)$$

7.2.2 Self-energy and magnetic dipole moment

The two-dimensional Fourier transforms of the magnetic induction and the vorticity due to the magnetic vortex loop, $\tilde{\mathbf{B}}_{\mathrm{V}}^{k_y,k_z}$ and $\tilde{\mathbf{V}}^{k_y,k_z}$, together with the two-dimensional Fourier transform of the scalar magnetic potential $\tilde{\psi}_{\mathrm{V}}^{k_y,k_z}$ allow for various key quantities to be derived, as shown in the following.

The self-energy of the magnetic vortex loop is given by

$$F_{\mathrm{V}} = \frac{1}{2\mu_0} \int dv \, (\mathbf{V} \cdot \mathbf{B}_V) \bigg|_{x>0} + \frac{1}{2} \int \mathrm{ds} \cdot (\mathbf{V}\,\psi_V) \bigg|_{x=0} \tag{7.50}$$

according to eqn. (B.18), the shielding effect of the semi-infinite paramagnet being contained in the actual delineation of $\mathbf{B}_{\mathrm{V}}$ and ψ_{V}. If the Cartesian components of the magnetic induction, $B_{\mathrm{V},x}(x,y,z)$ and $B_{\mathrm{V},y}(x,y,z)$, and those of the vorticity, $V_x(x,y,z)$ and $V_y(x,y,z)$, as well as the scalar magnetic potential $\psi_{\mathrm{V}}(x,y,z)$, are substituted for their respective two-dimensional Fourier transforms $\tilde{B}_{\mathrm{V},x}^{k_y,k_z}(x)$ and $\tilde{B}_{\mathrm{V},y}^{k_y,k_z}(x)$, $\tilde{V}_x^{k_y,k_z}(x)$ and $\tilde{\mathrm{V}}_y^{k_y,k_z}(x)$, as well as $\tilde{\psi}_{\mathrm{V}}^{k_y,k_z}(x)$ via eqns. (7.44), (7.46), and (7.48), eqn. (7.50) reduces to

$$\begin{aligned} F_{\mathrm{V}} = \frac{1}{4\pi\mu_0} \int_{-\infty}^{\infty} dk_y \, \frac{1}{\pi} \int_{0}^{\infty} dk_z \Bigg\{ \int_{0}^{\infty} dx \left(\tilde{V}_y^{k_y,k_z}(x) \tilde{B}_{\mathrm{V},y}^{k_y,k_z}(x) - \tilde{V}_x^{k_y,k_z}(x) \tilde{B}_{\mathrm{V},x}^{k_y,k_z}(x) \right) \\ + \mu_0 \tilde{V}_x^{k_y,k_z}(x) \tilde{\psi}_{\mathrm{V}}^{k_y,k_z}(x) \bigg|_{x=0} \Bigg\}. \end{aligned} \tag{7.51}$$

Evaluating eqn. (7.51) using eqns. (7.45), (7.47), and (7.49) upon implementation of the cut-off procedure regarding the vortex core introduced in section 2.1.3 and resorting to a limiting extreme admits the simple approximate form for the self-energy of the magnetic vortex loop,

$$F_{\mathrm{V}} \simeq \frac{a\Phi_0^2}{4\mu_0\lambda_{\mathrm{L}}^2} \left[\ln\left(\frac{a}{\xi_{\mathrm{GL}}}\right) - \left(\frac{\pi a^2}{8\lambda_{\mathrm{L}}^2}\right)\kappa \right]; \quad \xi_{\mathrm{GL}} < a \ll \lambda_{\mathrm{L}}, \tag{7.52}$$

with the image current strength $\kappa = (\mu - \mu_0)/(\mu + \mu_0)$, a result which aptly generalizes previous findings for a bare flat superconducting constituent (Galaiko 1966; Petukhov and Chechetkin 1973; Koshelev 1992; Samokhvalov 1995) by adding a correction due to the presence of the paramagnetic shield. The merely slight dependence on the magnetic characteristic of the latter obviously stems from the fact that the paramagnet domain's magnetic field induced by the vortex loop largely hinges upon quantizing the superconductor domain's magnetic flux.

The magnetic dipole moment of the vortex loop is defined by

$$\mathbf{m}_{\mathrm{V}} = \frac{1}{2} \int dv \, (\mathbf{r} \times \mathbf{j}) \bigg|_{x>0} \tag{7.53}$$

in terms of the density of the induced supercurrent $\mathbf{j}$ following eqn. (A.22). Thus, eqn. (A.35) provides the single nontrivial Cartesian component of the magnetic dipole moment of the loop,

$$m_{\mathrm{V},y} = (\pi a \Phi_0/\mu_0)\left[I_1(a/\lambda_{\mathrm{L}}) - L_1(a/\lambda_{\mathrm{L}})\right], \tag{7.54}$$

where I_1 and L_1 denote the modified Bessel function of the first kind of order one and the modified Struve function of order one, respectively (Abramowitz and Stegun 1972). The evidence mentioned earlier causes $m_{\mathrm{V},y}$ to lack information about the magnetic characteristic of the paramagnetic shield. A practically important approximation of eqn. (7.54) ensues,

$$m_{\mathrm{V},y} \simeq \pi a^2 \Phi_0/2\mu_0\lambda_{\mathrm{L}}; \quad a \ll \lambda_{\mathrm{L}}, \tag{7.55}$$

which reveals that, for small values of a, the magnetic dipole moment varies with the area covered by the loop.

7.2.3 Surface penetration field

The physical observables considered above make up the change in the Gibbs free energy of the superconducting constituent in the applied magnetic field relative to that of the flux-free Meissner state,

$$\Delta G = F_{\mathrm{V}} - \mathbf{m}_{\mathrm{V}} \cdot (\mu_0 \mathbf{H}_{\mathrm{a}}), \tag{7.56}$$

which furnishes the proper base for investigating the Bean–Livingston barrier against fluctuation penetration of magnetic flux. This barrier occurs due to a competition between the image force which attracts the vortex loop to the superconductor/paramagnet interface, accounted for by the self-energy term in eqn. (7.56), and the driving force due to the action of the applied magnetic field which pushes the vortex into the superconductor constituent, expressed by the potential energy term in eqn. (7.56). Thus, when H_{a} is fixed and the radius of the growing vortex loop has reached its critical size $a = a_{\mathrm{c}}$ defined by the condition $\partial\Delta G/\partial a = 0$, further expansion of the vortex loop becomes irreversible and its entry proceeds. Depending on the quality of the superconductor/paramagnet interface, this may happen at different values of the critical loop radius a_{c} throughout the range of validity of eqns. (7.52) and (7.55), and possibly also beyond. Yet, for the Bean–Livingston barrier to be completely suppressed, a_{c} must approach the scale of roughness σ of the superconductor/paramagnet interface. Whilst in the case of an *ideal* interface, with scale of roughness σ delineated by $0 \leq \sigma < \xi_{\mathrm{GL}}$, penetration of magnetic vortex loops takes place for $a_{\mathrm{c}} \simeq \xi_{\mathrm{GL}}$; in the case of a *real* interface, with scale of roughness σ obeying $\xi_{\mathrm{GL}} < \sigma \ll \lambda_{\mathrm{L}}$, penetration of magnetic vortex loops occurs for $a_{\mathrm{c}} \simeq \sigma$. Typifying H_{a} by H_{p}, when magnetic flux first enters across the flat superconductor/paramagnet interface, thus yields the surface penetration field

$$H_{\mathrm{p}} \simeq \frac{\Phi_0}{4\pi\mu_0\lambda_{\mathrm{L}}\sigma}\left[\ln\left(\frac{\mathrm{e}\sigma}{\xi_{\mathrm{GL}}}\right) - \left(\frac{3\pi\sigma^2}{8\lambda_{\mathrm{L}}^2}\right)\kappa\right]; \quad \xi_{\mathrm{GL}} < \sigma \ll \lambda_{\mathrm{L}} \tag{7.57}$$

with Euler's number $\mathrm{e} = 2.718\ldots$ This reveals that, even in the limit $\mu \to \infty$, the paramagnet only has a minute effect on the lowering of the Bean–Livingston barrier for the entry of magnetic flux, and hence on the reduction of the surface penetration current density $j_p = H_p/\lambda_L$, since in the infinite geometry under consideration here, the applied magnetic field remains almost completely unshielded by the paramagnetic constituent. Given the range of variability of σ indicated above, H_p adopts values between H_{c1} and H_{cth}, the lower critical field for the first appearance of a straight magnetic vortex in a homogeneous superconductor covering all space,

$$H_{c1} \simeq \frac{\Phi_0}{4\pi\mu_0\lambda_L^2} \ln\left(\frac{\lambda_L}{\xi_{GL}}\right); \quad \xi_{GL} \ll \lambda_L, \tag{7.58}$$

and the thermodynamic critical field

$$H_{cth} \simeq \frac{\Phi_0}{4\pi\mu_0\lambda_L\xi_{GL}}; \quad \xi_{GL} \ll \lambda_L, \tag{7.59}$$

respectively (Orlando and Delin 1991). Remarkably enough, the latter characteristic equals the surface penetration field for a straight magnetic vortex parallel to the ideally smooth flat surface of a semi-infinite homogeneous superconductor facing vacuum (de Gennes 1994), an outcome which also obtains from eqn. (7.57) when formally letting $\sigma \to \xi_{GL}$ and $\mu \to \mu_0$.

Figure 7.6(a) shows the variation of the normalized Gibbs free energy of the semi-circular magnetic vortex loop $\Delta G/E_V\lambda_L$ with the normalized loop radius a/λ_L for a range of values of the normalized amplitude of the applied magnetic field H_a/H_p, when the normalized scale of roughness σ/λ_L is fixed. Thus, the ordering of the

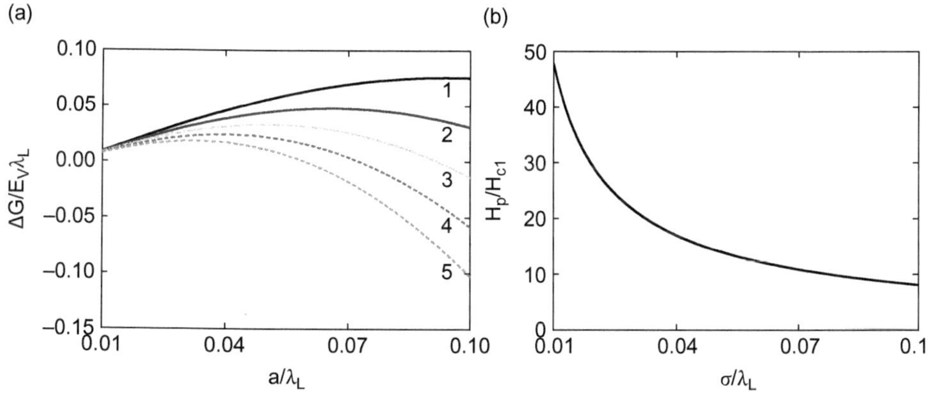

Fig. 7.6 (a) Dependence of the normalized Gibbs free energy of the semi-circular magnetic vortex loop $\Delta G/E_V\lambda_L$ on the normalized loop radius a/λ_L, when the normalized amplitude of the applied magnetic field $H_a/H_p = 0.6, 0.8, 1.0, 1.2, 1.4$, assuming the normalized scale of roughness $\sigma/\lambda_L = 0.05$; (b) variation of the normalized surface penetration field H_p/H_{c1} with the normalized scale of roughness σ/λ_L, when $\lambda_L/\xi_{GL} = 1000$, referring to the superconductor/paramagnet configuration of Fig. 7.5. The solid/dashed curves delineate physically stable and unstable regimes, respectively, of the flux-free Meissner state, separated by the marginally stable Meissner state (dash-dotted curve) in between.

curves unfolds that the height of the Bean–Livingston barrier shrinks, whilst the position of its maximum recedes, as the strength of the applied magnetic field augments. If $H_a/H_p < 1$, the energy barrier is too high for magnetic vortex loops of a radius exceeding the scale of roughness of the interface barrier to enter the superconducting constituent, whence the flux-free Meissner state remains preserved. Conversely, if $H_a/H_p > 1$, the energy barrier proves low enough for magnetic vortex loops of a radius falling short of the scale of roughness of the interface barrier to enter the superconducting constituent, whence a transition from the flux-free Meissner state to the partially flux-filled critical state occurs. The value $H_a/H_p = 1$ delimits the two regimes. Figure 7.6(b) illustrates the dependence of the normalized surface penetration field H_p/H_{c1} on the normalized scale of roughness σ/λ_L. Starting from a distinct maximum at the smallest value of σ/λ_L, there is a monotonic fall towards unity, the value prevailing for a straightened magnetic vortex loop, as σ/λ_L continues to grow. We note that the sensitivity to the permeability of the paramagnetic constituent μ is too weak for a resolution on the graphical scales of Fig. 7.6.

7.3 First entry of magnetic flux across a curved superconductor/paramagnet interface

In order to explore critical characteristics associated with the first entry of magnetic flux across a curved superconductor/paramagnet interface, we proceed from the derivations of section 7.1 for a magnetically sheathed superconductor filament. The distributions of the magnetic induction and the magnetic field around a magnetic vortex loop together with its self-energy and magnetic dipole moment admit evaluations of quantities like the transverse lower critical field, the surface penetration field, and the critical current density, given the loop's specific shape.

7.3.1 Magnetic vortex loop of arbitrary shape

Let us therefore go ahead by first addressing a planar magnetic vortex loop of arbitrary shape at the curved interface of an infinitely extended, cylindrical type-II superconductor filament of radius r_s in the marginal Meissner state, covered with a paramagnetic sheath of thickness d_m, of permeability μ, and subject to an applied magnetic field of strength H_a in the y-direction of a Cartesian coordinate system x, y, z, as Fig. 7.7 portrays. The theoretical analysis presented in the following draws upon the work of Yampolskii and Genenko (2005), based in turn on the previous study by Genenko et al. (1998).

7.3.1.1 Field distribution

In the superconductor domain $0 < r < r_s$, the magnetic induction due to the applied magnetic field, $\mathbf{B}$, obeys eqn. (7.1) subject to the second equation (7.2), with the solution given by eqn. (7.9), using cylindrical coordinates r, ϕ, z adapted to the magnetically sheathed filament. The magnetic induction due to the magnetic vortex loop, $\mathbf{B}_V$, supplements this making up the total magnetic induction

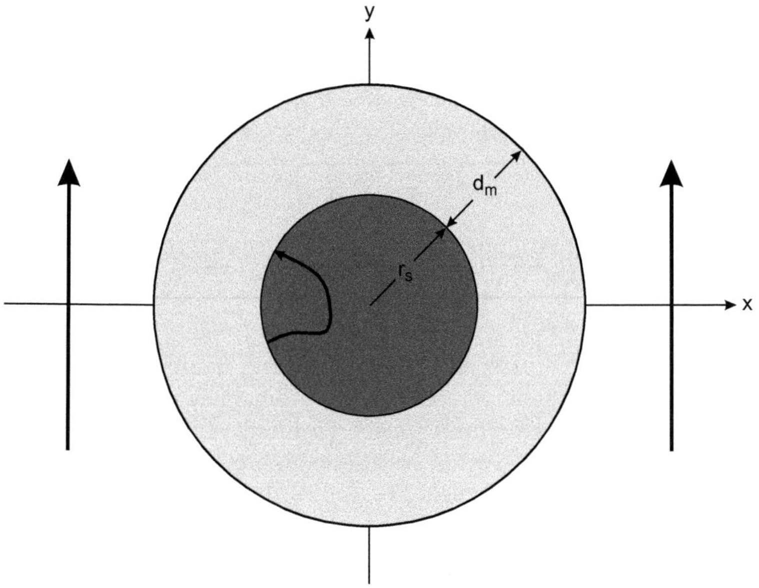

Fig. 7.7 Cross-sectional view of a magnetic vortex loop of arbitrary shape (full line) situated in the x, y-plane at the curved interface of a cylindrical superconductor filament (dark shading) covered with a coaxial paramagnetic sheath (light shading) extending infinitely in the z-direction of a Cartesian coordinate system x, y, z. The definition of the radius of the superconductor filament r_{s} and the definition of the thickness of the paramagnetic sheath d_{m} as well as the direction of the applied magnetic field $\mathbf{H}_{\mathrm{a}}$ (bold arrows) are marked.

$$\mathbf{B}_{\mathrm{tot}} = \mathbf{B} + \mathbf{B}_{\mathrm{V}}. \tag{7.60}$$

From eqns. (2.20) and (2.23), $\mathbf{B}_{\mathrm{V}}$ is determined by the generalized London equation

$$\mathbf{B}_{\mathrm{V}} + \lambda_{\mathrm{L}}^2 \, \nabla \times (\nabla \times \mathbf{B}_{\mathrm{V}}) = \mathbf{V} \tag{7.61}$$

with the vorticity

$$\mathbf{V} = \Phi_0 \int_C \mathrm{d}\mathbf{l} \, \delta(\mathbf{r} - \mathbf{l}), \tag{7.62}$$

the line integral extending along the core of the vortex loop C. The magnetic field $\mathbf{H}$ outside the superconductor domain $r \geq r_{\mathrm{s}}$ and the magnetic induction $\mathbf{B}$ in the entire space satisfy eqn. (7.2), with the respective solutions given by eqns. (7.11) and (7.13). The magnetic field due to the vortex loop, $\mathbf{H}_{\mathrm{V}}$, complements $\mathbf{H}$ making up the total magnetic field

$$\mathbf{H}_{\mathrm{tot}} = \mathbf{H} + \mathbf{H}_{\mathrm{V}}. \tag{7.63}$$

Focusing on the distributions of the vortex magnetic field $\mathbf{H}_{\mathrm{V}}$ outside the superconductor domain $r \geq r_{\mathrm{s}}$ and the vortex magnetic induction $\mathbf{B}_{\mathrm{V}}$ in the entire space, we proceed from the homogeneous Maxwell equations

$$\nabla \times \mathbf{H}_\mathrm{V} = 0 \quad \text{and} \quad \nabla \cdot \mathbf{B}_\mathrm{V} = 0, \tag{7.64}$$

where $\mathbf{B}_\mathrm{V} = \mu \mathbf{H}_\mathrm{V}$ in the paramagnet domain $r_\mathrm{s} \leq r < r_\mathrm{s} + d_\mathrm{m}$ and $\mathbf{B}_\mathrm{V} = \mu_0 \mathbf{H}_\mathrm{V}$ otherwise. This suggests the introduction of a scalar magnetic potential ψ_V to yield the representation of the magnetic field due to the vortex loop,

$$\mathbf{H}_\mathrm{V} = -\nabla \psi_\mathrm{V}, \tag{7.65}$$

identically satisfying the first equation (7.64), whereas the second equation (7.64) entails the Laplace equation

$$\Delta \psi_\mathrm{V} = 0. \tag{7.66}$$

Furthermore, we invoke the presence of an insulating nonmagnetic buffer layer of infinitesimal thickness at the superconductor/paramagnet interface to avoid complications arising from the proximity effect (Schmidt 1997). Continuity of both the normal component of the vortex magnetic induction $\mathbf{B}_\mathrm{V}$ and the tangential component of the vortex magnetic field $\mathbf{H}_\mathrm{V}$ across the superconductor/paramagnet interface as well as the paramagnet/vacuum interface hence demands that the boundary conditions

$$\lim_{r \to r_\mathrm{s}-} B_{\mathrm{V},r}(r,\phi) = -\mu \left.\frac{\partial \psi_\mathrm{V}}{\partial r}\right|_{r=r_\mathrm{s}+} \quad \text{and} \quad \lim_{r \to r_\mathrm{s}-} B_{\mathrm{V},\phi}(r,\phi) = -\mu_0 \left.\frac{1}{r}\frac{\partial \psi_\mathrm{V}}{\partial \phi}\right|_{r=r_\mathrm{s}+} \tag{7.67a}$$

$$\mu \left.\frac{\partial \psi_\mathrm{V}}{\partial r}\right|_{r=r_\mathrm{s}+d_\mathrm{m}-} = \mu_0 \left.\frac{\partial \psi_\mathrm{V}}{\partial r}\right|_{r=r_\mathrm{s}+d_\mathrm{m}+} \quad \text{and} \quad \left.\frac{1}{r}\frac{\partial \psi_\mathrm{V}}{\partial \phi}\right|_{r=r_\mathrm{s}+d_\mathrm{m}-} = \left.\frac{1}{r}\frac{\partial \psi_\mathrm{V}}{\partial \phi}\right|_{r=r_\mathrm{s}+d_\mathrm{m}+} \tag{7.67b}$$

hold. In addition, the magnetic induction and the scalar magnetic potential due to the vortex loop, $\mathbf{B}_\mathrm{V}$ and ψ_V, along with their derivatives must vanish asymptotically far away from the magnetic vortex loop. The required solutions thus can be constructed using the method of Fourier transforms.

The cylindrical components of the magnetic induction $\mathbf{B}_\mathrm{V}$ in the superconductor domain $0 \leq r < r_\mathrm{s}$, governed by eqn. (7.61) subject to the second equation (7.64), must fulfill the set of equations

$$\left.\begin{aligned} &B_{\mathrm{V},r} - \lambda_\mathrm{L}^2 \left(\Delta B_{\mathrm{V},r} - \frac{1}{r^2} B_{\mathrm{V},r} - \frac{2}{r^2}\frac{\partial B_{\mathrm{V},\phi}}{\partial \phi} \right) = V_r \\ &B_{\mathrm{V},\phi} - \lambda_\mathrm{L}^2 \left(\Delta B_{\mathrm{V},\phi} - \frac{1}{r^2} B_{\mathrm{V},\phi} + \frac{2}{r^2}\frac{\partial B_{\mathrm{V},r}}{\partial \phi} \right) = V_\phi \\ &B_{\mathrm{V},z} - \lambda_\mathrm{L}^2 \Delta B_{\mathrm{V},z} = 0. \end{aligned}\right\} \tag{7.68}$$

Analysis of eqn. (7.68) proceeds by expressing $\mathbf{B}_\mathrm{V}$ in terms of its two-dimensional Fourier transform $\tilde{\mathbf{B}}_\mathrm{V}^{n,k}$,

$$\mathbf{B}_\mathrm{V}(r,\phi,z) = \sum_{n=-\infty}^{\infty} \frac{1}{\pi} \int_0^\infty dk\, \tilde{\mathbf{B}}_\mathrm{V}^{n,k}(r) \exp(in\phi) \cos(kz), \tag{7.69}$$

and likewise expressing $\mathbf{V}$ in terms of its two-dimensional Fourier transform $\tilde{\mathbf{V}}^{n,k}$,

$$\mathbf{V}(r,\phi,z) = \sum_{n=-\infty}^{\infty} \frac{1}{\pi} \int_0^{\infty} dk\, \tilde{\mathbf{V}}^{n,k}(r) \exp(in\phi) \cos(kz). \tag{7.70}$$

Hence, substituting the representations, eqns. (7.69) and (7.70), in eqn. (7.68) upon introducing the ansatz

$$\tilde{h}^{-}_{n,k}(r) = \tilde{B}^{n,k}_{\mathrm{V},r}(r) - i\tilde{B}^{n,k}_{\mathrm{V},\phi}(r) \tag{7.71a}$$

$$\tilde{h}^{+}_{n,k}(r) = \tilde{B}^{n,k}_{\mathrm{V},r}(r) + i\tilde{B}^{n,k}_{\mathrm{V},\phi}(r) \tag{7.71b}$$

with auxiliary functions $\tilde{h}^{-}_{n,k}(r)$ and $\tilde{h}^{+}_{n,k}(r)$ admits separation of $\tilde{h}^{-}_{n,k}$ and $\tilde{h}^{+}_{n,k}$ in the resulting set of equations, while leaving $\tilde{B}^{n,k}_{\mathrm{V},z}$ preserved. The appropriate solution for the respective components of $\tilde{\mathbf{B}}^{n,k}_{\mathrm{V}}$ in eqn. (7.69) thus comes down to

$$\tilde{B}^{n,k}_{\mathrm{V},r}(r) = \frac{1}{2}[a^{-}_{n,k}(r)I_{n-1}(qr) + a^{+}_{n,k}(r)I_{n+1}(qr) + b^{-}_{n,k}(r)K_{n-1}(qr) + b^{+}_{n,k}(r)K_{n+1}(qr)] \tag{7.72a}$$

$$\tilde{B}^{n,k}_{\mathrm{V},\phi}(r) = \frac{i}{2}[a^{-}_{n,k}(r)I_{n-1}(qr) - a^{+}_{n,k}(r)I_{n+1}(qr) + b^{-}_{n,k}(r)K_{n-1}(qr) - b^{+}_{n,k}(r)K_{n+1}(qr)] \tag{7.72b}$$

$$\tilde{B}^{n,k}_{\mathrm{V},z}(r) = ic_{n,k}I_n(qr) \tag{7.72c}$$

with $q = \sqrt{k^2 + 1/\lambda_{\mathrm{L}}^2}$, where I_n and K_n denote modified Bessel functions of the first and second kinds, respectively, of order n (Abramowitz and Stegun 1972). The coefficients in eqns. (7.72a), (7.72b) read

$$a^{-}_{n,k}(r) = C^{-}_{n,k} + \int_r^{r_s} dr'r'\, \tilde{W}^{-}_{n,k}(r')K_{n-1}(qr') \tag{7.73a}$$

$$b^{-}_{n,k}(r) = \int_0^{r} dr'r'\, \tilde{W}^{-}_{n,k}(r')I_{n-1}(qr') \tag{7.73b}$$

$$a^{+}_{n,k}(r) = C^{+}_{n,k} + \int_r^{r_s} dr'r'\, \tilde{W}^{+}_{n,k}(r')K_{n+1}(qr') \tag{7.73c}$$

$$b^{+}_{n,k}(r) = \int_0^{r} dr'r'\, \tilde{W}^{+}_{n,k}(r')I_{n+1}(qr') \tag{7.73d}$$

with

$$C^-_{n,k} = S^-_{n,k} - \mu_0 k \alpha_{n,k} \left[I_{n-1}(kr_s) + (\mu/\mu_0 - 1) I'_n(kr_s) \right] / I_{n-1}(qr_s) + \mu_0 k \beta_{n,k} \left[K_{n-1}(kr_s) - (\mu/\mu_0 - 1) K'_n(kr_s) \right] / I_{n-1}(qr_s) \quad (7.74a)$$

$$C^+_{n,k} = S^+_{n,k} - \mu_0 k \alpha_{n,k} \left[I_{n+1}(kr_s) + (\mu/\mu_0 - 1) I'_n(kr_s) \right] / I_{n+1}(qr_s) + \mu_0 k \beta_{n,k} \left[K_{n+1}(kr_s) - (\mu/\mu_0 - 1) K'_n(kr_s) \right] / I_{n+1}(qr_s), \quad (7.74b)$$

I'_n and K'_n meaning derivatives of the modified Bessel functions of order n with respect to their arguments. Furthermore,

$$S^-_{n,k} = -\left[K_{n-1}(qr_s)/I_{n-1}(qr_s) \right] \int_0^{r_s} dr' r' \, \tilde{W}^-_{n,k}(r') I_{n-1}(qr') \quad (7.75a)$$

$$S^+_{n,k} = -\left[K_{n+1}(qr_s)/I_{n+1}(qr_s) \right] \int_0^{r_s} dr' r' \, \tilde{W}^+_{n,k}(r') I_{n+1}(qr'), \quad (7.75b)$$

where

$$\tilde{W}^-_{n,k}(r) = [\tilde{V}^{n,k}_r(r) - i\tilde{V}^{n,k}_\phi(r)]/\lambda^2_{\rm L} \quad (7.76a)$$

$$\tilde{W}^+_{n,k}(r) = [\tilde{V}^{n,k}_r(r) + i\tilde{V}^{n,k}_\phi(r)]/\lambda^2_{\rm L} \quad (7.76b)$$

as well as

$$\alpha_{n,k} = A^{(1)}_{n,k}(S^-_{n,k} + S^+_{n,k} - S_{n,k})/D_{n,k}(r_s, d_m) \quad (7.77a)$$

$$\beta_{n,k} = A^{(2)}_{n,k}(S^-_{n,k} + S^+_{n,k} - S_{n,k})/D_{n,k}(r_s, d_m) \quad (7.77b)$$

with

$$S_{n,k} = \frac{2r_s}{q\lambda^2_{\rm L}} \tilde{V}^{n,k}_r(r_s) K_n(qr_s) \quad (7.78)$$

and

$$A^{(1)}_{n,k} = (1/\mu - 1/\mu_0) K_n(k(r_s + d_m))/I_n(k(r_s + d_m)) \times K'_n(k(r_s + d_m))/I'_n(k(r_s + d_m)) \quad (7.79a)$$

$$A^{(2)}_{n,k} = (1/\mu_0) K_n(k(r_s + d_m))/I_n(k(r_s + d_m)) - (1/\mu) K'_n(k(r_s + d_m))/I'_n(k(r_s + d_m)), \quad (7.79b)$$

where

$$D_{n,k}(r_s, d_m) = A^{(1)}_{n,k}(u^-_{n,k} + u^+_{n,k}) - A^{(2)}_{n,k}(v^-_{n,k} + v^+_{n,k}) \quad (7.80)$$

with

$$u^-_{n,k} = \mu_0 k[I_{n-1}(kr_\mathrm{s})/I_{n-1}(qr_\mathrm{s}) + I_{n+1}(kr_\mathrm{s})/I_{n+1}(qr_\mathrm{s})] - (2\mu_0 k^2/q)I_n(kr_\mathrm{s})/I_n(qr_\mathrm{s}) \tag{7.81a}$$

$$v^-_{n,k} = \mu_0 k[K_{n-1}(kr_\mathrm{s})/K_{n-1}(qr_\mathrm{s}) + K_{n+1}(kr_\mathrm{s})/K_{n+1}(qr_\mathrm{s})] + (2\mu_0 k^2/q)K_n(kr_\mathrm{s})/I_n(qr_\mathrm{s}) \tag{7.81b}$$

$$u^+_{n,k} = (\mu - \mu_0)kI'_n(kr_\mathrm{s})[1/I_{n-1}(qr_\mathrm{s}) + 1/I_{n+1}(qr_\mathrm{s})] \tag{7.81c}$$

$$v^+_{n,k} = (\mu_0 - \mu)kK'_n(kr_\mathrm{s})[1/I_{n-1}(qr_\mathrm{s}) + 1/I_{n+1}(qr_\mathrm{s})]. \tag{7.81d}$$

Finally, the coefficient in eqn. (7.72c) reads

$$c_{n,k} = \mu_0 k\left[\alpha_{n,k}I_n(kr_\mathrm{s}) + \beta_{n,k}K_n(kr_\mathrm{s})\right]/I_n(qr_\mathrm{s}). \tag{7.82}$$

The appropriate solution of eqn. (7.66) for the scalar magnetic potential ψ_V, expressed through its two-dimensional Fourier transform $\tilde{\psi}_\mathrm{V}^{n,k}$, appears as

$$\psi_\mathrm{V}(r,\phi,z) = \sum_{n=-\infty}^{\infty} \frac{1}{\pi}\int_0^\infty dk\, \tilde{\psi}_\mathrm{V}^{n,k}(r)\exp(in\phi)\cos(kz), \tag{7.83}$$

where in the paramagnet domain $r_\mathrm{s} \le r < r_\mathrm{s} + d_\mathrm{m}$,

$$\tilde{\psi}_\mathrm{V}^{n,k}(r) = \alpha_{n,k}I_n(kr) + \beta_{n,k}K_n(kr) \tag{7.84}$$

and in the vacuum domain $r \ge r_\mathrm{s} + d_\mathrm{m}$,

$$\tilde{\psi}_\mathrm{V}^{n,k}(r) = \gamma_{n,k}K_n(kr) \tag{7.85}$$

with

$$\gamma_{n,k} = \frac{S^-_{n,k} + S^+_{n,k} - S_{n,k}}{\mu_0 k(r_\mathrm{s}+d_\mathrm{m})I_n(k(r_\mathrm{s}+d_\mathrm{m}))I'_n(k(r_\mathrm{s}+d_\mathrm{m}))D_{n,k}(r_\mathrm{s},d_\mathrm{m})}. \tag{7.86}$$

We note that, although derived for a magnetic vortex loop of arbitrary shape in the identified x, y-plane, the Fourier transforms, eqns. (7.72), (7.84), and (7.85), can be utilized for any single, or multiple, magnetic vortex configuration, duplicating the proper representations of an unshielded superconductor filament, if $\mu \to \mu_0$ (Genenko et al. 1998). By due choice of the vortex shape they particularly admit examinations of the transverse lower critical field as well as studies of the entry of magnetic flux across the curved interface of the magnetically shielded superconductor filament, subject to an applied magnetic field and/or carrying a transport current, issues dealt with in sections 7.3.2 and 7.3.3.

7.3.1.2 Self-energy and magnetic dipole moment

Further theoretical analysis simplifies considerably if the magnetic vortex loop is symmetric with respect to the x, z-plane, a premise supposed to apply to the sequel. Describing the irreducible part of the loop's specific shape by a smooth and monotonic function $\phi = \chi(r)$ therefore proves convenient. Hence, if the radial extent of the core of the vortex loop varies between $r_s - a$ and r_s, where $a \leq r_s$ represents the depth of penetration of the loop into the superconductor domain, $\chi(r) = \pi$ for $0 \leq r \leq r_s - a$ and $0 < \chi(r) < \pi$ for $r_s - a < r \leq r_s$. Given this approach, the two-dimensional Fourier transforms of the nontrivial cylindrical components of the vorticity $\mathbf{V}$, eqn. (7.62), turn out as

$$\tilde{V}_r^{n,k}(r) = \Phi_0\,\theta(r - r_s + a)\sin(n\chi(r))/i\pi r \tag{7.87a}$$

$$\tilde{V}_\phi^{n,k}(r) = \Phi_0\,\theta(r - r_s + a)\,\chi'(r)\cos(n\chi(r))/\pi, \tag{7.87b}$$

χ' meaning the derivative of the auxiliary function χ with respect to its argument. Use of eqn. (7.87) in eqns. (7.72) and (7.84) admits representations of physical observables like the self-energy and the magnetic dipole moment of the magnetic vortex loop in terms of χ.

The self-energy of the vortex loop adopts the form

$$F_V = \frac{1}{2\mu_0}\int dv\,(\mathbf{V}\cdot\mathbf{B}_V)\Big|_{0\leq r<r_s} + \frac{1}{2}\int d\mathbf{s}\cdot(\mathbf{V}\,\psi_V)\Big|_{r=r_s} \tag{7.88}$$

according to eqn. (B.18), as for a bare superconductor filament (Genenko et al. 1998), the shielding effect of the paramagnetic sheath being contained in the actual delineation of $\mathbf{B}_V$ and ψ_V. If the cylindrical components of the magnetic induction, $B_{V,r}(r,\phi,z)$ and $B_{V,\phi}(r,\phi,z)$, and those of the vorticity, $V_r(r,\phi,z)$ and $V_\phi(r,\phi,z)$, as well as the scalar magnetic potential $\psi_V(r,\phi,z)$ are substituted for their respective two-dimensional Fourier transforms $\tilde{B}_{V,r}^{n,k}(r)$ and $\tilde{B}_{V,\phi}^{n,k}(r)$, $\tilde{V}_r^{n,k}(r)$ and $\tilde{V}_\phi^{n,k}(r)$, as well as $\tilde{\psi}_V^{n,k}(r)$ via eqns. (7.69), (7.70), and (7.83), with inherent symmetry properties due to the assumed geometry of the loop taken into account, eqn. (7.88) reduces to

$$F_V = \frac{1}{2\mu_0}\sum_{n=-\infty}^{\infty}\int_0^{\infty} dk\left\{\int_0^{r_s} dr\,r\left(\tilde{V}_\phi^{n,k}(r)\tilde{B}_{V,\phi}^{n,k}(r) - \tilde{V}_r^{n,k}(r)\tilde{B}_{V,r}^{n,k}(r)\right)\right.$$
$$\left. -\mu_0 r\tilde{V}_r^{n,k}(r)\tilde{\psi}_V^{n,k}(r)\Big|_{r=r_s}\right\}. \tag{7.89}$$

The magnetic dipole moment of the vortex loop is defined by

$$\mathbf{m}_V = \frac{1}{2}\int dv\,(\mathbf{r}\times\mathbf{j})\Big|_{0\leq r<r_s} + \int dv\,\mathbf{M}\Big|_{r_s\leq r<r_s+d_m} \tag{7.90}$$

in terms of the density of the induced supercurrent $\mathbf{j}$ and the local magnetization $\mathbf{M}$ of the paramagnetic sheath following eqns. (A.38), (A.39), and (A.48). Thus, eqn. (A.56) furnishes the single nontrivial Cartesian component of the magnetic dipole moment of the loop,

$$m_{\mathrm{V},y} = \frac{2\pi i}{\mu_0} \lim_{k\to 0} \left\{ \int_0^{r_s} dr\, r \left(\tilde{B}^{1,k}_{\mathrm{V},r}(r) - i\tilde{B}^{1,k}_{\mathrm{V},\phi}(r) \right) + ir^2 \tilde{B}^{1,k}_{\mathrm{V},\phi}(r) \Big|_{r=r_\mathrm{s}} \right.$$
$$\left. -(\mu-\mu_0) r \tilde{\psi}^{1,k}_{\mathrm{V}}(r) \Big|_{r=r_\mathrm{s}}^{r=r_\mathrm{s}+d_\mathrm{m}} \right\}, \qquad (7.91)$$

the shielding effect of the paramagnetic sheath here being accounted for in $\tilde{B}^{1,k}_{\mathrm{V},r}(r)$ and $\tilde{B}^{1,k}_{\mathrm{V},\phi}(r)$ as well as $\tilde{\psi}^{1,k}_{\mathrm{V}}(r)$. Recourse to eqns. (7.72), (7.84), and (7.87) in eqn. (7.91) consequently yields the normalized result

$$m_{\mathrm{V},y}/m_{\mathrm{UV},y} = [2\mu/\mu_0 + (\mu^2/\mu_0^2+1)\eta_{\mathrm{sm}}]/[2\mu/\mu_0 + (\mu/\mu_0+1)(1+(\mu/\mu_0-1)\zeta_\mathrm{s})\eta_{\mathrm{sm}}], \qquad (7.92)$$

where

$$\zeta_\mathrm{s} = I_1'(r_\mathrm{s}/\lambda_\mathrm{L})/I_0(r_\mathrm{s}/\lambda_\mathrm{L}) \quad \text{and} \quad \eta_{\mathrm{sm}} = (d_\mathrm{m}/r_\mathrm{s}+2)d_\mathrm{m}/r_\mathrm{s}, \qquad (7.93)$$

with the limiting form of $m_{\mathrm{V},y}$ for $\mu \to \mu_0$ in the case of an unshielded superconductor filament,

$$m_{\mathrm{UV},y} = \frac{4\Phi_0}{\mu_0\lambda_\mathrm{L}} \int_{r_\mathrm{s}-a}^{r_\mathrm{s}} dr\, r\, (I_1(r/\lambda_\mathrm{L})/I_0(r_\mathrm{s}/\lambda_\mathrm{L})) \sin\chi(r), \qquad (7.94)$$

I_0 and I_1 denoting modified Bessel functions of the first kind of orders zero and one, respectively, and I_1' meaning the derivative of the latter with respect to its argument (Abramowitz and Stegun 1972). The conspicuous universality of the dependence of the normalized Cartesian component of the magnetic dipole moment $m_{\mathrm{V},y}/m_{\mathrm{UV},y}$ on the geometrical dimensions and materials characteristics of both the superconductor filament and the paramagnetic sheath, eqn. (7.92), holds regardless of the specific shape of the magnetic vortex loop. Sensitivity to the loop's shape merely concerns $m_{\mathrm{UV},y}$, eqn. (7.94), i.e. a bare superconductor filament.

The ratio $m_{\mathrm{V},y}/m_{\mathrm{UV},y}$, eqn. (7.92), as a function of μ/μ_0 for any fixed $r_\mathrm{s}/\lambda_\mathrm{L}$ and $d_\mathrm{m}/r_\mathrm{s}$, exhibits a *minimum* at the relative magnetic permeability

$$\mu/\mu_0 = \left[2\zeta_\mathrm{s} - 1 + \sqrt{\sigma^+_{\mathrm{sm}}}\sqrt{1-(\zeta_\mathrm{s}-1)\zeta_\mathrm{s}\,\sigma^-_{\mathrm{sm}}}\right] \Big/ \left[1-(\zeta_\mathrm{s}-1)(\sigma^+_{\mathrm{sm}}+\sigma^-_{\mathrm{sm}})/2\right], \qquad (7.95)$$

where

$$\sigma^+_{\mathrm{sm}} = 2(1/\eta_{\mathrm{sm}}+1) \quad \text{and} \quad \sigma^-_{\mathrm{sm}} = 2(1/\eta_{\mathrm{sm}}-1). \qquad (7.96)$$

Two important approximations of eqn. (7.92) deserve mentioning: firstly,

$$m_{\mathrm{V},y}/m_{\mathrm{UV},y} \simeq [4\mu/\mu_0 + 2(\mu^2/\mu_0^2+1)\eta_{\mathrm{sm}}]/[4\mu/\mu_0 + (\mu/\mu_0+1)^2\eta_{\mathrm{sm}}]; \quad r_\mathrm{s} \ll \lambda_\mathrm{L}, \qquad (7.97)$$

with the minimum of $m_{\mathrm{V},y}/m_{\mathrm{UV},y}$ located at the relative magnetic permeability

$$\mu/\mu_0 \simeq 1; \quad r_\mathrm{s} \ll \lambda_\mathrm{L}, \qquad (7.98)$$

where magnetic shielding is virtually absent; secondly,

$$m_{V,y}/m_{UV,y} \simeq [2 + (\mu/\mu_0 + \mu_0/\mu)\eta_{sm}]/[2 + (\mu/\mu_0 + 1)\eta_{sm}]; \quad r_s \gg \lambda_L, \qquad (7.99)$$

with the minimum of $m_{V,y}/m_{UV,y}$ located at the relative magnetic permeability

$$\mu/\mu_0 \simeq 1 + \sqrt{\sigma_{sm}^{+}}; \quad r_s \gg \lambda_L, \qquad (7.100)$$

where tangible magnetic shielding prevails. We note that the ratio $m_{V,y}/m_{UV,y}$ in both limits, eqns. (7.97) and (7.99), lacks information about the material characteristic of the superconductor filament, λ_L.

Figure 7.8 illustrates the dependence of the normalized magnetic dipole moment $m_{V,y}/m_{UV,y}$ on the normalized thickness of the paramagnetic sheath d_m/r_s for a range of values of its relative magnetic permeability μ/μ_0, and the dependence on the relative magnetic permeability of the paramagnetic sheath μ/μ_0 for a range of values of its normalized thickness d_m/r_s, addressing a *thin* superconductor filament. Keeping μ/μ_0 fixed and starting from unity at the smallest value of d_m/r_s, there is a monotonic rise followed by a transition to saturation, as d_m/r_s continues to grow. Accordingly, the curves approach a step-like shape, if μ/μ_0 augments (Fig. 7.8(a)). Keeping d_m/r_s fixed and starting from unity at the vacuum value of μ/μ_0, there is again a monotonic rise followed by a settlement at the level prevailing for an ideal paramagnetic shield, as μ/μ_0 continues to grow. The effect gets more pronounced as d_m/r_s augments and virtually saturates from $\mu/\mu_0 \simeq 200$ (Fig. 7.8(b)). Figure 7.9 elucidates the dependence of the normalized magnetic dipole moment $m_{V,y}/m_{UV,y}$

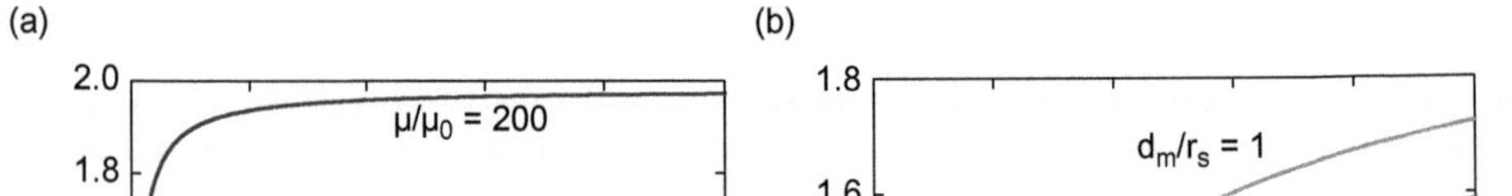

Fig. 7.8 Variation of the normalized magnetic dipole moment $m_{V,y}/m_{UV,y}$ with (a) the normalized thickness of the paramagnetic sheath d_m/r_s for three different values of its relative magnetic permeability μ/μ_0 identified on the curves, and (b) the relative magnetic permeability of the paramagnetic sheath μ/μ_0 for three different values of its normalized thickness d_m/r_s identified on the curves, when the normalized radius of the superconductor filament $r_s/\lambda_L = 0.1$, referring to the superconductor/paramagnet configuration of Fig. 7.7.

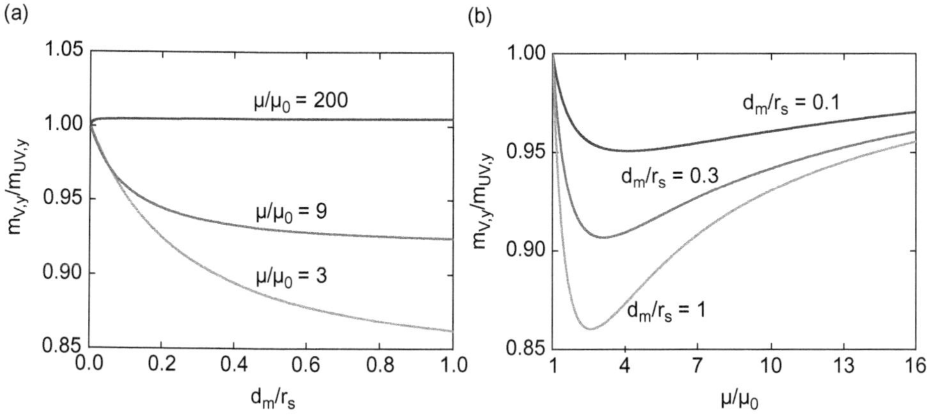

Fig. 7.9 Variation of the normalized magnetic dipole moment $m_{\mathrm{V},y}/m_{\mathrm{UV},y}$ with (a) the normalized thickness of the paramagnetic sheath $d_{\mathrm{m}}/r_{\mathrm{s}}$ for three different values of its relative magnetic permeability μ/μ_0 identified on the curves, and (b) the relative magnetic permeability of the paramagnetic sheath μ/μ_0 for three different values of its normalized thickness $d_{\mathrm{m}}/r_{\mathrm{s}}$ identified on the curves, when the normalized radius of the superconductor filament $r_{\mathrm{s}}/\lambda_{\mathrm{L}} = 100$, referring to the superconductor/paramagnet configuration of Fig. 7.7.

on the normalized thickness of the paramagnetic sheath $d_{\mathrm{m}}/r_{\mathrm{s}}$ for a range of values of its relative magnetic permeability μ/μ_0, and the dependence on the relative magnetic permeability of the paramagnetic sheath μ/μ_0 for a range of values of its normalized thickness $d_{\mathrm{m}}/r_{\mathrm{s}}$, addressing a *thick* superconductor filament. Keeping μ/μ_0 fixed and starting from unity at the smallest value of $d_{\mathrm{m}}/r_{\mathrm{s}}$, there is a drop in small values of μ/μ_0 and a rise in large values of μ/μ_0 followed by a transition to saturation, as $d_{\mathrm{m}}/r_{\mathrm{s}}$ continues to grow. Accordingly, the curves approach a step-like shape, if μ/μ_0 augments (Fig. 7.9(a)). Keeping $d_{\mathrm{m}}/r_{\mathrm{s}}$ fixed and starting from unity at the vacuum value of μ/μ_0, there is a drop towards a minimum for all values of $d_{\mathrm{m}}/r_{\mathrm{s}}$ and an ensuing rise followed by a settlement at the level prevailing for an ideal paramagnetic shield, as μ/μ_0 continues to grow. The effect gets more pronounced as $d_{\mathrm{m}}/r_{\mathrm{s}}$ augments and virtually saturates from $\mu/\mu_0 \simeq 200$ (Fig. 7.9(b)).

7.3.2 Straight magnetic vortex

Setting the scene with a basic illustration of the theoretical framework of section 7.3.1, we now appraise a straight magnetic vortex across an infinitely extended, cylindrical type-II superconductor filament of radius r_{s} in the marginal Meissner state, covered with a paramagnetic sheath of thickness d_{m}, of permeability μ, and subject to an applied magnetic field of strength H_{a} in the y-direction of a Cartesian coordinate system x, y, z, as Fig. 7.10 depicts. This represents a spatially confined analogue to the problem of an infinitely extended, straight magnetic vortex in an otherwise flux-free, unshielded superconductor that fills all space

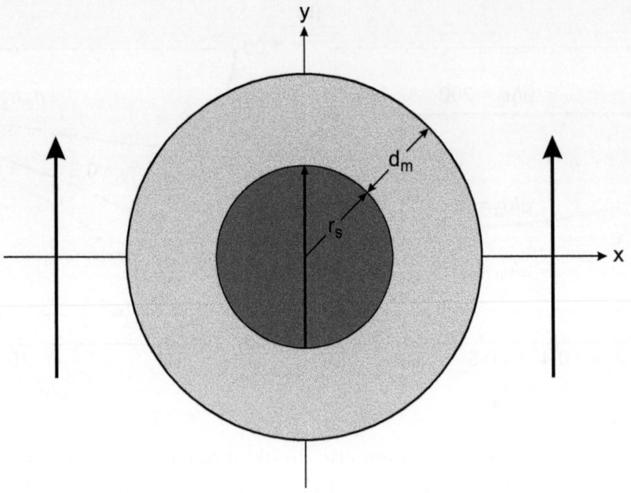

Fig. 7.10 Cross-sectional view of a central straight magnetic vortex (full line) situated in the x, y-plane at the curved interface of a cylindrical superconductor filament (dark shading) covered with a coaxial paramagnetic sheath (light shading) extending infinitely in the z-direction of a Cartesian coordinate system x, y, z. The definition of the radius of the superconductor filament r_s and the definition of the thickness of the paramagnetic sheath d_m as well as the direction of the applied magnetic field $\mathbf{H}_a$ (bold arrows) are marked.

(cf. section 2.1.3.1). Because of the alignment of the vortex parallel to the applied magnetic field, the two-dimensional Fourier transform of its vorticity only reveals a single nontrivial cylindrical component, $\tilde{V}_r^{n,k}(r)$, given by eqn. (7.87a) with $a = r_s$ and

$$\chi(r) \equiv \pi/2 \quad \text{for } 0 < r \leq r_s. \tag{7.101}$$

The account that follows enhances the work of Yampolskii and Genenko (2005).

7.3.2.1 *Self-energy and magnetic dipole moment*

The two-dimensional Fourier transforms of the nontrivial cylindrical components of the magnetic induction and the vorticity due to the magnetic vortex, $\tilde{B}_{V,r}^{n,k}$ and $\tilde{V}_r^{n,k}$, together with the two-dimensional Fourier transform of the scalar magnetic potential $\tilde{\psi}_V^{n,k}$ allow for the establishment of key quantities, as shown previously.

Simple approximate forms for the self-energy of the magnetic vortex F_V derive from eqn. (7.89) utilizing eqns. (7.72a), (7.84), and (7.87a) with eqn. (7.101) upon implementation of the cut-off procedure regarding the vortex core introduced in section 2.1.3, by resorting to limiting extremes. Thus, looking at a thin superconductor filament,

$$F_V \simeq \frac{r_s \Phi_0^2}{2\pi\mu_0\lambda_L^2} \ln\left(\frac{4er_s}{\xi_{GL}}\right); \quad \xi_{GL} \ll r_s \ll \lambda_L \tag{7.102}$$

with Euler's number $\mathrm{e} = 2.718...$ obtains; conversely, when addressing a thick superconductor filament,

$$F_{\mathrm{V}} \simeq \frac{r_{\mathrm{s}}\Phi_0^2}{2\pi\mu_0\lambda_{\mathrm{L}}^2} \ln\left(\frac{\lambda_{\mathrm{L}}}{\mathrm{e}^{\gamma}\xi_{\mathrm{GL}}}\right); \quad r_{\mathrm{s}} \gg \lambda_{\mathrm{L}} \gg \xi_{\mathrm{GL}} \tag{7.103}$$

with Euler's constant $\gamma = 0.577\ldots$ prevails. We note that F_{V} in both limits, eqns. (7.102) and (7.103), lacks information about the geometrical dimension and material characteristic of the paramagnetic sheath, d_{m} and μ, since perceptible shielding effects on the self-energy of the vortex here are virtually absent (Genenko, Rauh, and Yampolskii 2005).

The single nontrivial Cartesian component of the magnetic dipole moment of the magnetic vortex, $m_{\mathrm{V},y}$, with its distinct sensitivity to the paramagnetic environment, obeys eqn. (7.91), and hence eqn. (7.92), the limiting form of $m_{\mathrm{V},y}$ for $\mu \to \mu_0$ in the case of an unshielded superconductor filament, $m_{\mathrm{UV},y}$, being determined by eqn. (7.94) with eqn. (7.101). This gives

$$m_{\mathrm{UV},y} = (2\pi r_{\mathrm{s}}\Phi_0/\mu_0)\left[I_1(r_{\mathrm{s}}/\lambda_{\mathrm{L}})L_0(r_{\mathrm{s}}/\lambda_{\mathrm{L}}) - L_1(r_{\mathrm{s}}/\lambda_{\mathrm{L}})I_0(r_{\mathrm{s}}/\lambda_{\mathrm{L}})\right]/I_0(r_{\mathrm{s}}/\lambda_{\mathrm{L}}) \tag{7.104}$$

(Gradshteyn and Ryzhik 1994). I_n and L_n denote the modified Bessel function of the first kind of order n and the modified Struve function of order n, respectively (Abramowitz and Stegun 1972). A useful approximation of eqn. (7.104) reads

$$m_{\mathrm{UV},y} \simeq 2r_{\mathrm{s}}^3\Phi_0/3\mu_0\lambda_{\mathrm{L}}^2; \quad r_{\mathrm{s}} \ll \lambda_{\mathrm{L}}, \tag{7.105}$$

which reveals that, for small values of r_{s}, the magnetic dipole moment $m_{\mathrm{UV},y}$ is controlled by the volume of part of the superconductor filament with length r_{s}. Conversely, an asymptotic form of eqn. (7.104) is

$$m_{\mathrm{UV},y} \simeq 4r_{\mathrm{s}}\Phi_0/\mu_0; \quad r_{\mathrm{s}} \gg \lambda_{\mathrm{L}}, \tag{7.106}$$

which links the magnetic dipole moment $m_{\mathrm{UV},y}$ for large values of r_{s} to the diameter of the superconductor filament.

7.3.2.2 *Transverse lower critical field*

The self-energy of the straight magnetic vortex and its magnetic dipole moment establish the change of the Gibbs free energy of the superconducting constituent in the applied magnetic field relative to that of the flux-free Meissner state,

$$\Delta G = F_{\mathrm{V}} - \mathbf{m}_{\mathrm{V}} \cdot (\mu_0 \mathbf{H}_{\mathrm{a}}), \tag{7.107}$$

which furnishes the proper base for investigations of the thermodynamic stability of the vortex across the superconductor filament. Two antagonistic effects are manifest, viz. the introduction of the vortex into the superconducting constituent, accounted for by the self-energy term in eqn. (7.107) which contributes positively to ΔG, and the work on the vortex done by the applied magnetic field, accounted for by the potential energy term in eqn. (7.107) which contributes negatively to ΔG. Thus, $\Delta G > 0$ at zero applied magnetic field, $H_{\mathrm{a}} = 0$. As H_{a} assumes finite strengths, ΔG wanes, reaching the value $\Delta G = 0$ at the transverse lower critical field $H_{\mathrm{a}} = H_{\mathrm{c1}}$, the limit to the regime of thermodynamic stability, where the vortex across the superconductor filament appears first. An increase of H_a beyond H_{c1} entails $\Delta G < 0$, reinforcing the thermodynamic stability of the vortex state. Equation (7.107) yields $H_{c1} = F_V/\mu_0 m_{V,y}$. When reverting to eqn. (7.92), we therefore obtain the normalized result for the transverse lower critical field,

$$H_{c1}/H_{c1,UV} = [2\mu/\mu_0 + (\mu/\mu_0 + 1)(1 + (\mu/\mu_0 - 1)\zeta_s)\eta_{sm}]/[2\mu/\mu_0 + (\mu^2/\mu_0^2 + 1)\eta_{sm}], \tag{7.108}$$

with the limiting form of H_{c1} for $\mu \to \mu_0$ in the case of an unshielded superconductor filament, $H_{c1,UV} = F_V/\mu_0 m_{UV,y}$. The ratio $H_{c1}/H_{c1,UV}$, eqn. (7.108), as a function of μ/μ_0 for any fixed r_s/λ_L and d_m/r_s, exhibits a *maximum* at the relative magnetic permeability μ/μ_0 identified by eqn. (7.95).

Two important approximations of eqn. (7.108) deserve stating: firstly,

$$H_{c1}/H_{c1,UV} \simeq [4\mu/\mu_0 + (\mu/\mu_0 + 1)^2\eta_{sm}]/[4\mu/\mu_0 + 2(\mu^2/\mu_0^2 + 1)\eta_{sm}]; \quad r_s \ll \lambda_L \tag{7.109}$$

with

$$H_{c1,UV} \simeq \frac{3\Phi_0}{4\pi\mu_0 r_s^2}\ln\left(\frac{4er_s}{\xi_{GL}}\right); \quad \xi_{GL} \ll r_s \ll \lambda_L \tag{7.110}$$

according to eqns. (7.102) and (7.105), the maximum of $H_{c1}/H_{c1,UV}$ being located close to the vacuum value of the relative magnetic permeability μ/μ_0, eqn. (7.98); secondly,

$$H_{c1}/H_{c1,UV} \simeq [2 + (\mu/\mu_0 + 1)\eta_{sm}]/[2 + (\mu/\mu_0 + \mu_0/\mu)\eta_{sm}]; \quad r_s \gg \lambda_L \tag{7.111}$$

with

$$H_{c1,UV} \simeq \frac{\Phi_0}{8\pi\mu_0\lambda_L^2}\ln\left(\frac{\lambda_L}{e^\gamma\xi_{GL}}\right); \quad r_s \gg \lambda_L \gg \xi_{GL} \tag{7.112}$$

thanks to eqns. (7.103) and (7.106), the maximum of $H_{c1}/H_{c1,UV}$ being located at an increased value of the relative magnetic permeability μ/μ_0, eqn. (7.100). We note that the ratio $H_{c1}/H_{c1,UV}$ in both limits, eqns. (7.109) and (7.111), contains no information about the material characteristic of the superconductor filament, λ_L.

Figure 7.11 illustrates the dependence of the normalized transverse lower critical field $H_{c1}/H_{c1,UV}$ on the normalized thickness of the paramagnetic sheath d_m/r_s for a range of values of its relative magnetic permeability μ/μ_0, and the dependence on the relative magnetic permeability of the paramagnetic sheath μ/μ_0 for a range of values of its normalized thickness d_m/r_s, addressing a *thin* superconductor filament. Keeping μ/μ_0 fixed and starting from unity at the smallest value of d_m/r_s, there is a monotonic drop followed by a transition to saturation, as d_m/r_s continues to grow. Accordingly, the curves approach a step-like shape, if μ/μ_0 augments (Fig. 7.11(a)). Keeping d_m/r_s fixed and starting from unity at the vacuum value of μ/μ_0, there is again a monotonic drop followed by a settlement at the level prevailing for an ideal paramagnetic shield, as μ/μ_0 continues to grow. The effect gets more pronounced as d_m/r_s augments and virtually saturates from $\mu/\mu_0 \simeq 200$ (Fig. 7.11(b)). It thus

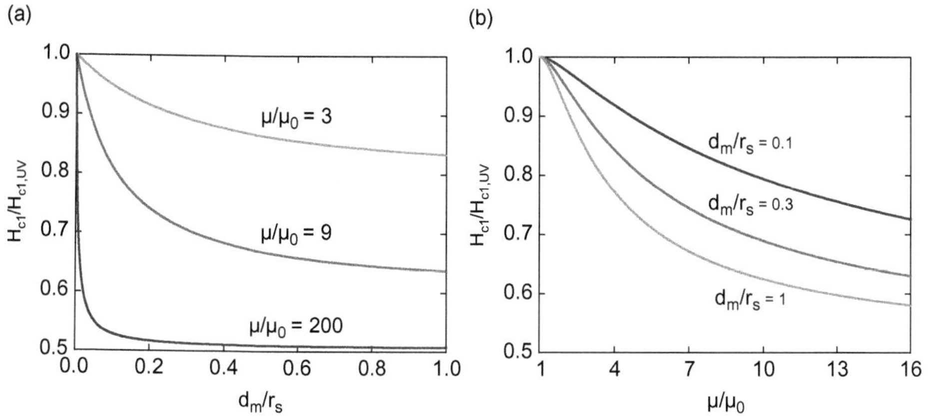

Fig. 7.11 Variation of the normalized transverse lower critical field $H_{c1}/H_{c1,UV}$ with (a) the normalized thickness of the paramagnetic sheath d_m/r_s for three different values of its relative magnetic permeability μ/μ_0 identified on the curves, and (b) the relative magnetic permeability of the paramagnetic sheath μ/μ_0 for three different values of its normalized thickness d_m/r_s identified on the curves, when the normalized radius of the superconductor filament $r_s/\lambda_L = 0.1$, referring to the superconductor/paramagnet configuration of Fig. 7.10.

appears that magnetic shielding of thin superconductor filaments has a detrimental effect on superconductivity, promoting the vortex state at *reduced* transverse lower critical fields. Figure 7.12 elucidates the dependence of the normalized transverse lower critical field $H_{c1}/H_{c1,UV}$ on the normalized thickness of the paramagnetic sheath d_m/r_s for a range of values of its relative magnetic permeability μ/μ_0, and the dependence on the relative magnetic permeability of the paramagnetic sheath μ/μ_0 for a range of values of its normalized thickness d_m/r_s, addressing a *thick* superconductor filament. Keeping μ/μ_0 fixed and starting from unity at the smallest value of d_m/r_s, there is a rise for small values of μ/μ_0 and a drop for large values of μ/μ_0 followed by a transition to saturation, as d_m/r_s continues to grow. Accordingly, the curves approach a step-like shape, if μ/μ_0 augments (Fig. 7.12(a)). Keeping d_m/r_s fixed and starting from unity at the vacuum value of μ/μ_0, there is a rise towards a maximum for all values of d_m/r_s and an ensuing drop followed by a settlement at the level prevailing for an ideal paramagnetic shield, as μ/μ_0 continues to grow. The effect gets more pronounced as d_m/r_s augments and virtually saturates from $\mu/\mu_0 \simeq 200$ (Fig. 7.12(b)). It thus appears that magnetic shielding of thick superconductor filaments can have a beneficial effect on superconductivity, preserving the flux-free Meissner state up to moderately *increased* transverse lower critical fields.

A comparison with the lower critical field of an infinitely extended, bare superconductor covering all space warrants reporting at last. Despite counteracting protection of the flux-free Meissner state, magnetic shielding of *thin* superconductor filaments still offers scope for increasing the transverse lower critical field H_{c1},

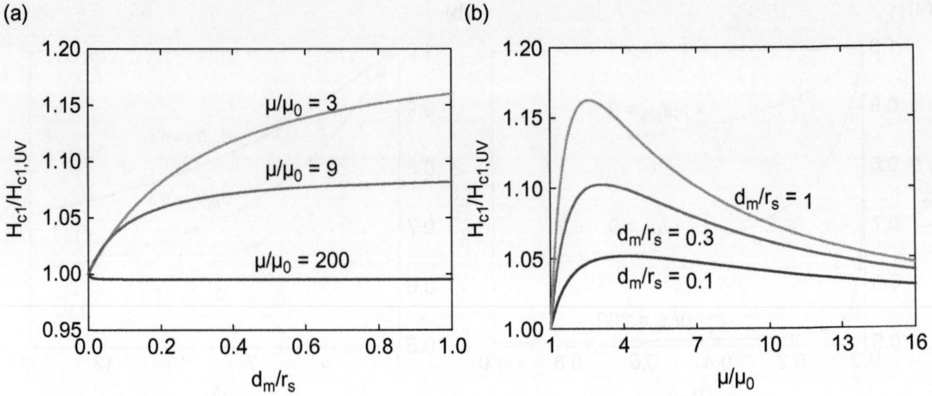

Fig. 7.12 Variation of the normalized transverse lower critical field $H_{c1}/H_{c1,UV}$ with (a) the normalized thickness of the paramagnetic sheath d_m/r_s for three different values of its relative magnetic permeability μ/μ_0 identified on the curves, and (b) the relative magnetic permeability of the paramagnetic sheath μ/μ_0 for three different values of its normalized thickness d_m/r_s identified on the curves, when the normalized radius of the superconductor filament $r_s/\lambda_L = 100$, referring to the superconductor/paramagnet configuration of Fig. 7.10.

eqn. (7.109) with eqn. (7.110), above the lower critical field given by eqn. (7.58) through a judicious choice of small values of d_m/r_s and μ/μ_0. The transverse lower critical field H_{c1} for *thick* superconductor filaments, eqn. (7.111) with eqn. (7.112), only surpasses the lower critical field defined by eqn. (7.58) if magnetic shielding is strong, as for large values of d_m/r_s and small values of μ/μ_0.

7.3.3 Regular arc-shaped magnetic vortex loop

Continuing with a more sophisticated illustration of the theoretical framework of section 7.3.1, we now study a regular arc-shaped magnetic vortex loop at the curved interface of an infinitely extended, cylindrical type-II superconductor filament of radius r_s in the marginal Meissner state, covered with a paramagnetic sheath of thickness d_m, of permeability μ, and subject to an applied magnetic field of strength H_a in the y-direction of a Cartesian coordinate system x, y, z, as Fig. 7.13 depicts. This builds on the problem of fluctuation penetration of a semi-circular magnetic vortex loop in the presence of the Bean–Livingston barrier at the flat interface between a superconductor and a paramagnet, each filling semi-infinite space (cf. section 7.2). Because of the planar orientation of the vortex loop, the two-dimensional Fourier transform of its vorticity reveals two nontrivial cylindrical components, $\tilde{V}_r^{n,k}(r)$ and $\tilde{V}_\phi^{n,k}(r)$, given by eqs. (7.87a), (7.87b) with $a < r_s$ and

$$\chi(r) = \begin{cases} \arccos\left((a^2 - r^2 - r_s^2)/2rr_s\right) & \text{for } r_s - a < r \leq \sqrt{r_s^2 - a^2} \\ \arccos\left(-\sqrt{r_s^2 - a^2}/r_s\right) & \text{for } \sqrt{r_s^2 - a^2} < r \leq r_s. \end{cases} \tag{7.113}$$

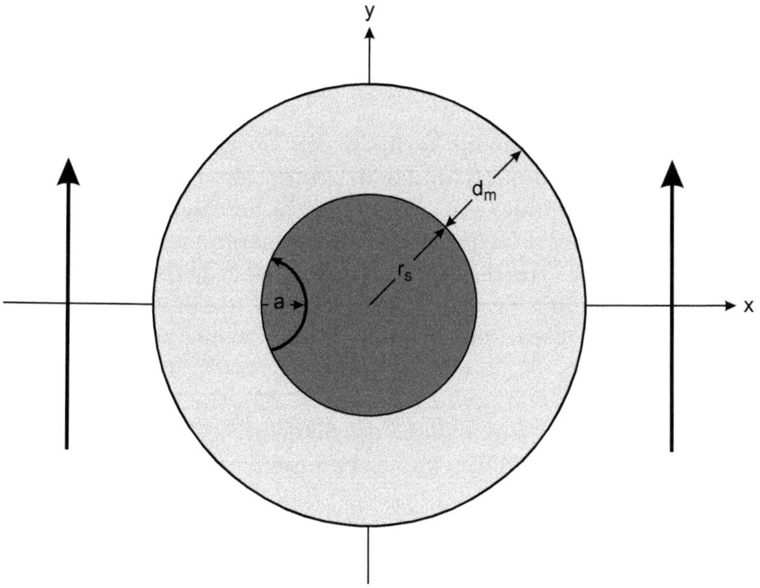

Fig. 7.13 Cross-sectional view of a regular arc-shaped magnetic vortex loop (full line) situated in the x, y-plane at the curved interface of a cylindrical superconductor filament (dark shading) covered with a coaxial paramagnetic sheath (light shading) extending infinitely in the z-direction of a Cartesian coordinate system x, y, z. The definition of the radius of the superconductor filament r_{s}, the definition of the depth of penetration of the loop a, and the definition of the thickness of the paramagnetic sheath d_{m} as well as the direction of the applied magnetic field $\mathbf{H}_{\mathrm{a}}$ (bold arrows) are marked.

Equation (7.113) relates to a circular arc with added straight parts, ensuring orthogonal alignment of both ends of the loop at the superconductor/paramagnet interface. The presentation that follows complements the work of Genenko and Yampolskii (2005) and Genenko, Rauh, and Yampolskii (2005).

7.3.3.1 Self-energy and magnetic dipole moment

The two-dimensional Fourier transforms of the nontrivial cylindrical components of the magnetic induction, $\tilde{B}_{\mathrm{V},r}^{n,k}$ and $\tilde{B}_{\mathrm{V},\phi}^{n,k}$, and those of the vorticity, $\tilde{V}_r^{n,k}$ and $\tilde{V}_\phi^{n,k}$, together with the two-dimensional Fourier transform of the scalar magnetic potential $\tilde{\psi}_{\mathrm{V}}^{n,k}$ again allow for the evaluation of key quantities, as shown before.

Practically important simplifications obtain by focusing on small penetration depths, $a \ll r_{\mathrm{s}}$, when the loop virtually adopts the shape of a circular arc with radius a alone, a premise adhered to in the following. Approximate forms for the self-energy of the magnetic vortex loop F_{V} then derive from eqn. (7.89) utilizing eqns. (7.72), (7.84), and (7.87) with eqn. (7.113) upon implementation of the cut-off procedure regarding the vortex core introduced in section 2.1.3, by resorting to limiting extremes. Thus, looking at a thin or thick superconductor filament,

$$F_{\rm V} \simeq \frac{a\Phi_0^2}{4\mu_0\lambda_{\rm L}^2} \ln\left(\frac{a}{\xi_{\rm GL}}\right); \quad \xi_{\rm GL} < a \ll r_{\rm s} \ll \lambda_{\rm L} \quad \text{or} \quad r_{\rm s} \gg \lambda_{\rm L} \gg a > \xi_{\rm GL} \tag{7.114}$$

obtains, duplicating the dominant term in eqn. (7.52) for the flat interface of the semi-infinite superconductor/paramagnet composite examined in section 7.2.2. Remarkably, $F_{\rm V}$ in either limit, eqn. (7.114), lacks information about the geometrical dimension and material characteristic of the paramagnetic sheath, $d_{\rm m}$ and μ, since the correction due to the magnetic shielding effect here is even less significant than that for the semi-infinite configuration referred to above.

The single nontrivial Cartesian component of the magnetic dipole moment of the magnetic vortex loop, $m_{{\rm V},y}$, with its distinct sensitivity to the paramagnetic environment, obeys eqn. (7.91), and hence eqn. (7.92), the limiting form of $m_{{\rm V},y}$ for $\mu \to \mu_0$ as to an unshielded superconductor filament, $m_{{\rm UV},y}$, being determined by eqn. (7.94) with eqn. (7.113). This enables two useful approximations: firstly,

$$m_{{\rm UV},y} \simeq 3a^2 r_{\rm s}\Phi_0/2\mu_0\lambda_{\rm L}^2; \quad a \ll r_{\rm s} \ll \lambda_{\rm L}, \tag{7.115}$$

which reveals that, for small values of $r_{\rm s}$, the magnetic dipole moment $m_{{\rm UV},y}$ is controlled by both the area covered by the magnetic vortex loop and the radial size of the superconductor filament; secondly,

$$m_{{\rm UV},y} \simeq 3a^2\Phi_0/\mu_0\lambda_{\rm L}; \quad r_{\rm s} \gg \lambda_{\rm L} \gg a, \tag{7.116}$$

which links the magnetic dipole moment $m_{{\rm UV},y}$ for large values of $r_{\rm s}$ to the area covered by the magnetic vortex loop alone.

7.3.3.2 Surface penetration field

The self-energy of the regular arc-shaped magnetic vortex loop and its magnetic dipole moment establish the change of the Gibbs free energy of the superconducting constituent in the applied magnetic field relative to that of the flux-free Meissner state,

$$\Delta G = F_{\rm V} - \mathbf{m}_{\rm V} \cdot (\mu_0 \mathbf{H}_{\rm a}), \tag{7.117}$$

which again furnishes the proper base for investigations of the Bean–Livingston barrier against fluctuation penetration of magnetic flux. This barrier occurs due to a competition between the image force which attracts the vortex loop to the superconductor/paramagnet interface, accounted for by the self-energy term in eqn. (7.117), and the driving force due to the action of the applied magnetic field which pushes the vortex into the superconductor constituent, expressed by the potential energy term in eqn. (7.117). Thus, when $H_{\rm a}$ is fixed and the radius of the growing vortex loop has reached its critical size $a = a_{\rm c}$ defined by the condition $\partial\Delta G/\partial a = 0$, further expansion of the vortex loop becomes irreversible and its entry proceeds. As with a flat superconductor/paramagnet interface, this may happen at different values of the critical loop radius $a_{\rm c}$ throughout the range of validity of eqns. (7.114) to (7.116), and possibly also beyond. Yet, for the Bean–Livingston barrier to be completely suppressed, $a_{\rm c}$ must approach the scale of roughness

σ of the superconductor/paramagnet interface. Whilst in the case of an *ideal* interface, with σ delineated by $0 \leq \sigma < \xi_{\rm GL}$, penetration of magnetic vortex loops takes place for $a_{\rm c} \simeq \xi_{\rm GL}$; in the case of a *real* interface, with scale of roughness σ obeying $\xi_{\rm GL} < \sigma \ll \lambda_{\rm L}$, penetration of magnetic vortex loops occurs for $a_{\rm c} \simeq \sigma$. Typifying $H_{\rm a}$ by $H_{\rm p}$, when magnetic flux first enters across the curved superconductor/paramagnet interface, and using eqns. (7.97), (7.114), and (7.115) thus yields the normalized surface penetration field

$$H_{\rm p}/H_{\rm p,UV} \simeq [4\mu/\mu_0 + (\mu/\mu_0+1)^2\eta_{\rm sm}]/[4\mu/\mu_0 + 2(\mu^2/\mu_0^2+1)\eta_{\rm sm}]; \quad r_{\rm s} \ll \lambda_{\rm L}, \tag{7.118}$$

duplicating eqn. (7.109), with reference to an unshielded thin superconductor filament, if $\mu \to \mu_0$,

$$H_{\rm p,UV} \simeq \frac{\Phi_0}{12\mu_0 r_{\rm s}\sigma}\ln\left(\frac{{\rm e}\sigma}{\xi_{\rm GL}}\right); \quad \xi_{\rm GL} < \sigma \ll r_{\rm s} \ll \lambda_{\rm L}. \tag{7.119}$$

Conversely, observing eqns. (7.99), (7.114), and (7.116) gives

$$H_{\rm p}/H_{\rm p,UV} \simeq [2 + (\mu/\mu_0+1)\eta_{\rm sm}]/[2 + (\mu/\mu_0 + \mu_0/\mu)\eta_{\rm sm}]; \quad r_{\rm s} \gg \lambda_{\rm L}, \tag{7.120}$$

duplicating eqn. (7.111), with reference to an unshielded thick superconductor filament, if $\mu \to \mu_0$,

$$H_{\rm p,UV} \simeq \frac{\Phi_0}{24\mu_0 \lambda_{\rm L}\sigma}\ln\left(\frac{{\rm e}\sigma}{\xi_{\rm GL}}\right); \quad r_{\rm s} \gg \lambda_{\rm L} \gg \sigma > \xi_{\rm GL}. \tag{7.121}$$

Normalizing the surface penetration field $H_{\rm p}$ with respect to the transverse lower critical field for the first appearance of a central straight magnetic vortex across the magnetically shielded superconductor filament, $H_{\rm c1}$, defined in eqns. (7.109) to (7.112) thus affords the result for a thin superconductor filament,

$$H_{\rm p}/H_{\rm c1} \simeq \frac{\pi r_{\rm s}}{9\sigma}\ln\left(\frac{{\rm e}\sigma}{\xi_{\rm GL}}\right)\Big/\ln\left(\frac{4{\rm e}r_{\rm s}}{\xi_{\rm GL}}\right); \quad \xi_{\rm GL} < \sigma \ll r_{\rm s} \ll \lambda_{\rm L}, \tag{7.122}$$

and the result for a thick superconductor filament,

$$H_{\rm p}/H_{\rm c1} \simeq \frac{\pi \lambda_{\rm L}}{3\sigma}\ln\left(\frac{{\rm e}\sigma}{\xi_{\rm GL}}\right)\Big/\ln\left(\frac{\lambda_{\rm L}}{{\rm e}^{\gamma}\xi_{\rm GL}}\right); \quad r_{\rm s} \gg \lambda_{\rm L} \gg \sigma > \xi_{\rm GL}. \tag{7.123}$$

These findings, together with eqns. (7.109) and (7.111), reveal that the paramagnet can have a substantial effect on the Bean–Livingston barrier for entry of magnetic flux, since in the restricted geometry under consideration here, the applied magnetic field is prone to considerable shielding by the paramagnetic constituent. Thus, taking a *thin* superconductor filament, eqn. (7.118) unfolds a strong monotonic *depression* of the surface penetration field $H_{\rm p}$ with increasing permeability μ down to even half $H_{\rm p,UV}$ as $\mu \to \infty$. Conversely, taking a *thick* superconductor filament, eqn. (7.120) displays a moderate nonmonotonic *enhancement* of the surface penetration field $H_{\rm p}$ with increasing permeability μ distinguished by a maximum for μ according to eqn. (7.100) followed by an approach towards $H_{\rm p,UV}$ as $\mu \to \infty$.

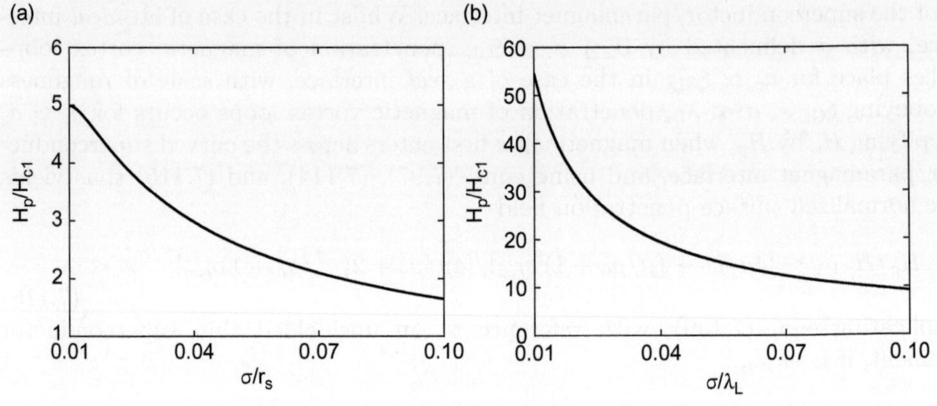

Fig. 7.14 Variation of the normalized surface penetration field H_p/H_{c1} with (a) the normalized scale of roughness σ/r_s for the normalized radius of the superconductor filament $r_s/\lambda_L = 0.1$ and (b) the normalized scale of roughness σ/λ_L for the normalized radius of the superconductor filament $r_s/\lambda_L = 100$, when $\lambda_L/\xi_{GL} = 1000$, referring to the superconductor/paramagnet configuration of Fig. 7.13.

Given the range of variability of σ indicated above, H_p adopts values between H_{c1} and H_{cth}, the transverse lower critical field and the thermodynamic critical field, respectively, according to eqns. (7.102) and (7.103),

$$H_{cth} \simeq \frac{\Phi_0}{2\pi\mu_0\lambda_L\xi_{GL}}; \quad \xi_{GL} \ll r_s \ll \lambda_L \quad \text{or} \quad r_s \gg \lambda_L \gg \xi_{GL} \tag{7.124}$$

(Orlando and Delin 1991), irrespective of the geometrical dimension and material characteristic of the paramagnetic sheath, d_m and μ. Remarkably enough, H_{cth} from eqn. (7.124) turns out to be twice the thermodynamic critical field for a straight magnetic vortex in a homogeneous superconductor filling all space, eqn. (7.59).

Figure 7.14 illustrates the dependence of the normalized surface penetration field H_p/H_{c1} on the scale of roughness σ in different normalizations: σ/r_s for a thin superconductor filament (Fig. 7.14 (a)) and σ/λ_L for a thick superconductor filament (Fig. 7.14 (b)). Starting from a distinct maximum at the smallest value of σ/r_s or σ/λ_L, there is a monotonic fall in either case; as the scale of roughness continues to grow, the range covered by H_p/H_{c1}, however, turns out to be far greater for the thick superconductor filament than for the thin superconductor filament. Entry of magnetic flux thus is much further delayed for a magnetically shielded *thick* superconductor filament, to an extent already found for the flat interface of a semi-infinite superconductor/paramagnet composite (cf. Fig. 7.6(b)).

7.3.3.3 Critical current density

If, in addition to the applied magnetic field, a transport current is imposed on the superconducting constituent, the Gibbs free energy is changed due to the contribution of the work done by the transport current. The first magnetic flux entry into the unshielded current-carrying cylindrical type-II superconductor was studied in the works of Genenko (1994, 1995) and Genenko et al. (1998). In the case of the magnetically shielded superconductor considered here, the change of the Gibbs free energy relative to that of the flux-free Meissner state without a transport current now reads

$$\Delta G = F_{\mathrm{V}} - \mathbf{m}_{\mathrm{V}} \cdot (\mu_0 \mathbf{H}_{\mathrm{a}}) + \Delta U_{\mathrm{V},I}. \tag{7.125}$$

The first two terms on the right-hand side of eqn. (7.125) duplicate those in eqn. (7.117), and the third term accounts for the change of the potential energy due to the work done on the magnetic vortex loop during growth up to size a in the presence of the transport current, starting from a nucleus at the superconductor/paramagnet interface. We have

$$\Delta U_{\mathrm{V},I} = -\Phi_0 \int_0^a dc \int_C \mathrm{d}\mathbf{l} \cdot (\mathbf{e}_n \times \mathbf{j}), \tag{7.126}$$

where $\mathbf{e}_n$ represents the outward normal unit vector along the core of the magnetic vortex loop C and $\mathbf{j}$ marks the density of the imposed transport current I, with its single nontrivial Cartesian component j_z. Provided that $a << r_{\mathrm{s}}$, the latter virtually remains constant within the extent of the magnetic vortex loop, which is why in eqn. (7.126) the isotropic distribution of a vortex-free superconductor filament

$$j_z(r) = (I/2\pi r_{\mathrm{s}}\lambda_{\mathrm{L}}) I_0(r/\lambda_{\mathrm{L}})/I_1(r_{\mathrm{s}}/\lambda_{\mathrm{L}}) \quad \text{in } 0 \le r \le r_{\mathrm{s}} \tag{7.127}$$

(Schafroth 1960) may be assumed, I_0 and I_1 denoting modified Bessel functions of the first kind of orders zero and one, respectively (Abramowitz and Stegun 1972). This allows for two important extremes to be identified: firstly, transport current flow extends over the entire cross section of the superconductor filament, admitting

$$\Delta U_{\mathrm{V},I} \simeq -a^2 \Phi_0 I/2r_{\mathrm{s}}^2; \quad a \ll r_{\mathrm{s}} \ll \lambda_{\mathrm{L}}, \tag{7.128}$$

which shows that, for small values of r_{s}, the change of the potential energy $\Delta U_{\mathrm{V},I}$ is controlled by the area covered by the magnetic vortex loop and the radial size of the superconductor filament; secondly, transport current flow is essentially confined to a layer of thickness λ_{L} beneath the surface of the superconductor filament, rendering

$$\Delta U_{\mathrm{V},I} \simeq -a^2 \Phi_0 I/4r_{\mathrm{s}}\lambda_{\mathrm{L}}; \quad r_{\mathrm{s}} \gg \lambda_{\mathrm{L}} \gg a, \tag{7.129}$$

which links the change of the potential energy $\Delta U_{\mathrm{V},I}$ for large values of r_{s} to both the area covered by the magnetic vortex loop, the radial size of the superconductor filament and the London penetration depth.

When H_{a} is fixed and the radius of the growing vortex loop has reached its critical size $a = a_{\mathrm{c}}$ defined by the condition $\partial \Delta G/\partial a = 0$, further expansion of the vortex loop becomes irreversible and its entry proceeds. This may happen at different values of the critical loop radius a_{c} throughout the range of validity of eqns. (7.114) to (7.116), (7.128), and (7.129), and possibly also beyond. Yet, for the Bean–Livingston barrier to be completely suppressed, a_{c} must approach the scale of roughness σ of the superconductor/paramagnet interface. Whilst in the case of an *ideal* interface, with scale of roughness σ delineated by $0 \le \sigma < \xi_{\mathrm{GL}}$, penetration of magnetic vortex loops takes place for $a_c \simeq \xi_{GL}$, in the case of a *real* interface, with scale of roughness σ obeying $\xi_{GL} < \sigma \ll \lambda_L$, penetration of magnetic vortex loops occurs for $a_c \simeq \sigma$. Typifying H_a by H_p, when magnetic flux first enters across the curved superconductor/paramagnet interface whereby a transport current is absent, expressing furthermore the critical transport current $I = I_c$ through j_c via $I_c = \pi r_s^2 j_c$ and using eqns. (7.97), (7.99), (7.114) to (7.116) as well as (7.128), (7.129) thus yields the field-dependent critical current density

$$j_c/j_{c,\max} = 1 - H_a/H_p \quad \text{for } 0 \leq H_a \leq H_p, \tag{7.130}$$

calling upon eqns. (7.118) to (7.121). The result, eqn. (7.130), displays a universal linear decrease in the critical current density j_c with the strength of the applied magnetic field from its maximum at zero field, $H_a = 0$,

$$j_{c,\max} \simeq \frac{\Phi_0}{4\pi\mu_0\lambda_L^2\sigma} \ln\left(\frac{e\sigma}{\xi_{GL}}\right); \quad \xi_{GL} < \sigma \ll r_s \ll \lambda_L \tag{7.131}$$

or

$$j_{c,\max} \simeq \frac{\Phi_0}{2\pi\mu_0 r_s\lambda_L\sigma} \ln\left(\frac{e\sigma}{\xi_{GL}}\right); \quad r_s \gg \lambda_L \gg \sigma > \xi_{GL}, \tag{7.132}$$

down to 0 at $H_a = H_p$, irrespective of the geometrical dimension and material characteristic of the paramagnetic sheath, in accord with experiments (Soltanian et al. 2001). Remarkably enough, the sensitivity of $j_{c,\max}$ to σ equals that of H_p for both extremes, as juxtapositions of eqn. (7.131) with eqns. (7.118), (7.119) and eqn. (7.132) with eqns. (7.120), (7.121) tell.

7.3.3.4 Numerical estimates

Let us conclude by reverting to the example of a cylindrical iron-sheathed MgB_2 filament at the temperature T = 20 K, with geometrical dimensions $r_s = 500\ \mu$m, $d_m = 250\ \mu$m and material characteristics $\lambda_L = 140$ nm, $\xi_{GL} = 5$ nm as well as $\mu/\mu_0 = 46$ (Canfield, Bud'ko, and Finnemore 2003). Roughly probing the theoretically admissible range, we posit that the scale of roughness varies between its lower limit $\sigma = \xi_{GL}$ and $\sigma = 10\,\xi_{GL}$. Equations (7.120), (7.121) then predict a monotonic fall of the surface penetration field from $H_p \simeq 1.0 \times 10^5$ A/m down to $H_p \simeq 3.3 \times 10^4$ A/m, with a concomitant enhancement of H_p due to paramagnetic shielding of the superconductor filament of about 2%, a relatively small effect owing to the large value of μ/μ_0. Equation (7.132) adheres to a monotonic drop of the maximum critical current density from $j_{c,\max} \simeq 7.5 \times 10^8$ A/m^2 down to $j_{c,\max} \simeq 2.5 \times 10^8$ A/m^2, given the same premise. An evaluation of the semi-empirical derivations of critical characteristics in section 7.1.3.2 on the basis of the present results thus suggests that omitting the phenomenon of penetration of magnetic flux in the theoretical frame apparently leads to an overestimate of the surface penetration field and to an underestimate of the current density that the superconductor filament is at most capable to sustain.

8

Meissner state and critical state of thin superconductors in arbitrary paramagnetic environments: finite-element simulations

We now revisit the issue of magnetic shielding of thin type-II superconductor strips and tubes which recommend themselves as most promising constituents for realizations of the beneficial effects offered by superconductor/paramagnet heterostructures. In doing so, we build on the results of Chapters 3 to 5 concerning both the flux-free Meissner state and the partially flux-filled critical state, that the sheet current in superconductor strips may be strongly redistributed when exposed to ideal, paramagnetic or diamagnetic, shields, bringing about substantial enhancements of the total loss-free current that can be carried by the strips. Likewise, we note that significant redistribution of the sheet current in superconductors of flat, annular, and cylindrical shapes may also occur under the influence of nonideal paramagnetic environments, as the outcomes of Chapters 6 and 7 show. However, while these observations mainly establish principal traits based on certain limitations of the geometrical dimensions and material characteristics of the various constituents involved which largely enable the use of analytic techniques, we here aim at relaxing constraints to some extent and taking a step towards investigating superconductor/paramagnet heterostructures of practical relevance too. This, admittedly, necessitates turning to computational techniques right from the start.

To implement them, we call upon ANSYS and COMSOL, commercial finite-element programs described elsewhere (Pryor 2011; Gadala 2020) for computing scalar, vector, and tensor fields associated with diverse material states, though not originally designed for modelling superconducting states, by exploiting magnetostatic–electrostatic analogues. The method in its present form, assuming that critical sheet currents are independent of the local magnetic field, is easy to handle and readily accessible, without requiring adaptations as in more sophisticated treatments based on other multiphysics packages (Gömöry 2006; Campbell 2007; Grilli, Ashworth, and Civale 2007; Hong et al. 2007; Wolsky and Campbell 2008; Nguyen, Ashworth and Willis 2009; Badía-Majós and López 2014; Grilli et al.

Electromagnetics of Superconductor/Paramagnet Heterostructures. Yuri Genenko and Hermann Rauh, Oxford University Press.
 DOI: 10.1093/9780191782855.003.0008

2014; Majoros, Sumption, Collings, and van der Laan 2014; Grilli 2016; Shen, Grilli, and Coombs 2020; Statra, Menana, and Douine 2022). Although a detailed outline of the method will be given for a superconductor strip inside a paramagnetic cavity, adaptations for, e.g., a superconductor strip deposited on a paramagnetic substrate or a magnetically coated superconducting tubular wire are readily available. The conditions of both an imposed transport current and an applied magnetic field can be examined and the flux-free Meissner state as well as the partially flux-filled critical state taken into account.

We propose two essentially different shielding configurations for a superconductor strip inside a paramagnetic cavity. Results cover the distribution of the magnetic field around the strip, the distribution of the sheet current in the strip, and the maximum total transport current available to the strip, when addressing the flux-free Meissner state. Selecting the paramagnet configuration with its superior performance of the two, we extend our finite-element analysis to the partially flux-filled critical state and also calculate the hysteretic AC loss suffered by the strip. This makes it possible to specify criteria for the design of superconductor/paramagnet heterostructures with efficient paramagnetic shields, complementing previous numerical work on hysteretic AC losses in planar superconductor/soft-magnet heterostructures with simple, albeit realistic shielding geometries and arbitrary permeabilities of their magnetic constituents (Glowacki and Majoros 2000; Majoros, Glowacki, and Campbell 2000b; Sumption et al. 2002; Gömöry et al. 2007; Gömöry, Vojenčiak, Pardo, and Šouc 2009; Ma and Rauh 2013). Emphasis on hysteretic AC losses is laid as well when assessing the intricate shielding effects of substrate magnetism on the properties of superconducting tapes and tubes in the presence of an applied magnetic field, tying in with a host of previous literature (Amemiya and Nakahata 2007; Claassen 2007; Grilli, Ashworth, and Civale 2007; Nguyen, Sastry, and Schwartz 2007; Claassen and Thieme 2008; Nguyen, Ashworth, and Willis 2009; Amemiya, Nakahata, Fujiwara, and Shiohara 2010; Gömöry et al. 2010; Lousberg et al. 2010; Mikitik, Mawatari, Wan, and Sirois 2013; Fareed, Robert, and Ruiz 2019; Fareed and Ruiz 2021).

8.1 Fundamentals of finite-element simulations of a magnetically shielded superconductor strip

Let us first give an exposition of the concepts and premises underlying our finite-element simulations of physical states of a magnetically shielded superconductor strip. This includes formulations of the original magnetostatic problem and the corresponding electrostatic problem as well as elaborations of magnetostatic–electrostatic analogues, which pave the way for computations with the help of ANSYS or COMSOL relating to individually chosen paramagnetic environments. The account that follows essentially follows along the lines of Genenko, Rauh, Krüger, and Narayanan (2009).

8.1.1 Magnetostatic problem

We consider an infinitely extended type-II superconductor strip of width $2w$ and thickness $d \ll w$ limited by the range $-w \leq x \leq w$ and located between two infinitely extended paramagnets with permeabilities μ_1 and μ_2, the direction of translational invariance of this heterostructure being parallel to the z-axis of a Cartesian coordinate system x, y, z, as depicted in Fig. 8.1. A cross section distinguishes the domain D occupied by the strip from the respective domains $D^{(1)}$ and $D^{(2)}$ occupied by the paramagnets, leaving the domain $D^{(0)}$ as a vacuum with permeability μ_0. The strip is supposed to carry a transport current of magnitude I in the absence of an applied magnetic field. By virtue of the restriction concerning the dimensions of the strip, changes in this current over the thickness of the strip may be ignored and, for mathematical convenience, the strip is regarded as infinitesimally thin, enabling the physical state of the strip to be characterized by the sheet current $J(x)$ alone (Brandt 1992). The latter measures the discontinuity of the tangential component of the magnetic field $H_{\mathrm{t}}^{(0)}$ across the thickness of the strip.

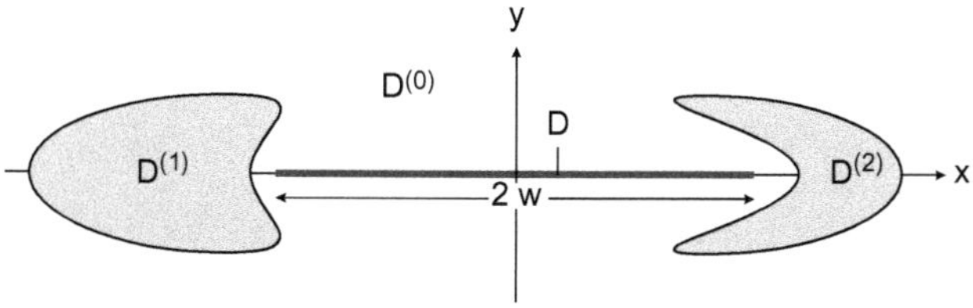

Fig. 8.1 Cross-sectional view of a superconductor strip of width $2w$ (dark shading) located between two paramagnets of arbitrary shape (light shading) extending infinitely in the z-direction of a Cartesian coordinate system x, y, z. The domain D occupied by the strip, the domains $D^{(1)}$ and $D^{(2)}$ occupied by the paramagnets, and the domain $D^{(0)}$ left for a vacuum are marked.

The physical states of the superconductor strip are crucially defined through the boundary conditions imposed on the surface of the strip. Thus, for the strip to exhibit the flux-free Meissner state, the normal component of the static magnetic field $\mathbf{H}^{(0)}$ in the vacuum must vanish on the whole surface of the strip: $H_{\mathrm{n}}^{(0)} = 0$ along ∂D. If, by comparison, the strip were to reveal the partially flux-filled critical state due to the first flux penetration from the strip edges, the normal component of this field must vanish only on the central, flux-free part of the surface of the strip: $H_{\mathrm{n}}^{(0)} = 0$ along ∂D_{ff}. Implementing Bean's model of the critical state duly adapted to the geometry of the strip (Bean 1962, 1964; Norris 1970; Brandt and Indenbom 1993), we assume that the sheet current J within the strip, and hence the tangential component of the static magnetic field $\mathbf{H}^{(0)}$ in the vacuum, remain constant on the marginal, flux-penetrated parts of the surface of the strip characterized by the permeability μ_0, of depths $s_1, s_2 > 0$ measured from the left or, right end of the strip respectively: $H_{\mathrm{t}}^{(0)} = \mathrm{const}$ along $\partial D_{s_1,s_2}$.

In studying the variation in the sheet current $J(x)$ over the width of the strip, when modelling the Meissner state for individual paramagnetic shields, we invoke strip protection against entry of magnetic flux owing to an edge barrier typified by the sheet current value J_b, as noted in section 3.1.2; this ensures that the strip remains flux free provided that $J < J_\mathrm{b}$ at both edges of the strip, $x = \pm w$. The maximum total transport current possible, I_max, obtains if $J = J_\mathrm{b}$ at least at one edge of the strip, irrespective of a shielding paramagnetic environment. On the other hand, when modelling the critical state for given paramagnetic shields, we resort to the (field-independent) critical sheet current value J_c prevailing in the marginal, flux-penetrated parts of the strip, of depths $s_1, s_2 > 0$, where $J = J_\mathrm{c}$. Compliance with the Maxwell equations requires that the sheet current $J(x)$ be continuous over the whole width of the strip adopting, in particular, the value J_c at the outer boundaries $x = -w + s_1$ and $x = w - s_2$ of the central, flux-free part of the strip adjacent to the marginal, flux-penetrated parts. In the central part itself, the sheet current $J(x)$ must satisfy the condition $J < J_\mathrm{c}$ in the immediate vicinity of these boundaries since, with Lorentz-like forces acting upon magnetic vortices, magnetic flux would otherwise also penetrate this part until the full critical state with $J = J_\mathrm{c}$ throughout the strip is reached. If proceeding from a total transport current of fixed magnitude I, the parameters s_1 and s_2 must be determined self-consistently such that $J = J_\mathrm{c}$ at the interfaces between the flux-free and the flux-filled parts, resulting in a continuous distribution of the sheet current over the whole width of the strip (Norris 1970; Brandt and Indenbom 1993).

Additional conditions relating to the tangential and normal components of the static magnetic fields $\mathbf{H}^{(i)}$ in the domains $D^{(i)}$, when their respective boundaries are traversed, demand that

$$H_\mathrm{t}^{(i)} = H_\mathrm{t}^{(k)} \quad \text{and} \quad \mu_i H_\mathrm{n}^{(i)} = \mu_k H_\mathrm{n}^{(k)} \quad \text{on } \partial D^{(i,k)}; \quad i,k = 0,1,2 \ (i \neq k) \tag{8.1}$$

(Jackson 1998). These requirements include the cases of the paramagnets facing each other or a vacuum, but exclude direct contact between the paramagnets and the superconductor strip to avoid complications arising from the proximity effect (Schmidt 1997). Needless to say that, on general physical grounds, the static magnetic field $\mathbf{H}^{(0)}$ in the vacuum must vanish asymptotically far away from the heterostructure.

The magnetic field $\mathbf{H}$ in the entire space outside the domain D satisfies the homogeneous Maxwell equations

$$\nabla \times \mathbf{H} = 0 \quad \text{and} \quad \nabla \cdot \mathbf{H} = 0, \tag{8.2}$$

which suggest the introduction of a magnetic vector potential $\mathbf{A}$ to yield the representation of the magnetic field,

$$\mu_0 \mathbf{H} = \nabla \times \mathbf{A}, \tag{8.3}$$

identically satisfying the second equation (8.2). With the geometry of the heterostructure at hand, the current density $\mathbf{j}$ has a nonzero component j_z only, and thus allows the choice of a single nontrivial component A_z of $\mathbf{A}$, both depending on

x and y. By reference to the Coulomb gauge $\nabla \cdot \mathbf{A} = 0$, the first equation (8.2) then generates the Laplace equation

$$\triangle A_z = 0. \tag{8.4}$$

Turning to Ampère's law in the domain D,

$$\nabla \times \mathbf{H} = \mathbf{j}, \tag{8.5}$$

and applying Stokes' theorem, the total transport current permeating the cross section of the superconductor strip, I, can be expressed in terms of the sheet current J or the tangential component of the magnetic field $H_\mathrm{t}^{(0)}$ on the surface of this strip:

$$I = \int_{-w}^{w} dx\, J(x) = \lim_{d \to 0} \oint_{\partial D} dl\, H_\mathrm{t}^{(0)}, \tag{8.6}$$

i.e. through the solution of eqn. (8.2) or (8.4) under consideration of the boundary and far-field conditions given above, with the component of the magnetic vector potential A_z prescribed on the whole surface of the strip: $A_z =$ const on ∂D, if the flux-free Meissner state were addressed. By contrast, studies of the partially flux-filled critical state require the component of the magnetic vector potential A_z only to be prescribed on the surface of the central, flux-free part of the strip, $A_z =$ const on ∂D_ff, in conjunction with the sheet current J in the marginal, flux-penetrated parts of the strip determined self-consistently, $J = J_\mathrm{c}$ along $\partial D_{s_1,s_2}$, so that a continuous distribution of $J(x)$ over the whole width of the strip is obtained which, according to eqn. (8.6), itself determines I.

8.1.2 Electrostatic problem

The desired simulation of the magnetic field can be attained by proper reinterpretation of the solution of a corresponding electrostatic problem, as will be shown. Let us therefore imagine that the heterostructure portrayed in Fig. 8.1 consisted of an infinitely extended metal or composite strip of width $2w$ and thickness $d \ll w$ limited by the range $-w \leq x \leq w$ and located between two infinitely extended dielectrics with permittivities ε_1 and ε_2. A cross section distinguishes the domain D occupied by the strip from the domains $D^{(1)}$ and $D^{(2)}$ occupied by the dielectrics, leaving the domain $D^{(0)}$ as a vacuum with permittivity ε_0. The strip is supposed to carry a static electric charge per unit length of the strip, Q, in the absence of an applied electric field. By virtue of the restriction concerning the dimensions of the strip, the physical state of the strip can be characterized by the surface charge density $\sigma(x)$ alone. The latter measures the discontinuity of the normal component of the electric field $E_\mathrm{n}^{(0)}$ across the thickness of the strip (Jackson 1998).

If we wish to emulate the flux-free Meissner state, the strip is understood to exhibit metallic conductivity, whence the tangential component of the static electric field $\mathbf{E}^{(0)}$ in the vacuum must vanish on the whole surface of the strip: $E_\mathrm{t}^{(0)} = 0$ along ∂D. If, however, the strip were addressed in the partially flux-filled critical state, it is supposed to consist of a central, metallic part and marginal, dielectric parts.

The tangential component of this field then must vanish only on the central part of the surface of the strip: $E_{\mathrm{t}}^{(0)} = 0$ along ∂D_{ff}. In conformity with Bean's model of the critical state duly adapted to the geometry of the strip, we assume that the surface charge density σ at the boundary of the strip, and hence the normal component of the static electric field $\mathbf{E}^{(0)}$ in the vacuum, remain constant on the dielectric parts of the surface of the strip characterized by the permittivity ε_0, of widths $s_1, s_2 > 0$ measured from the left or right end of the strip: $E_{\mathrm{n}}^{(0)} = \text{const}$ along $\partial D_{s_1,s_2}$.

In studying the variation in the surface charge density $\sigma(x)$ over the width of the strip, when addressing the Meissner state, we invoke a surface charge density value σ_b which ensures protection of this state as long as $\sigma < \sigma_{\mathrm{b}}$ at both edges of the strip, $x = \pm w$. The maximum total electric charge possible, per unit length of the strip, Q_{max}, then obtains if $\sigma = \sigma_{\mathrm{b}}$ at least at one edge of the strip. On the other hand, when considering the critical state, we refer to a (field-independent) critical surface charge density value σ_{c} prevailing in the marginal, dielectric parts of the strip, of widths $s_1, s_2 > 0$, where $\sigma = \sigma_{\mathrm{c}}$. The surface charge density $\sigma(x)$ must be continuous over the whole width of the strip adopting, in particular, the value $\sigma = \sigma_{\mathrm{c}}$ at the outer boundaries $x = -w + s_1$ and $x = w - s_2$ of the central, metallic part of the strip adjacent to the marginal, dielectric parts. In the central part itself, the surface charge density $\sigma(x)$ must satisfy the condition $\sigma < \sigma_{\mathrm{c}}$ in the immediate vicinity of these boundaries so as to ensure stability of the critical state. If proceeding from a total electric charge per unit length of fixed magnitude Q, the parameters s_1 and s_2 must be determined self-consistently such that $\sigma = \sigma_{\mathrm{c}}$ at the interfaces between the metallic and the dielectric parts, resulting in a continuous distribution of the surface charge density over the whole width of the strip.

Additional conditions relating to the tangential and normal components of the static electric fields $\mathbf{E}^{(i)}$ in the domains $D^{(i)}$, when their respective boundaries are traversed, demand that

$$E_{\mathrm{t}}^{(i)} = E_{\mathrm{t}}^{(k)} \quad \text{and} \quad \varepsilon_i E_{\mathrm{n}}^{(i)} = \varepsilon_k E_{\mathrm{n}}^{(k)} \quad \text{on } \partial D^{(i,k)}; \quad i,k = 0,1,2 \ (i \neq k) \tag{8.7}$$

(Jackson 1998). Needless to say that, on general physical grounds, the static electric field $\mathbf{E}^{(0)}$ in the vacuum must vanish asymptotically far away from the heterostructure.

The electric field $\mathbf{E}$ in the entire space outside the domain D satisfies the homogeneous Maxwell equations

$$\nabla \times \mathbf{E} = 0 \quad \text{and} \quad \nabla \cdot \mathbf{E} = 0, \tag{8.8}$$

which permit the introduction of a scalar electric potential φ depending on x and y to yield the representation of the electric field,

$$\mathbf{E} = -\nabla\varphi, \tag{8.9}$$

identically satisfying the first equation (8.8), whereas the second equation (8.8) entails the Laplace equation

$$\triangle\varphi = 0. \tag{8.10}$$

Making recourse to Gauss's law in the domain D,

$$\nabla \cdot (\varepsilon_0 \mathbf{E}) = \rho, \tag{8.11}$$

with the volume charge density ρ and applying the divergence theorem, the total electric charge per unit length of the metal, or composite, strip located on its surface, Q, can be expressed in terms of the surface charge density σ or the normal component of the electric field $E_\mathrm{n}^{(0)}$ on the surface of this strip:

$$Q = \int_{-w}^{w} dx\, \sigma(x) = \varepsilon_0 \lim_{d \to 0} \oint_{\partial D} dl\, E_\mathrm{n}^{(0)}, \tag{8.12}$$

referring to the solution of eqn. (8.8) or (8.10) under consideration of the boundary and far-field conditions defined earlier, with the scalar electric potential φ prescribed on the whole surface of the strip: $\varphi =$ const on ∂D, if the flux-free Meissner state were addressed. By contrast, emulating the partially flux-filled critical state requires the potential φ to be only prescribed on the surface of the central, metallic part of the strip, $\varphi =$ const on ∂D_ff, in conjunction with the charge density σ on the dielectric parts of the surface of the strip determined self-consistently, $\sigma = \sigma_\mathrm{c}$ along $\partial D_{s_1,s_2}$, so that a continuous distribution of $\sigma(x)$ over the whole width of the strip is obtained which, according to eqn. (8.12), itself determines Q.

8.1.3 Magnetostatic–electrostatic analogues

We now relate the component of the magnetic vector potential A_z as a solution of eqn. (8.4) and the scalar electric potential φ as a solution of eqn. (8.10) in the domains $D^{(i)}$ by setting

$$A_z^{(i)} = \alpha \mu_0 \varepsilon_i \varphi^{(i)}; \quad i = 0, 1, 2 \tag{8.13}$$

with an arbitrary constant, α. Then, in the entire space outside the domain D, the electric field and the magnetic field derived from these potentials are mutually orthogonal: $\mathbf{E}^{(i)} \cdot \mathbf{H}^{(i)} = 0$, since

$$H_x^{(i)} = -\alpha \varepsilon_i E_y^{(i)} \quad \text{and} \quad H_y^{(i)} = \alpha \varepsilon_i E_x^{(i)}; \quad i = 0, 1, 2, \tag{8.14}$$

revealing a coincidence of the lines of the magnetic field generated by the transport current and the equipotentials of the electric field created by the surface charge. In particular, the assignments

$$H_\mathrm{t}^{(i)} = \alpha \varepsilon_i E_\mathrm{n}^{(i)} \quad \text{and} \quad H_\mathrm{n}^{(i)} = -\alpha \varepsilon_i E_\mathrm{t}^{(i)} \quad \text{on } \partial D^{(i)}; \quad i = 0, 1, 2 \tag{8.15}$$

hold. This, in turn, means that, for a metal strip, the tangential component of the static electric field $\mathbf{E}^{(0)}$ in the vacuum, and hence the normal component of the static magnetic field $\mathbf{H}^{(0)}$ in the vacuum, vanish on the surface of the strip: $E_\mathrm{t}^{(0)} = 0$ together with $H_\mathrm{n}^{(0)} = 0$ along ∂D. The domain D thus is devoid of magnetic flux and therefore represents a superconductor strip in the Meissner state. On the other hand, for a composite strip, the tangential component of the static electric field $\mathbf{E}^{(0)}$ in the vacuum, and hence the normal component of the static magnetic field $\mathbf{H}^{(0)}$ in the vacuum, only vanish on the central, metallic part of the surface of the strip: $E_\mathrm{t}^{(0)} = 0$ together with $H_\mathrm{n}^{(0)} = 0$ along ∂D_ff. Over and above, the normal component of the static electric field $\mathbf{E}^{(0)}$ in the vacuum, and hence the tangential component of the static magnetic field $\mathbf{H}^{(0)}$ in the vacuum, remain constant on the dielectric parts of the surface of the strip: $E_\mathrm{n}^{(0)} = \mathrm{const}$ together with $H_\mathrm{t}^{(0)} = \mathrm{const}$ along $\partial D_{s_1,s_2}$. The domain D thus is just partially filled with magnetic flux and represents a superconductor strip in the critical state. By substituting eqn. (8.15) into eqn. (8.7), additional conditions relating to the tangential and normal components of the static magnetic fields $\mathbf{H}^{(i)}$ in the domains $D^{(i)}$, when their respective boundaries are traversed, emerge:

$$\varepsilon_i^{-1} H_\mathrm{n}^{(i)} = \varepsilon_k^{-1} H_\mathrm{n}^{(k)} \quad \text{and} \quad H_\mathrm{t}^{(i)} = H_\mathrm{t}^{(k)} \quad \text{on } \partial D^{(i,k)}; \quad i,k = 0,1,2 \ (i \neq k). \tag{8.16}$$

Equation (8.16) in view of eqn. (8.1) defines the correct boundary conditions for the magnetic field of the heterostructure illustrated in Fig. 8.1, with the domains $D^{(i)}$ characterized by the permeabilities ε_i^{-1}.

Since the sheet current J is proportional to the tangential component of the magnetic field $H_\mathrm{t}^{(0)}$ or, equivalently, to the normal component of the electric field $E_\mathrm{n}^{(0)}$, the relation

$$I = \lim_{d\to 0} \oint_{\partial D} dl\, H_\mathrm{t}^{(0)} = \alpha\varepsilon_0 \lim_{d\to 0} \oint_{\partial D} dl\, E_\mathrm{n}^{(0)} = \alpha Q \tag{8.17}$$

follows from eqns. (8.6), (8.12), and (8.15), demonstrating a correspondence between the total transport current flowing through the strip, in the case of the magnetostatic problem, and the total electric charge per unit length carried by the surface of the strip, in the case of the electrostatic analogue. Thus, once the distribution of the electric field $\mathbf{E}$, obtained from the scalar electric potential φ and, if applicable, the surface charge density σ, is known, the distributions of both the magnetic field $\mathbf{H}$ and the sheet current J as well as the total transport current I can be established from eqns. (8.14) and (8.17), except for the common factor α.

8.1.4 Quasistatic extension

The preceding analysis not only applies to static quantities, but also some way beyond. Thus, for the superconductor strip in the partially flux-filled critical state, hysteretic AC losses caused by a transport current that varies periodically with time, at fixed magnitude I, can be ascertained, resorting to the quasistatic approach. The scenario of entry and exit of magnetic flux in the presence of magnetic shields is basically the same as for an isolated strip (cf. section 4.1.2); in particular, the distribution of the magnetic field across the width of the strip, obtained by the gradual increase in the transport current from the virgin state of the strip up to the state associated with the maximum value of the current, changes sign as the current alternates between I and $-I$. Accordingly, the energy dissipated during a steady-state cycle of the AC transport current, per unit length of the strip, $U_{\rm ac}$, is determined by the normal component of the magnetic field $H_{\rm n}^{(0)}$ for a given value of I or, equivalently, by the tangential component of the electric field $E_{\rm t}^{(0)}$ for a chosen value of Q,

$$U_{\rm ac} = 4\mu_0 J_{\rm c} \lim_{d\to 0} \left\{ \int_{w-s_2}^{w} dx \int_{w-s_2}^{x} dx'\, H_{\rm n}^{(0)}(x') - \int_{-w}^{-w+s_1} dx \int_{x}^{-w+s_1} dx'\, H_{\rm n}^{(0)}(x') \right\}$$

$$= 4\alpha^2 \varepsilon_0 \mu_0 \sigma_{\rm c} \lim_{d\to 0} \left\{ \int_{-w}^{-w+s_1} dx \int_{x}^{-w+s_1} dx'\, E_{\rm t}^{(0)}(x') - \int_{w-s_2}^{w} dx \int_{w-s_2}^{x} dx'\, E_{\rm t}^{(0)}(x') \right\}, \quad (8.18)$$

making use of eqns. (8.6), (8.12), and (8.15). We note that in the limit of vanishing penetration of magnetic flux, $s_1, s_2 \to 0$, when the strip reverts to the flux-free Meissner state, eqn. (8.18) yields the result $U_{\rm ac} = 0$, as it should.

It is interesting to note that although characteristics such as the sheet current J, the total transport current I, and the hysteretic AC loss $U_{\rm ac}$ still hinge upon the free constant α when exploiting magnetostatic–electrostatic analogues, all quantities of interest here are uniquely defined through this technique: the ratios $J/J_{\rm b}$ and $I/I_{\rm b}$ for the strip to exhibit the Meissner state, with the maximum total transport current in the case of an isolated strip $I_{\rm b} \simeq \pi(2wd)^{1/2} J_{\rm b}$ (Genenko, Usoskin, and Freyhardt 1999) or the ratios $J/J_{\rm c}$ and $I/I_{\rm c}$ for the strip to reveal the critical state, with the critical total transport current in the case of an isolated strip $I_{\rm c} = 2wJ_{\rm c}$ (Norris 1970) as well as the ratio $U_{\rm ac}/U_0$ for the strip to unfold the latter state.

8.1.5 Practical implementation

We now refer to paramagnetic or dielectric configurations of complete, twofold mirror symmetry.

8.1.5.1 Analytic frame

For such configurations, the permeabilities of the paramagnets $\mu_1, \mu_2 \to \mu > \mu_0$ and, in the sense of proper reinterpretation, therefore the permittivities of the corresponding dielectrics $\varepsilon_1, \varepsilon_2 \to \varepsilon < \varepsilon_0$ coincide; likewise, the depths of penetration of magnetic flux into the superconductor strip and the widths of the marginal, dielectric parts of the composite strip agree: $s_1, s_2 \to s \geq 0$. The hysteretic AC loss per unit length of the strip, from the second part of eqn. (8.18), then simply reads as

$$U_{\rm ac} = 8\alpha^2 \varepsilon_0 \mu_0 \sigma_{\rm c} \lim_{d\to 0} \int_{-w}^{-w+s} dx \int_{x}^{-w+s} dx'\, E_{\rm t}^{(0)}(x'). \quad (8.19)$$

If, for given paramagnet shapes, the width of the magnetically shielded strip $2w$ and its thickness d are fixed, the distance between the edges of the strip and the paramagnets remains as the only variable geometrical quantity. On the other hand, when letting the permeability of the paramagnets $\mu \to \mu_0$ and, hence, the permittivity of the dielectrics $\varepsilon \to \varepsilon_0$, to mimic a vacuum environment, the correspondence between the depth of penetration of magnetic flux and the total transport current becomes

$$s = w\left[1 - \sqrt{1 - I^2/I_c^2}\right], \tag{8.20}$$

whereas the scaled form for the electromagnetic energy loss during a steady-state cycle of the AC transport current, per unit length of the isolated strip,

$$U_{ac} = U_0 f\left(I/I_c\right) \quad \text{with} \quad f(i) = (1-i)\ln(1-i) + (1+i)\ln(1+i) - i^2, \tag{8.21}$$

obtains, duplicating a result established in section 4.1.2 by different means, which exposes the critical transport current I_c in $U_0 = 2\mu_0 I_c^2/\pi$ (Brandt and Indenbom 1993)

8.1.5.2 Computational scheme

Finite-element simulations with the help of ANSYS or COMSOL and magnetostatic–electrostatic analogues, for which the substitution $\mu/\mu_0 \to \varepsilon_0/\varepsilon$ applies, run as follows. The first step includes a geometrical definition of the domains occupied by the metal, or composite, strip and by the dielectrics making up the entire heterostructure; this is complemented by assigning a chosen value of the permittivity to the domains occupied by the dielectrics and of the due value of the permittivity to the domain left as a vacuum. The domain envisaged for numerical analysis—inwardly delimited by the domain of the strip—is outwardly delineated by a quadratic frame much larger than the domain covered by the heterostructure. The second step involves generating an adapted seed of mesh points of a triangular grid, fostered by the flexibility which the software at hand allows in putting computational elements wherever they seem fit. Thus, a substantially refined mesh size m near the surface and the edges of the strip, i.e. in regions where pronounced field inhomogeneities are bound to occur, recommends itself; a practically still manageable lower limit is $m/w = 0.001$ in conjunction with $d = 2m$. The third step concerns the algorithmic solution of eqn. (8.10) governing the scalar electric potential φ, from which, via eqn. (8.9), the static electric field $\mathbf{E}$ is derived. This assumes that the boundary conditions, eqn. (8.7), and the far-field condition apply. Satisfaction of the latter, asymptotic requirement for a quadratic frame of large, but *finite* extent is approximately ensured by invoking a continuously distributed surface charge on its outer boundary, balancing the total charge per unit length Q carried by the strip. A convenient realization, in view of the gauge invariance of eqn. (8.10), consists of setting the scalar electric potential $\varphi = 0$ on this boundary.

For the metal strip, a nonzero electric potential value $\varphi = \varphi_0$ is assigned to the whole surface of the strip and adjusted such that the peaks of the ensuing distribution of the surface charge density σ at both edges of the strip, $x = \pm w$, match a pre-set surface charge density value σ_b. In this way, emulation of the edge-barrier-protected, flux-free Meissner state carrying the maximum transport current possible is secured. For the composite strip, a surface charge density value σ_c is prescribed on the marginal, dielectric parts of the strip of width s; the latter is supplied, rather than self-consistently fixed, to facilitate practical computations, implying that the total charge per unit length Q carried by the strip follows as a deduced quantity. A nonzero electric potential value $\varphi = \varphi_0$ assigned to the central, metallic part of the strip is adjusted such that the peaks of the resulting distribution of the surface charge density σ at the interfaces between the metallic part and the dielectric parts, $x = -w + s$ and $x = w - s$, match the surface charge density value σ_c. Again, in this way, emulation of the partially flux-filled critical state carrying the maximum transport current possible is ensured.

Naturally, all finite-element simulations are subject to tests for convergence of the predicted results with enlargements of the quadratic frame underlying the numerical analysis and refinements of the size of the meshes setting up the triangular grid; they enjoy a reduction to a quarter of the true numerical expenditure by taking the mirror symmetry of the respective problems into account.

8.2 Superconductor strip inside a paramagnetic cavity with an imposed transport current

We now investigate the magnetic shielding effect on transport current flow and possible hysteretic AC loss in the superconductor strip with the help of the ANSYS finite-element program, referring to both the flux-free Meissner state and the partially flux-filled critical state, for two essentially different paramagnetic environments illustrated in Fig. 8.2: a horseshoe-shaped paramagnet configuration and a wedge-shaped paramagnet configuration representing examples of concave and convex, respectively, open magnetic cavities. The choice of their geometrical design is motivated by the findings of sections 3.4, 3.7, and 4.2 that open paramagnetic environments yield distinct redistributions of the sheet current towards the centre of the strip (associated with significant enhancements of the total loss-free current

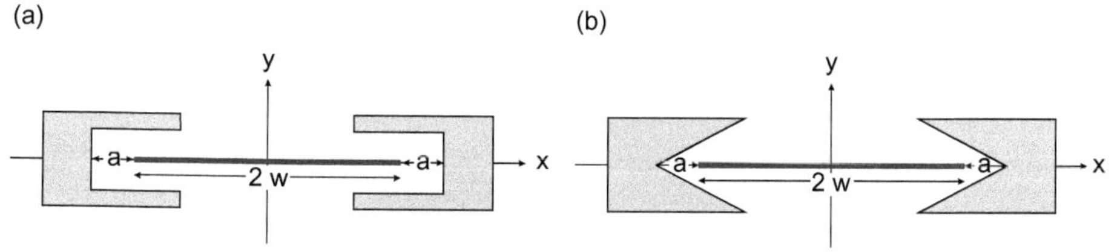

Fig. 8.2 Cross-sectional view of a superconductor strip (dark shading) located between two paramagnets (light shading) extending infinitely in the z-direction of a Cartesian coordinate system x, y, z for (a) a horseshoe-shaped paramagnet configuration representing a concave open cavity and (b) a wedge-shaped paramagnet configuration representing a convex open cavity. The definition of the distance between the edges of the strip and the paramagnets a is marked. Reproduced with permission from Genenko, Rauh, Krüger, and Narayanan (2009), *Supercond. Sci. Technol.* 22, pg 055001 © IOP Publishing. All rights reserved. DOI: 10.1088/0953-2048/22/5/055001.

that can be carried by the strip) and by the expectation of particularly strong shielding effects. In the finite-element simulations, the distance between the edges of the strip and the paramagnets a, the magnetic permeability of the shields μ, and the depth of penetration of magnetic flux into the superconductor strip s are regarded as constitutive parameters. The following report adheres to the work of Genenko, Rauh, Krüger, and Narayanan (2009).

8.2.1 Horseshoe-shaped paramagnet configuration: Meissner state

The influence of the paramagnetic environment shows up in the static magnetic field associated with the particular superconductor/paramagnet heterostructure as well as in the sheet current and the maximum loss-free current that can be carried by the superconductor strip. Both the geometrical and the magnetic characteristics of the shields play crucial roles thereby.

8.2.1.1 Field distribution

Figure 8.3 presents an illustrative example of the distribution of the static magnetic field **H** around a magnetically shielded superconductor strip in the flux-free Meissner state, for the horseshoe-shaped paramagnet configuration of Fig. 8.2(a), with the normalized distance between the edges of the strip and the paramagnets $a/w = 0.2$ and the relative magnetic permeability of the shields $\mu/\mu_0 = 1000$. Whereas the

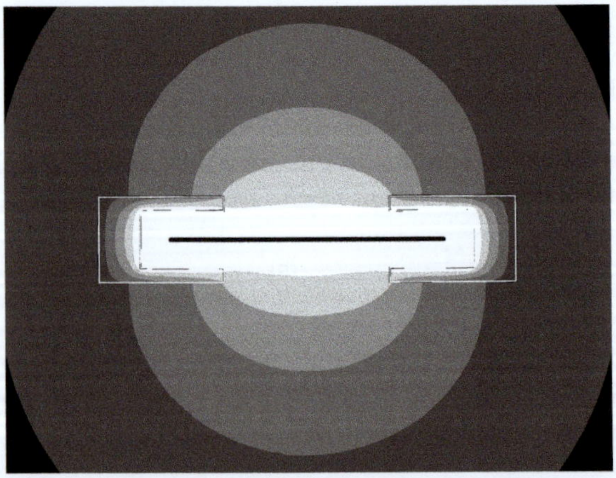

Fig. 8.3 Cross-sectional view of the distribution of the static magnetic field **H** around a magnetically shielded superconductor strip (dark shading) for the relative magnetic permeability of the shields $\mu/\mu_0 = 1000$, referring to the configuration of Fig. 8.2(a). The strength of the magnetic field is characterized stepwise by regions of assorted depths of tone, whose boundaries define lines of the magnetic field; the paramagnets are indicated by white contour double lines. Reproduced with permission from Genenko, Rauh, Krüger, and Narayanan (2009), *Supercond. Sci. Technol.* 22, pg 055001 © IOP Publishing. All rights reserved. DOI: 10.1088/0953-2048/22/5/055001.

lines of the magnetic field at either side of the strip take on virtually circular shapes far away from the heterostructure, like for an isolated strip, they show refraction at the almost orthogonally intersected surfaces of the paramagnets and enhanced concentrations inside. The diversion of the lines of the magnetic field away from the edges of the strip (not graphically resolved) in this concave open paramagnetic cavity comes about by a compromise between paramagnetic plate elements parallel to the superconductor strip, tending to reduce the maximum total current in the strip, and paramagnetic plate elements perpendicular to the superconductor strip, tending to increase the maximum total current in the strip. The present analysis makes it possible to examine the interplay of both effects.

8.2.1.2 Current distribution

Addressing this paramagnet configuration, with the relative magnetic permeability of the shields $\mu/\mu_0 = 1000$, Fig. 8.4(a) manifests the variation in the normalized sheet current J/J_b over the width of the strip for a range of values of the normalized distance between the edges of the strip and the paramagnets a/w. A general trait due to sheet current redistribution in the strip is the appearance of a central, elevated plateau, whose height abates, whilst its width extends, for increasing values of a/w, until it ceases to exist. Figure 8.4(b) presents the variation of the corresponding normalized maximum total transport current I_{max}/I_b with the normalized

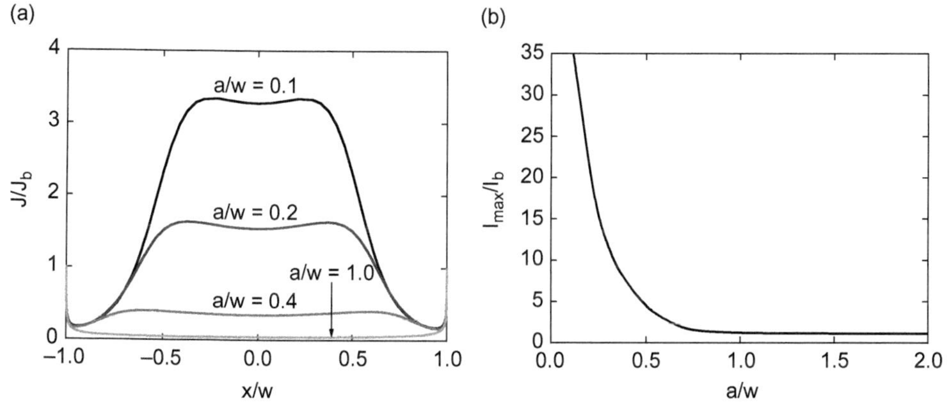

Fig. 8.4 (a) Distribution of the normalized sheet current J/J_b as a function of the normalized coordinate x/w over the width of the magnetically shielded superconductor strip in the flux-free Meissner state for four different values of the normalized distance between the edges of the strip and the paramagnets a/w identified on the curves; (b) variation of the corresponding normalized maximum total transport current I_{max}/I_b with the normalized distance between the edges of the strip and the paramagnets a/w, when the relative magnetic permeability of the shields $\mu/\mu_0 = 1000$, referring to the horseshoe-shaped paramagnet configuration of Fig. 8.2(a). Reproduced with permission from Genenko, Rauh, Krüger, and Narayanan (2009), *Supercond. Sci. Technol.* 22, pg 055001 © IOP Publishing. All rights reserved. DOI: 10.1088/0953-2048/22/5/055001.

distance between the edges of the strip and the paramagnets a/w. Starting from a distinct extreme at the smallest value of a/w, there is a monotonic drop followed by an approach towards unity, the value prevailing for an isolated strip, as a/w continues to grow. For this paramagnet configuration, with the normalized distance between the edges of the strip and the paramagnets $a/w = 0.2$, Fig. 8.5(a) presents the variation of the normalized sheet current J/J_b over the width of the strip for a range of values of the relative magnetic permeability of the shields μ/μ_0. As before, a general trait due to sheet current redistribution in the strip is the appearance of a central, elevated plateau, whose height abates, whilst its width extends, though with decreasing values of μ/μ_0, until it ceases to exist. Figure 8.5(b) shows the variation in the corresponding normalized maximum total transport current I_{max}/I_b with the relative magnetic permeability of the shields μ/μ_0. There is a monotonic rise starting from unity, the value appropriate to an isolated strip adopted for the vacuum value of μ/μ_0, followed by a transition to saturation at large values of μ/μ_0.

In summary, the results for the horseshoe-shaped paramagnet configuration of Fig. 8.2(a) establish the general tendency of the normalized maximum total transport current I_{max}/I_b to rise above unity when the normalized distance between the edges of the strip and the paramagnets a/w shrinks or when the relative magnetic permeability of the shields μ/μ_0 grows. We note explicitly the enhanced ratio $I_{max}/I_b \simeq 20.8$ attained at $a/w = 0.2$ for $\mu/\mu_0 = 1000$, according to Figs. 8.4(b) and 8.5(b).

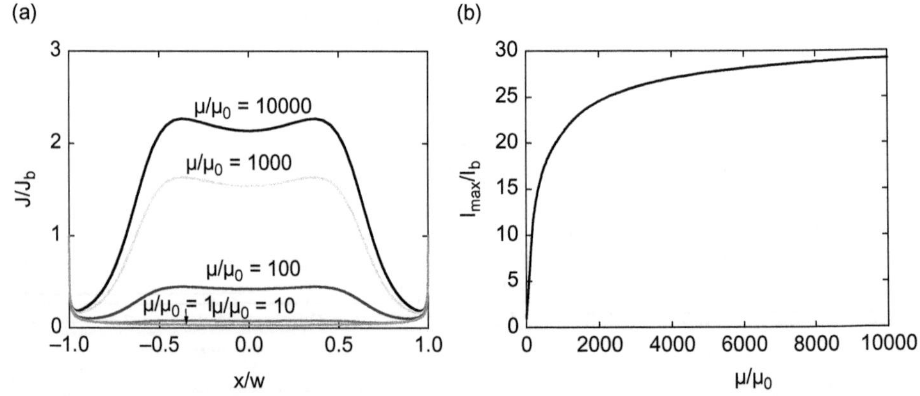

Fig. 8.5 (a) Distribution of the normalized sheet current J/J_b as a function of the normalized coordinate x/w over the width of the magnetically shielded superconductor strip in the flux-free Meissner state for five different values of the relative magnetic permeability of the shields μ/μ_0 identified on the curves; (b) variation of the corresponding normalized maximum total transport current I_{max}/I_b with the relative magnetic permeability of the shields μ/μ_0, when the normalized distance between the edges of the strip and the paramagnets $a/w = 0.2$, referring to the horseshoe-shaped paramagnet configuration of Fig. 8.2(a). Reproduced with permission from Genenko, Rauh, Krüger, and Narayanan (2009), *Supercond. Sci. Technol.* 22, pg 055001 © IOP Publishing. All rights reserved. DOI: 10.1088/0953-2048/22/5/055001.

8.2.2 Wedge-shaped paramagnet configuration: Meissner state

Again, the influence of the paramagnetic environment shows up in the static magnetic field associated with the particular superconductor/paramagnet heterostructure as well as in the sheet current and the maximum loss-free current that can be carried by the superconductor strip. Both the geometrical and the magnetic characteristics of the shields play crucial roles thereby.

8.2.2.1 Field distribution

Figure 8.6 presents an illustrative example of the distribution of the static magnetic field **H** around the magnetically shielded superconductor strip in the flux-free Meissner state, for the wedge-shaped paramagnet configuration of Fig. 8.2(b), with the normalized distance between the edges of the strip and the paramagnets $a/w = 0.2$ and the relative magnetic permeability of the shields $\mu/\mu_0 = 1000$. Again, whereas the lines of constant magnetic field at either side of the strip take on virtually circular shapes far away from the heterostructure, like for an isolated strip, they show refraction at the almost orthogonally intersected surfaces of the paramagnets and enhanced concentrations inside. The diversion of the lines of the magnetic field away from the edges of the strip (not graphically resolved) comes about by convex open paramagnetic cavity elements inclined with respect to the superconductor strip,

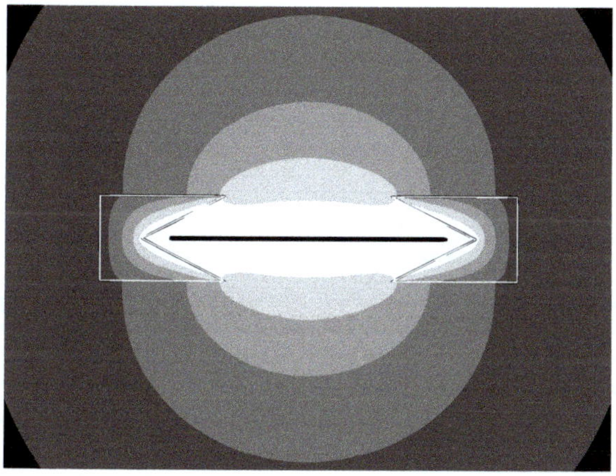

Fig. 8.6 Cross-sectional view of the distribution of the static magnetic field **H** around the magnetically shielded superconductor strip (dark shading) for the relative magnetic permeability of the shields $\mu/\mu_0 = 1000$, referring to the configuration of Fig. 8.2(b). The strength of the magnetic field is characterized stepwise by regions of assorted depths of tone, whose boundaries define lines of the magnetic field; the paramagnets are indicated by white contour double lines. Reproduced with permission from Genenko, Rauh, Krüger, and Narayanan (2009), *Supercond. Sci. Technol.* 22, pg 055001 © IOP Publishing. All rights reserved. DOI: 10.1088/0953-2048/22/5/055001.

which favour an increase of the maximum total current in the strip. The present analysis makes it possible to quantify the significance of the latter effect.

8.2.2.2 Current distribution

Examining this paramagnet configuration, with the relative magnetic permeability of the shields $\mu/\mu_0 = 1000$, Fig. 8.7(a) unfolds the variation in the normalized sheet current J/J_b over the width of the strip for a range of values of the normalized distance between the edges of the strip and the paramagnets a/w. A general trait due to sheet current redistribution in the strip now is the appearance of a central, substantially elevated plateau, whose height abates, whilst its width extends, for increasing values of a/w, until it ceases to exist. Figure 8.7(b) presents the variation in the corresponding normalized maximum total transport current I_{max}/I_b with the normalized distance between the edges of the strip and the paramagnets a/w. Starting from a pronounced extreme at the smallest value of a/w, there is a sharp drop followed by a gradual approach to unity, the value prevailing for an isolated strip, as a/w continues to grow. For this paramagnet configuration, with the normalized distance between the edges of the strip and the paramagnets $a/w = 0.2$, Fig. 8.8(a) presents the variation in the normalized sheet current J/J_b over the width of the strip for a range of values of the relative magnetic permeability of the

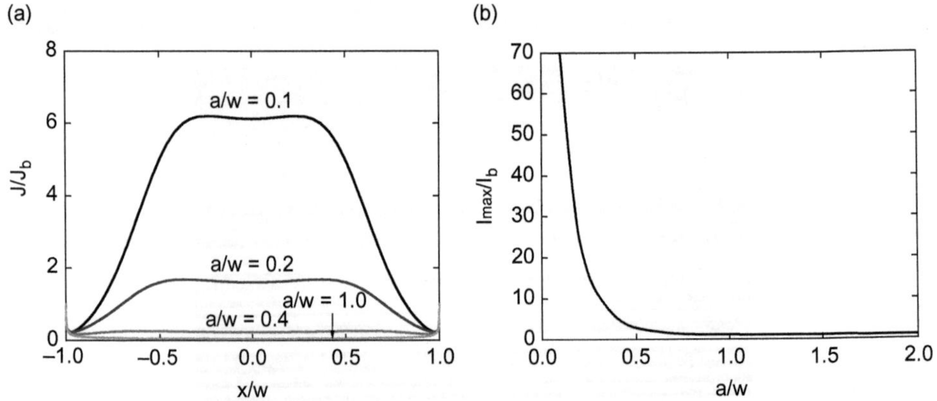

Fig. 8.7 (a) Distribution of the normalized sheet current J/J_b as a function of the normalized coordinate x/w over the width of the magnetically shielded superconductor strip in the flux-free Meissner state for four different values of the normalized distance between the edges of the strip and the paramagnets a/w identified on the curves; (b) variation of the corresponding normalized maximum total transport current I_{max}/I_b with the normalized distance between the edges of the strip and the paramagnets a/w, when the relative magnetic permeability of the shields $\mu/\mu_0 = 1000$, referring to the wedge-shaped paramagnet configuration of Fig. 8.2(b). Reproduced with permission from Genenko, Rauh, Krüger, and Narayanan (2009), *Supercond. Sci. Technol.* 22, pg 055001 DOI: 10.1088/0953-2048/22/5/055001.

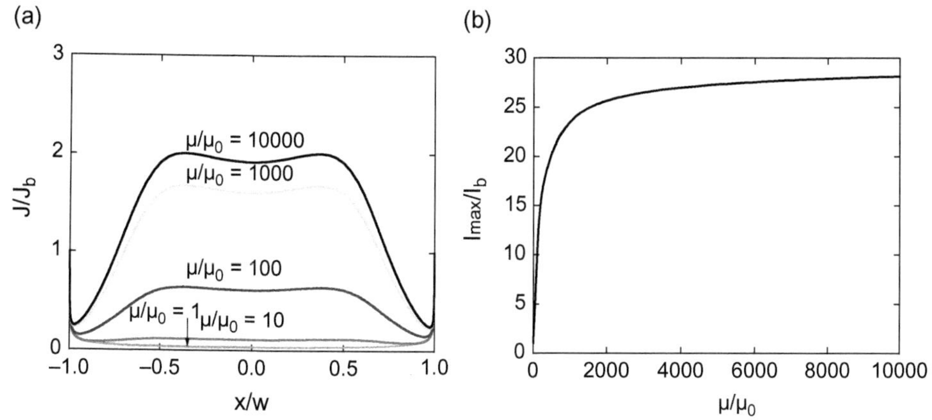

Fig. 8.8 (a) Distribution of the normalized sheet current J/J_{b} as a function of the normalized coordinate x/w over the width of the magnetically shielded superconductor strip in the flux-free Meissner state for five different values of the relative magnetic permeability of the shields μ/μ_0 identified on the curves; (b) variation in the corresponding normalized maximum total transport current $I_{\max}/I_{\mathrm{b}}$ with the relative magnetic permeability of the shields μ/μ_0, when the normalized distance between the edges of the strip and the paramagnets $a/w = 0.2$, referring to the wedge-shaped paramagnet configuration of Fig. 8.2(b). Reproduced with permission from Genenko, Rauh, Krüger, and Narayanan (2009), *Supercond. Sci. Technol.* 22, pg 055001 © IOP Publishing. All rights reserved. DOI: 10.1088/0953-2048/22/5/055001.

shields μ/μ_0. As before, a general trait due to sheet current redistribution in the strip is the appearance of a central, elevated plateau, whose height abates, whilst its width extends, though with decreasing values of μ/μ_0, until it ceases to exist. Figure 8.8(b) shows the variation in the corresponding normalized maximum total transport current $I_{\max}/I_{\mathrm{b}}$ with the relative magnetic permeability of the shields μ/μ_0. Starting from unity, the value appropriate to an isolated strip adopted for the vacuum value of μ/μ_0, there is a sharp monotonic rise followed by a transition to saturation at already moderate values of μ/μ_0.

In summary, the results for the wedge-shaped paramagnet configuration of Fig. 8.2(b) establish the general tendency of the normalized maximum total transport current $I_{\max}/I_{\mathrm{b}}$ to rise above unity when the normalized distance between the edges of the strip and the paramagnets a/w shrinks or when the relative magnetic permeability of the shields μ/μ_0 grows. We note explicitly the enhanced ratio $I_{\max}/I_{\mathrm{b}} \simeq 23.5$ attained at $a/w = 0.2$ for $\mu/\mu_0 = 1000$, according to Figs. 8.7(b) and 8.8(b).

8.2.3 Wedge-shaped paramagnet configuration: overcritical states

Having realized the fact that the horseshoe-shaped paramagnet configuration of Fig. 8.2(a) is outperformed by the wedge-shaped paramagnet configuration of

Fig. 8.2(b), as far as the enhancement of the maximum total transport current in the flux-free Meissner state is concerned, we confine ourselves to the latter configuration for studying the partially flux-filled critical state. Since physically stable overcritical current transport can only go along with small depths of penetration of magnetic flux into the superconductor strip in the whole range of the constitutive parameters discussed, the distribution of the static magnetic field around the magnetically shielded superconductor strip in the partially flux-filled critical state does not differ markedly from the corresponding distribution in the flux-free Meissner state, for which reason Fig. 8.6 is representative here too.

8.2.3.1 Current distribution

Examining the wedge-shaped magnet configuration of Fig. 8.2(b), with the normalized distance between the edges of the strip and the paramagnets $a/w = 0.2$, Fig. 8.9 displays the variation in the normalized sheet current J/J_c over the width of the strip for a range of values of the normalized flux penetration depth s/w, when the relative magnetic permeability of the shields μ/μ_0 is fixed. For the lowest value of the latter quantity, $\mu/\mu_0 = 10$, Fig. 8.9(a) illustrates the emergence of a central, subcritical plateau with $J/J_c < 1$ in the physically stable sheet current distribution, once magnetic flux penetrates the marginal parts of the strip, where $J/J_c = 1$; the height of this plateau grows, whilst its width persists, for increasing values of s/w, until the marginally stable, overcritical state associated with the largest value of s/w at hand is reached. In the latter state, the flux-filled marginal parts of the strip border on a zone of overcritical sheet current with $J/J_c > 1$ for the first time, which renders the whole sheet current distribution unstable. The fronts of the magnetic flux then move towards the centre of the strip, and the partially flux-filled overcritical state collapses into the fully flux-filled critical state with $J/J_c = 1$ prevailing all over the strip. For the intermediate value of the relative magnetic permeability of the shields, $\mu/\mu_0 = 100$, Fig. 8.9(b) illustrates the appearance of a central, elevated overcritical plateau with $J/J_c > 1$ in the physically stable sheet current distribution, once magnetic flux penetrates the marginal parts of the strip, where $J/J_c = 1$. Stability here is ensured by transitional zones of subcritical sheet current with $J/J_c < 1$ joining the overcritical plateau to the flux-filled marginal parts of the strip. The height of the central plateau grows, whilst its width persists and the transitional zones shrink, when s/w augments, until the marginally stable, overcritical state associated with the evanescence of these zones for the largest value of s/w at hand is reached. For the highest value of the relative magnetic permeability of the shields, $\mu/\mu_0 = 1000$, Fig. 8.9(c) reveals sheet current distributions to which the observations reported above *a fortiori* apply: partially flux-filled, strongly overcritical states of the strip characterized by the appearance of a central, substantially elevated overcritical plateau together with reduced transitional zones of subcritical sheet current in the physically stable sheet current distribution, once magnetic flux penetrates the marginal parts of the strip. Again, the height of this plateau grows, whilst its width persists and the transitional zones

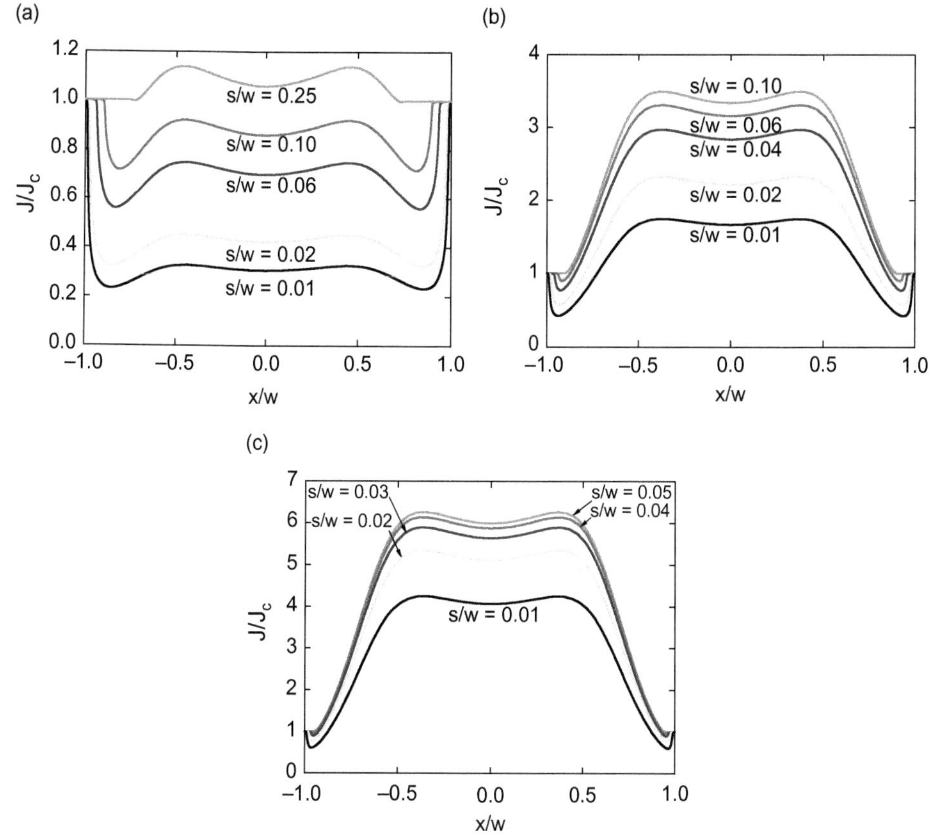

Fig. 8.9 Distribution of the normalized sheet current J/J_c as a function of the normalized coordinate x/w over the width of the magnetically shielded superconductor strip in the partially flux-filled critical state for five different values of the normalized flux penetration depth s/w identified on the curves, when the normalized distance between the edges of the strip and the paramagnets $a/w = 0.2$ and (a) the relative magnetic permeability of the shields $\mu/\mu_0 = 10$, (b) $\mu/\mu_0 = 100$, and (c) $\mu/\mu_0 = 1000$, referring to the wedge-shaped paramagnet configuration of Fig. 8.2(b). Reproduced with permission from Genenko, Rauh, Krüger, and Narayanan (2009), *Supercond. Sci. Technol.* 22, pg 055001 © IOP Publishing. All rights reserved. DOI: 10.1088/0953-2048/22/5/055001.

shrink, when s/w augments, until the marginally stable, overcritical state associated with the evanescence of these zones for the largest value of s/w at hand is reached.

8.2.3.2 Hysteretic AC loss

To demonstrate the implications of the behaviour of the sheet current for observables of practical interest such as the current-carrying capability and the quality of AC current transport of the magnetically shielded superconductor strip, we once more address the normalized distance between the edges of the strip and the paramagnets $a/w = 0.2$. Figure 8.10(a) displays the variation in the normalized flux penetration depth s/w with the normalized total transport current I/I_c for a range of values of the relative magnetic permeability of the shields μ/μ_0. At each value of μ/μ_0, there is a monotonic rise starting from zero in the physically stable regime of the partially flux-filled critical state up to the respective limit of stability, beyond which the collapse of this state into the fully flux-filled critical state with $I/I_c = 1$ ensues. It remains to add that at the vacuum value of μ/μ_0 in particular, these results meet the theoretical prediction of eqn. (8.20) appropriate to an isolated strip. The enhancement of the normalized maximum total transport current strongly builds up with increasing values of the relative magnetic permeability of the shields μ/μ_0. We note explicitly the maximum enhanced ratio $I_{\max}/I_c \simeq 4.7$ attained at $a/w = 0.2$ for $\mu/\mu_0 = 1000$. Figure 8.10(b) shows the variation in the corresponding normalized hysteretic AC loss suffered by the strip U_{ac}/U_0 with the normalized total transport current I/I_c, for the same range of values of the relative magnetic permeability of the shields μ/μ_0. Like before, at each value of μ/μ_0, there

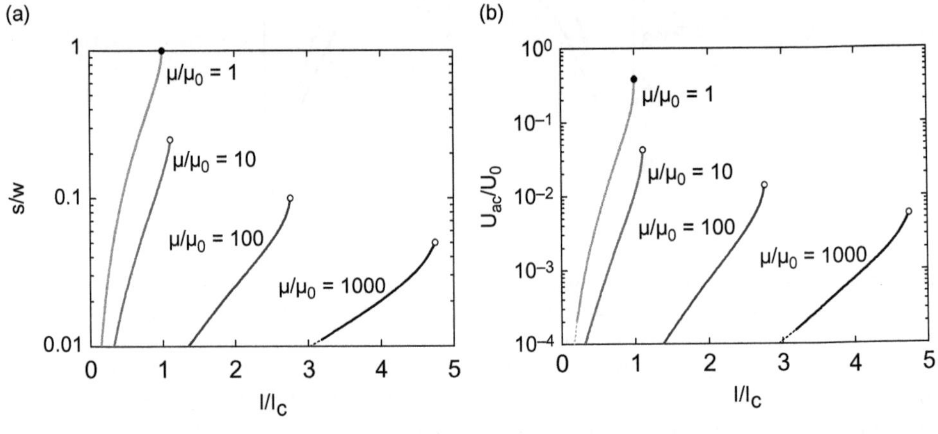

Fig. 8.10 (a) Dependence of the normalized flux penetration depth s/w on the normalized total transport current I/I_c for the magnetically shielded superconductor strip in the partially flux-filled critical state and (b) variation of the corresponding normalized hysteretic AC loss per cycle suffered by the strip U_{ac}/U_0 with the normalized total transport current I/I_c, when the normalized distance between the edges of the strip and the paramagnets $a/w = 0.2$, for four different values of the relative magnetic permeability of the shields μ/μ_0 identified on the curves, referring to the wedge-shaped paramagnet configuration of Fig. 8.2(b). The empty circles terminating some of the curves separate physically stable and unstable regimes of the partially flux-filled critical state, whereas the solid circles terminating the residual curves denote the onset of the full critical state; the dashed lines represent extrapolations into computationally inaccessible regimes and merely serve as guides to the eye. Reproduced with permission from Genenko, Rauh, Krüger, and Narayanan (2009), *Supercond. Sci. Technol.* 22, pg 055001 © IOP Publishing. All rights reserved. DOI: 10.1088/0953-2048/22/5/055001.

is a monotonic rise in the physically stable regime of the partially flux-filled critical state up to the respective limit of stability, beyond which the collapse of this state into the fully flux-filled critical state with $U_{\text{ac}}/U_0 = 2\ln 2 - 1$ ensues. At the vacuum value of μ/μ_0 in particular, these results meet the theoretical prediction of eqn. (8.21) appropriate to an isolated strip. The normalized hysteretic AC loss dramatically wanes with increasing values of the relative magnetic permeability of the shields μ/μ_0, even for normalized total transport currents in the overcritical regime, a fact owed to the reduction of the depth of penetration of magnetic flux into the superconductor strip in conjunction with a weakening of the strength of the associated magnetic field.

8.2.4 Discussion and conclusion

Numerical simulations of the flux-free Meissner state and the partially flux-filled critical state of a thin, current-carrying type-II superconductor strip surrounded by paramagnetic shields of restricted geometry and finite permeability have been performed by exploiting magnetostatic–electrostatic analogues. For a horseshoe-shaped paramagnetic environment representing a concave open cavity and a wedge-shaped paramagnetic environment representing a convex open cavity, the distribution of the static magnetic field around the superconductor strip, the distribution of the sheet current in the strip, and the maximum total transport current carried by the strip have been established in the flux-free Meissner state on the premise of strip protection against entry of magnetic flux owing to an edge barrier typified by the sheet current value J_{b}. Constitutive parameters used were the normalized distance between the edges of the strip and the paramagnets a/w and the relative permeability of the shields μ/μ_0. The results demonstrate substantial redistributions of the sheet current towards the centre of the strip associated with significant enhancements of the total loss-free current that can be carried by the strip for both the concave and the convex paramagnetic cavity, with a bias in favour of the latter type of environment, however. The effect of sheet current redistribution, which shows up in the maximum total transport current carried by the strip, is highly sensitive to variations of a/w and exhibits stabilization at large values of a/w or μ/μ_0, bearing resemblance to the predictions of numerical studies in sections 6.1 and 6.5 addressing a thin superconductor strip surrounded by semi-infinite paramagnetic environments of finite permeability.

For the wedge-shaped paramagnetic environment, the distribution of the sheet current in the strip, the total transport current carried by the strip, and, within a quasistatic approach, hysteretic AC losses suffered by the strip have been revealed in the partially flux-filled critical state, using the normalized distance between the edges of the strip and the paramagnets a/w, the relative permeability of the shields μ/μ_0, and the normalized flux penetration depth s/w as constitutive parameters. Like in the flux-free Meissner state, when a/w is fixed, substantial redistributions of the sheet current towards the centre of the strip associated with significant enhancements of the total loss-free current that can be carried by the strip occur here too, again bearing resemblance to the predictions of numerical studies in section 6.1 for a thin superconductor strip adjacent to a semi-infinite paramagnetic environment of finite permeability. This effect gets more pronounced with increasing values of μ/μ_0, while the range of permissible values of s/w for physically stable current transport narrows down and the maximum total current possible builds up at the same time. On the other hand, the hysteretic AC loss suffered by the strip drastically wanes when μ/μ_0 augments.

8.2.4.1 Contingencies for overcritical states

In the nondissipative overcritical states with the highest values of the total current, zones of merely infinitesimal extent of the marginal parts of the strip show penetration of magnetic flux, while the major part of the strip remains untouched. This obviously marks a situation where penetration of magnetic flux is inhibited by pinning magnetic vortices at internal defects, so that the (field-independent) critical sheet current value J_c takes over the role of the characteristic barrier sheet current identified for the Meissner state. If, by contrast, J_c were dependent on the local magnetic field, the profiles of the magnetic flux in the flux-filled marginal parts of the strip might change (Campbell and Evetts 1972). Such a dependence would not, however, preclude the existence of a wide, central flux-free zone as emerging from Bean's model of the critical state (Norris 1970; Brandt and Indenbom 1993; Zeldov et al. 1994), a prerequisite crucial to the appearance of nondissipative overcritical states, which has been detected empirically by magneto-optical means, though only in the case of an applied magnetic field (Johansen et al. 1996; Jooss et al. 1996). It is for this reason that non-dissipative overcritical states should also be observable experimentally when passing a transport current through a magnetically shielded superconductor strip in its zero-field cooled virgin state, where the total current is increased below its possible maximum. A procedure of this sort might bring about substantially enhanced loss-free currents, as compared to conventionally attainable critical currents, and significantly reduced hysteretic AC losses, as compared to those suffered by an unshielded strip.

The appearance of high values of the sheet current in the central, flux-free part of the magnetically shielded superconductor strip, for nondissipative overcritical states, may enable horizontal magnetic vortices to penetrate from the surface of the strip. If the latter is randomly rough, rather than ideally smooth, as the theoretical model implies, the Bean–Livingston barrier against entry of magnetic flux proves insignificant, and magnetic flux penetrates when the lower critical field H_{c1} is reached on the surface of the strip, whatever the microscopic details of its geometrical structure may be.

Numerical estimates. For an $YBa_2Cu_3O_7$-coated tape (Neumüller et al. 1997; Usoskin et al. 2005), the lower critical field amounts to $H_{c1} = 3.3 \times 10^4$ A/m at the liquid nitrogen temperature $T = 77$ K. Since, from eqn. (8.6) and symmetry with respect to the plane of the strip implied, the sheet current J is related to the tangential component of the magnetic field $H_t^{(0)}$ on the surface of the strip by $J = 2H_t^{(0)}$, the value of the characteristic sheet current for entry of magnetic flux $J_e = 6.6 \times 10^4$ A/m obtains. If this tape is of the highest available quality, with representative thickness $d_e = 1.0\,\mu m$ and critical current density $j_c = 1.0 \times 10^{10}\,A/m^2$ at $T = 77$ K (Neumüller et al. 1997; Usoskin et al. 2005), the sheet current in the critical state equals $J = 1.0 \times 10^4$ A/m and therefore remains far below J_e. If, on the other hand, the sheet current J is substantially enhanced due to magnetic shielding of the $YBa_2Cu_3O_7$- coated tape, it may in fact locally exceed the limiting value J_e. Such circumstances necessitate the use of superconductor coatings with a smaller thickness d_e to ensure that J stays below J_e throughout the central, flux-free part of the coating of the tape. Assuming, e.g., an enhancement effect of up to the factor of 10, the condition on the thickness $d_e \leq H_{c1}/5j_c$ must be satisfied for the central part of the coating of the tape so as to guarantee protection against entry of horizontal magnetic vortices. This constraint yields the upper bound for the thickness of the superconductor coating $d_e = 0.7\,\mu m$, a value slightly falling short of typical values of the thickness of commercially available coated tapes (Neumüller et al. 1997; Usoskin et al. 2005).

Inherent subtleties. We note in passing that, for a thin superconducting tape and a horseshoe-shaped paramagnetic environment with oscillating transport currents imposed or magnetic fields applied, some recent experimental (Krüger et al. 2013) and computational (Grilli et al. 2014) studies failed to reveal nondissipative overcritical states. This might be due to the following reasons. The central, flux-free zone of the magnetically shielded superconducting tape, as a precondition for the appearance of overcritical states, distinguishes itself by decoupling the local current density and the normal component of the magnetic field. Thus, a transport current could flow in the presence of a zero normal component of the magnetic field. The above numerical simulations use a power-law relationship between the current density and the electric field to represent the critical state, rather than taking Bean's model as a basis for this state. Such an ansatz prevents a decoupling of the local current density and the normal component of the magnetic field from the start. A finite current, hence, will always be accompanied by a finite electric field, and therefore by a nonzero normal component of the magnetic field, a fact which obliterates any possibility of a theoretical identification of overcritical states. Experimental observations of overcritical states, on the other hand, require the highest care in preparing the zero-field cooled flux-free states before a transport current is imposed, and this may be hard to attain.

8.2.4.2 Traits in sheet current distributions

Apart from obvious similarities regarding the performance of superconductor/paramagnet heterostructures made up of a thin superconductor strip and semi-infinite or infinite paramagnetic shields, the finite-element simulations at hand, with their emphasis on restricted geometries and finite permeabilities, disclose some new, essential traits due to sheet current redistribution in the strip. Thus, the enhancement in the flux-free Meissner state of the maximum total transport current for the horseshoe-shaped paramagnetic environment, which can be distinct at short normalized distances between the edges of the strip and the paramagnets a/w, very slowly saturates with increasing values of the relative permeability of the shields,

$\mu/\mu_0 \simeq 10000$ representing a typical estimate at $a/w = 0.2$. The enhancement of the maximum total transport current for the wedge-shaped magnetic environment, which can be even more distinct at short distances a/w, still slowly saturates with increasing values of the relative permeability of the shields, $\mu/\mu_0 \simeq 5000$ representing a typical estimate at $a/w = 0.2$. This is in stark contrast to the case of infinite paramagnetic shields analysed in section 6.1, where saturation occurs for $\mu/\mu_0 \simeq 200$ no less. A reduction in hysteretic AC losses by orders of magnitude, even with overcritical transport currents involved, can take place at already moderate values of the relative magnetic permeability of the shields in the partially flux-filled critical state and for the wedge-shaped magnetic environment. Thus, although a relative magnetic permeability as high as possible is to be preferred, the value $\mu/\mu_0 \simeq 50$ representative of an iron environment at low temperature T (Pan and Dou 2004) is sufficient to anticipate a considerable shielding effect. The sensitivity of these simulations to the particular shape of the paramagnetic environment opens up new vistas of research directed at optimizing the paramagnet configuration surrounding a thin superconductor strip with respect to dimension and geometry, so as to achieve the maximum possible enhancement of the total transport current and the maximum possible reduction of the hysteretic AC loss. Even without knowing in detail the results which such endeavours will produce, it seems clear that, for a significant shielding effect to persist, the transverse dimension of the shields (normal to the plane of the strip) should be shorter than their lateral extent (parallel to the plane of the strip) which, in turn, should be much smaller than the width of the strip.

In conclusion, exploiting magnetostatic–electrostatic analogues offers the possibility of studying superconductor/paramagnet heterostructures made up of a thin superconductor strip and finite paramagnetic shields. This approach admits convenient evaluations of the static magnetic field, the distribution of the sheet current, the variation in the maximum total transport current, and hysteretic AC losses suffered by the strip.

8.3 Hysteretic AC loss in a bilayer superconductor/paramagnet heterostructure with an applied magnetic field

Second-generation high-temperature superconductor wires rely on a coated-conductor technology which, because of manufacturing needs, often employs Ni-alloy tapes as supports (Malozemoff et al. 2008). The role of these soft-magnetic constituents has been debated for a long time since magnetic shielding can, in principle, assist or inhibit the penetration of magnetic flux, and hence the dissipation of electromagnetic energy, depending on the geometrical and material characteristics of the magnetic environment (Genenko, Snezhko, and Freyhardt 2000; Yampolskii, Yampolskaya, and Rauh 2006, 2007; Aladyshkin, Silhanek, Gillijns, and Moshchalkov 2009; Genenko and Rauh 2010).

For coated conductors in bilayer or multi-layer form, there is still controversy about experimental and theoretical results (Duckworth et al. 2003; Suenaga and Li 2006; Amemiya and Nakahata 2007; Grilli, Ashworth, and Civale 2007;

Claassen and Thieme 2008; Mawatari 2008; Suenaga et al. 2008; Gömöry et al. 2010). Some experiments demonstrate a monotonic rise of hysteretic AC losses with increasing permeability of soft-magnetic supports (Claassen and Thieme 2008), while others show a contingent fall in similar events (Duckworth et al. 2003; Grilli, Ashworth, and Civale 2007). Finite-element simulations of transport currents vindicate the build-up of hysteretic AC losses and their growth with the permeability of the supports (Amemiya and Nakahata 2007; Gömöry et al. 2010). Given an oscillating applied magnetic field, hysteretic AC losses may increase for low, but decrease for moderate, amplitudes H_a of such a field with reference to the particulars of nonmagnetic supports (Gömöry et al. 2010), whereas magnetization losses seem to only weakly hinge upon magnetic supports (Amemiya and Nakahata 2007; Sanchez, Del-Valle, Navau, and Chen 2010). In keeping with numerical studies (Gömöry et al. 2010), theoretical predictions emerge from analytical investigations of a planar superconductor/paramagnet heterostructure made up of an infinitesimally thin superconductor strip and an infinitesimally thin paramagnetic support, the two constituents being described by the sheet current J, with a (field-independent) critical value J_c, and an infinite magnetic permeability, $\mu \to \infty$ (Mawatari 2008). A similar model with infinitesimally thin superconductor and paramagnetic constituents, however, with a finite permeability μ was recently elaborated (Prigozhin and Sokolovsky 2023) where the paramagnetic support was fully characterized by a finite dimensionless parameter $(\mu/\mu_0)d_m/w$ with the support thickness d_m and the bilayer structure width $2w$. Accordingly, penetration of magnetic flux into, and hysteretic AC losses suffered by, the superconductor strip should be slightly enhanced for low values of H_a, but substantially reduced otherwise, as measured against the conditions of a nonmagnetic support. Bearing the above assumption for J_c in mind, this peculiar dependence of hysteretic AC losses cannot entirely be explained by allowing for a reduction of J_c at high values of H_a, despite suggestions of this sort (Gömöry et al. 2010). It is for these reasons that we take a fresh look at the principal effects of a paramagnetic support on hysteretic AC losses in a superconductor strip by making recourse to Bean's model of the critical state and exploiting magnetostatic-electrostatic analogues, the procedure introduced in section 8.1. We thereby extend the work of Genenko, Rauh, and Krüger (2011).

8.3.1 Theoretical model

Let us define the magnetostatic problem first by considering a planar superconductor/paramagnet heterostructure of bilayer geometry: an infinitely extended type-II superconductor strip of width $2w$ and thickness d limited by the range $-w \leq x \leq w$, $-d/2 \leq y \leq d/2$ of a Cartesian coordinate system x, y, z and deposited on a paramagnetic support of respective thickness d_m, with an insulating nonmagnetic buffer layer of infinitesimal thickness in between mitigating the proximity effect (Schmidt 1997), this structure being subject to a transverse applied magnetic field of strength H_a with no imposed transport current, as illustrated in Fig. 8.11. We choose materials and dimensions that second-generation coated conductors typically display (Claassen and Thieme 2008; Malozemoff et al. 2008; Suenaga et al. 2008),

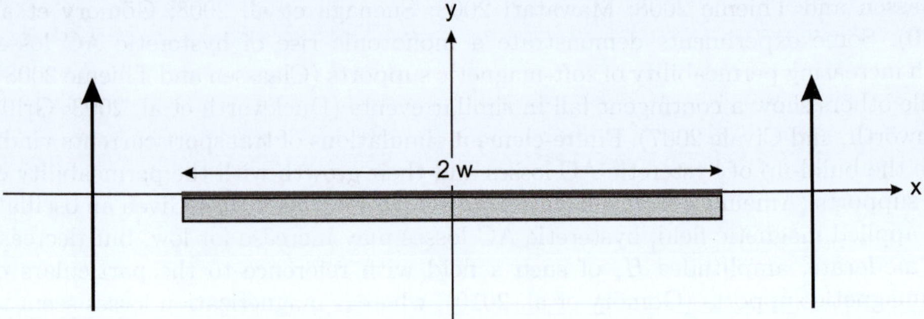

Fig. 8.11 Cross-sectional view of a superconductor strip (dark shading) deposited on a paramagnetic support (light shading) extending infinitely in the z-direction of a Cartesian coordinate system x, y, z. The direction of the applied magnetic field $\mathbf{H}_a$ (bold arrows) is marked.

viz. $YBa_2Cu_3O_7$ for the superconductor strip and Ni$-$5 at.% W for the paramagnetic support delineated by a finite permeability μ, with $w \simeq 4\,mm$, $d = 2\,\mu m$, and $d_\mathrm{m} = 100\,\mu m$, at the liquid nitrogen temperature $T = 77$ K. Since $d \ll w$, we again ignore spatial variations of the induced current on a length scale less than d and, for mathematical convenience, regard the strip as infinitesimally thin, so that its physical state can be characterized by the sheet current $J(x)$ alone. In conformity with Bean's model of the critical state duly adapted to the geometry of the strip for the chosen orientation of the applied magnetic field, magnetic flux penetrates from both edges of the strip; a flux-free region prevails in the central part of the strip, where the normal component of the magnetic field $H_\mathrm{n}^{(0)}$ disappears, whereas in the marginal, flux-penetrated parts of the strip, of depth s, the sheet current $J(x)$ assumes the (field-independent) critical value J_c.

8.3.1.1 *Magnetostatic–electrostatic analogues*

The physical state of the strip in the corresponding electrostatic problem is characterized by the surface charge density $\sigma(x)$ alone, invoking a central, metallic part of the strip at zero electrostatic potential φ, contiguous to marginal, dielectric parts of the strip, of depth s, where the surface charge density $\sigma(x)$ adopts the (field-independent) critical value σ_c; the support is represented by a dielectric with a finite permittivity ε. The heterostructure itself is subject to an electric field parallel to the strip, as eqn. (8.14) tells, with strength E_a, generated by finite potential values $\varphi = \pm\varphi_\mathrm{a}$ at opposite (left and right) sides of the quadratic computation frame by far encompassing the domain covered by a cross section of the heterostructure.

Therefore, the following analogues between the magnetostatic problem and the electrostatic problem hold: $J_\mathrm{c} = \alpha\sigma_\mathrm{c}$ with a free constant α from eqns. (8.6), (8.12), and (8.15), relating the critical sheet current to the critical surface charge density, $\mu/\mu_0 = \varepsilon_0/\varepsilon$ with the vacuum permeability μ_0 and the vacuum permittivity ε_0, linking the permeability of the paramagnetic support to the permittivity of the dielectric support, and $H_a = \alpha\varepsilon_0 E_a$ from eqn. (8.14), relating the strength of the applied magnetic field to the strength of the applied electric field. It is thus again clear that lines of the magnetic field **H** and equipotentials of the electric field **E** coincide. Hence, for an oscillating applied magnetic field with amplitude H_a, the penetration of magnetic flux and the consequential dissipation of electromagnetic energy can be determined by resorting to the quasistatic approach.

8.3.1.2 Computational base

Computations using the ANSYS finite-element program run as follows. Given ε and s for an arbitrarily chosen nonzero potential value φ_a, the critical value σ_c of the surface charge density is varied until continuous profiles of both normal components of the electric field $E_n^{(0)}$ over the surface of the strip are reached. Upon making recourse to the characteristic magnetic field $H_c = J_c/\pi$, the normalized flux penetration depth s/w then appears in terms of the ratio $H_a/H_c = \pi\varepsilon_0 E_a/\sigma_c$, independent of the constant α and the potential value φ_a. From eqn. (8.19) reinterpreted for an oscillating applied magnetic field and generalized for an asymmetric magnetic environment by the two sides of the plane of the strip, the normalized hysteretic AC loss per unit length of the heterostructure,

$$U_{ac}/H_a^2 = \left(8\mu_0\sigma_c/\varepsilon_0 E_a^2\right) \lim_{d\to 0} \int_{-w}^{-w+s} dx \int_{x}^{-w+s} dx'\, \bar{E}_t^{(0)}(x'), \tag{8.22}$$

with the average tangential component of the electric field on the surface of the strip $\bar{E}_t^{(0)}$, turns out to be independent of α and φ_a too, since the electric quantities σ_c, E_a and $\bar{E}_t^{(0)}$ all scale with the magnitude of φ_a. We comment that when letting the permeability of the paramagnet $\mu \to \mu_0$ and, hence the permittivity of the dielectric $\varepsilon \to \varepsilon_0$, to mimic a vacuum environment, the correspondence between the flux penetration depth and the amplitude of the applied magnetic field becomes

$$s = w\left[1 - \operatorname{sech}\left(H_a/H_c\right)\right], \tag{8.23}$$

whereas the scaled form for the normalized electromagnetic energy dissipated during a steady-state cycle of the applied magnetic field,

$$U_{ac}/H_a^2 = 4\pi\mu_0 w^2 f\left(H_a/H_c\right) \quad \text{with } f(h) = \left(2/h^2\right)\ln\cosh h - (1/h)\tanh h, \tag{8.24}$$

obtains, which exposes the half-width of the strip w (Brandt and Indenbom 1993).

8.3.2 Numerical results

Figure 8.12 reveals the distribution of the magnetic field **H** around the magnetically shielded superconductor strip for a fixed relative permeability of the paramagnetic support μ/μ_0. Two antagonistic effects come to the fore: on the one hand, the support tends to concentrate the magnetic field both in the centre and near the edges of the strip, facilitating penetration of magnetic flux; on the other hand, it

Fig. 8.12 Cross-sectional view of the distribution of the magnetic field **H** around the magnetically shielded superconductor strip, when $\mu/\mu_0 = 20$. The strength of the magnetic field is characterized stepwise by regions of assorted depths of tone, whose boundaries define lines of the magnetic field; the support together with the strip is indicated by black contour lines. Reproduced with permission from Genenko, Rauh, and Krüger (2011), *Appl. Phys. Lett.* 98, 152508, AIP Publishing LLC. DOI: https://doi.org/10.1063/1.3560461.

guides the magnetic field around the superconductor strip, impairing penetration of magnetic flux, as in closed geometries (cf. sections 7.1 to 7.3).

The variation in the normalized flux penetration depth s/w with the normalized amplitude of the applied magnetic field H_a/H_c, depicted in Fig. 8.13(a) for a range of values of the relative permeability of the paramagnetic support μ/μ_0, reveals ambiguous traits. Above $H_a/H_c \simeq 0.4$, penetration of magnetic flux is seen to abate monotonically with increasing μ/μ_0 from its extreme for $\mu/\mu_0 = 1$, the latter duplicating the theoretical expectation, eqn. (8.23), for the isolated superconductor strip; around $H_a/H_c = 0.4$, however, curves pertaining to different values of μ/μ_0 intersect and exchange positions below $H_a/H_c \simeq 0.4$, disclosing the tendency of the curve for $\mu/\mu_0 = 100$, the highest relative permeability taken into account, to even cross the curve for $\mu/\mu_0 = 1$ (the intercept lying outside the computationally accessible regime), in line with a similar observation under idealized conditions reported previously (Mawatari 2008). Figure 8.13(b) shows the variation in the normalized hysteretic AC loss U_{ac}/H_a^2 with the normalized amplitude of the applied magnetic field H_a/H_c for the same range of values of the relative permeability of the paramagnetic support μ/μ_0. Above $H_a/H_c \simeq 0.6$, hysteretic AC losses obviously wane monotonically with increasing μ/μ_0 from their extreme for $\mu/\mu_0 = 1$, the latter strictly matching the theoretical prediction of eqn. (8.24) for an isolated superconductor strip in the whole range of H_a/H_c; below $H_a/H_c \simeq 0.6$, however, curves for different values of μ/μ_0 cross each other and, eventually, also intersect the curve for $\mu/\mu_0 = 1$. This behaviour may be explained by the following rationale. At small values of H_a/H_c, and consequently low penetration of magnetic flux, the major contribution to U_{ac}/H_a^2 arises from narrow regions around the edges of the strip where the magnetic field is oriented mainly normal to the plane of the strip, in keeping with the Meissner effect. The paramagnetic support serves to enhance

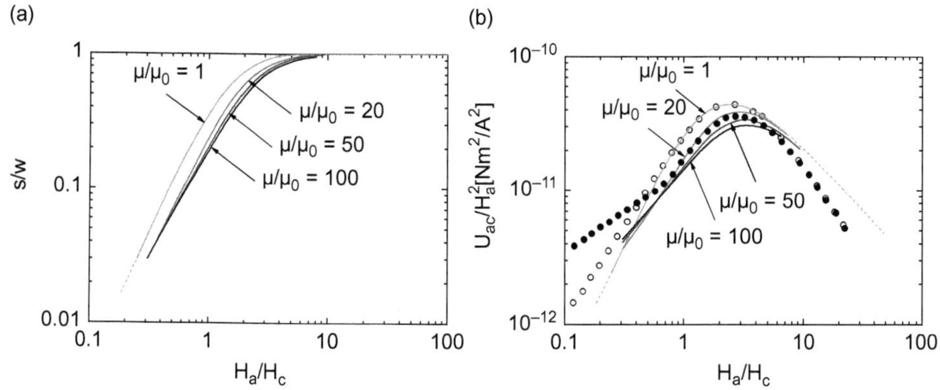

Fig. 8.13 Dependence of (a) the normalized flux penetration depth s/w and (b) the normalized hysteretic AC loss $U_{\rm ac}/H_{\rm a}^2$ on $H_{\rm a}/H_{\rm c}$, the normalized amplitude of the magnetic field applied to the magnetically shielded superconductor strip, in computationally accessible regimes, for four different values of the relative permeability of the support μ/μ_0 identified on the curves. The dashed lines represent analytical results for the isolated superconductor strip; the full and open circles bear witness to the experimental database for the magnetically shielded and isolated superconductor strip , respectively (Suenaga et al. 2008). Reproduced with permission from Genenko, Rauh, and Krüger (2011), *Appl. Phys. Lett.* 98, 152508, AIP Publishing LLC. DOI: https://doi.org/10.1063/1.3560461.

the strength of the magnetic field by attracting magnetic flux, thereby increasing the hysteretic AC loss. For large values of $H_{\rm a}/H_{\rm c}$, much deeper penetration of magnetic flux into the strip occurs, so that a substantial share of $U_{ac}/H_{\rm a}^2$ is due to the motion of magnetic flux over wide margins of the strip. The primary effect of the paramagnetic support here consists of tilting the magnetic field against the plane of the strip, reducing its normal component and thus curtailing the hysteretic AC loss.

8.3.3 Discussion and conclusion

A comparison with the presentation of measured data of the normalized hysteretic AC loss (Suenaga et al. 2008)—brought up by the factor $\pi/2$, thus forcing the maximum of $U_{ac}/H_{\rm a}^2$ in the case of an isolated superconductor strip to satisfy the theoretical test for $\mu/\mu_0 = 1$– reveals that empirical values obtained for a magnetically shielded superconductor strip, around the maximum of $U_{ac}/H_{\rm a}^2$, lie between the simulated results for $\mu/\mu_0 = 20$ and $\mu/\mu_0 = 50$, a fact which correlates satisfactorily with the estimate $\mu/\mu_0 \simeq 30$, noting that $H_{\rm c} \simeq 8\,{\rm A}/mm$, from experiments.

Nevertheless, remarkable differences between numerical results and empirical findings for the normalized hysteretic AC loss in the low- and high-value regimes of $H_{\rm a}/H_{\rm c}$ persist. Thus, the curve for $\mu/\mu_0 = 1$, whilst manifesting perfect coincidence with the analytical result, eqn. (8.24), in the whole range of $H_{\rm a}/H_{\rm c}$ and fair accord with experiment in the medium-value regime of H_a/H_c, deviates from the experiment below $H_a/H_c \simeq 0.6$ to an increasing extent. This discrepancy, for one thing, is due to the assumption of an infinitesimally thin superconductor strip, both in the analytical theory and in the numerical analysis, a simplification known to distinctly underestimate hysteretic AC losses in superconductor strips of rectangular cross section with high values of the aspect ratio $2w/d$ (Clem 2009; Ma and Rauh 2013). For another, it is caused by the idealization of a discontinuous, rather than smooth, transition between opposite current states during cycling of the applied magnetic field, inherent in Bean's model of the critical state (Rauh and Ma 2013). The difference in the asymptotic behaviour above $H_a/H_c \simeq 5$, exhibiting an overestimate by theory of hysteretic AC losses in superconductor strips on a nonmagnetic support, can be attributed to the decrease of the critical sheet current at high fields (Ma and Rauh 2013; Rauh and Ma 2013). By contrast, the growing underestimate of simulated hysteretic AC losses in a magnetically shielded superconductor strip below $H_a/H_c \simeq 0.8$ as compared to those found experimentally seems to demonstrate the importance of additional, ferromagnetic AC losses in the support, not accounted for here (Grilli, Ashworth, and Civale 2007; Claassen and Thieme 2008; Gömöry et al. 2010; Nguyen et al. 2010; Krüger, Grilli, and Farinon 2011).

In conclusion, the finite-element simulations explained underpin an equivocal effect of the paramagnetic support on hysteretic AC losses in a planar bilayer superconductor/paramagnet heterostructure subject to an oscillating transverse magnetic field, such losses being enhanced in the low-field regime and reduced in the high-field regime, given those for a nonmagnetic support. This phenomenon corresponds with theoretical predictions by Mawatari based on the simplified assumptions of an infinite permeability and an infinitesimally thin paramagnetic support.

8.4 Hysteretic AC loss in a magnetically coated superconducting tubular wire with an applied magnetic field

Cylindrical heterostructures made up of superconducting and paramagnetic constituents find use in various technological applications, power transmission cables involving second-generation high-temperature superconductor multistrand wires and electromagnetic coils based on coated conductors (Clem and Malozemoff 2010) or bulk MgB_2/Fe filaments (Sumption et al. 2002) playing a prominent role thereby. The combined magnetic shielding properties of these kinds of constituents render beneficial effects like a reduction of hysteretic AC losses (Majoros, Glowacki, and Campbell 2000; Sumption et al. 2002; Yampolskii, Genenko, Rauh, and Snezhko 2006) or an enhancement of the (field-dependent) critical current (Horvat, Wang, Soltanian, and Dou 2002; Genenko, Yampolskii, and Pan 2004; Genenko, Rauh, and Yampolskii 2005; Yampolskii and Genenko 2005), to name just a few. Given specific material characteristics and geometries, coaxial superconductor/paramagnet heterostructures may even disclose the exceptional hallmarks of magnetic cloaks. (cf. section 7.1.2.3).

Hysteretic AC losses in unshielded superconductors of cylindrical shapes have been established for the cases of an imposed oscillating transport current or an applied oscillating transverse magnetic field (Ashkin 1979; Majoros, Sumption, Collings, and van der Laan 2014) as well as for scenarios with both types of excitations in operation (Ruiz et al. 2012; Ruiz and Badía-Majós 2013). By comparison, hysteretic AC losses in superconductor strips on magnetic supports were until recently discussed controversially; details can be found in section 8.3. However, as shown there, by taking advantage of magnetostatic–electrostatic analogues, satisfactory agreement between numerically simulated (Genenko, Rauh, and Krüger 2011) and experimentally determined (Suenaga et al. 2008) hysteretic AC losses in a planar bilayer superconductor/paramagnet heterostructure subject to an oscillating transverse magnetic field could be achieved, with proper dependences on the permeability of the magnetic support μ and the amplitude of the applied magnetic field H_a. Thorough accounts of both analytical and numerical methods for modelling electromagnetic properties of high-temperature superconductors, including hysteretic AC losses in the presence of paramagnetic supports, can be found in the literature (Mikitik, Mawatari, Wan, and Sirois 2013; Grilli et al. 2014; Shen, Grilli, and Coombs 2020; Berrospe-Juarez, Trillaud, Zermeño, and Grilli 2021).

Regarding tubular superconductor/paramagnet heterostructures forged by polygonal assemblies of coated superconductor strips, great effort has been spent on finite-element simulations of hysteretic AC losses for an imposed oscillating transport current (Amemiya and Nakahata 2007; Miyagi, Umabuchi, Takahashi, and Tsukamoto 2007; Amemiya, Nakahata, Fujiwara, and Shiohara 2010). Accordingly, when assuming a nonlinear current–voltage characteristic of the superconductors and a constant permeability μ of the magnetic constituents, the transport current hysteretic AC loss with the magnetic supports placed inside turns out to be far less than with these supports placed outside (Miyagi, Umabuchi, Takahashi, and Tsukamoto 2007). Increasing μ in configurations with inner magnetic supports leads to enhancements of the hysteretic AC loss (Amemiya and Nakahata 2007), an effect which saturates for a relative magnetic permeability near $\mu/\mu_0 \simeq 100$, as anticipated from previous analysis (Genenko, Snezhko, and Freyhardt 2000; Genenko and Snezhko 2002). Incorporation of a nonlinear, but reversible, field dependence of the magnetic permeability μ has merely little bearing on the saturated magnetic property, as comparative modellings using fixed values of μ show (Amemiya and Nakahata 2007), very much like for a single magnetically shielded superconductor strip (Ma and Rauh 2013). Advanced studies of a power transmission cable involving two coaxial sets of coated superconductor strips—with hysteretic magnetization losses suffered by the magnetic constituents allowed for too—reveal that these losses control the dissipation of electromagnetic energy of the entire heterostructure (Amemiya, Nakahata, Fujiwara, and Shiohara 2010).

Although multistrand wires and electromagnetic coils are obviously exposed to transverse magnetic fields, only few works seem to have addressed the electromagnetic behaviour of tubular superconductor/paramagnet heterostructures in oscillating applied magnetic fields. An elaborate numerical study of the shielding properties of such heterostructures, for example, assumes a nonlinear current–voltage characteristic of the superconductors and a reversible field dependence of the permeability of the magnetic constituents μ, apart from including the field effect on the critical current density j_c (Lousberg et al. 2010). Hence, configurations with the paramagnet layers placed outside the superconductor constituents exhibit much stronger shieldings than those with the respective layers placed inside. Based on an elegant analytical approach (Mawatari 2008, 2009), the follow-up work probes the effect of outer magnetic supports on the hysteretic AC loss in a power transmission cable assembled from curved superconductor/soft-magnet tapes conforming to a cylindrical shape (He, Xue, Yong, and Zhou 2014). Both types of constituents are deemed infinitesimally thin, the magnetic supports being characterized by an infinite permeability, $\mu \to \infty$. A realistic enhancement of the hysteretic AC loss due to the magnetic supports obtains for an imposed transport current, consistent with previous investigations (Amemiya and Nakahata 2007; Miyagi, Umabuchi, Takahashi, and Tsukamoto 2007); results derived for an applied magnetic field of radial symmetry, however, scarcely hit, considering the fact that an external magnetic field must be source free. Significant progress exemplifies a theoretical analysis of the electromagnetic response of a cylindrical tubular wire represented by an infinitesimally thin superconductor constituent subject to an oscillating transverse magnetic field (Mawatari 2011). In a description where the superconductor is delineated by the sheet current J, with a (field-independent) critical value J_c, the profiles of the magnetic field, the surface penetration field, and the hysteretic AC losses that ensue pave the way towards research on the electromagnetic behaviour of tubular heterostructures embracing superconductor as well as paramagnet constituents. We extend this ansatz here for a coated superconducting tubular wire, with a coaxial paramagnetic support, by making recourse to Bean's model of the critical state and exploiting magnetostatic–electrostatic analogues, the procedure introduced in section 8.1. Our outlines largely adhere to the work of Genenko, Rauh, and Kurdi (2015).

8.4.1 Theoretical model

Let us define the magnetostatic problem first by considering a cylindrical superconductor/paramagnet heterostructure of bilayer geometry: an infinitely extended type-II superconducting tubular wire of radius r_s and thickness d, whose symmetry axis coincides with the z-axis of both a Cartesian coordinate system x, y, z and a cylindrical coordinate system r, ϕ, z. The tube is supposed to be deposited on an inner, or outer, paramagnetic support of respective thickness d_m, buffered by an insulating nonmagnetic layer of infinitesimal thickness in between mitigating the proximity effect (Schmidt 1997), and subject to a transverse applied magnetic field of strength H_a with no imposed transport current, as illustrated in Fig. 8.14. We choose dimensions that second-generation coated conductors typically display (Claassen and Thieme 2008; Suenaga et al. 2008), viz. $r_s = 5\,mm$, $d = 2\,\mu m$, and $d_m = 250\,\mu m$, understanding that the paramagnetic support is delineated by a

magnetically coated superconducting tubular wire portrayed in Fig. 8.17, again for the relative permeability of the support μ/μ_0 set and three progressive values of the normalized amplitude of the applied magnetic field H_a/H_c. At the lowest value, $H_a/H_c = 2.0$, the interior of the wire is still completely shielded from the applied magnetic field by the paramagnetic support on top of the superconducting tube: the lines of the magnetic field are refracted at the wire's outer surface and guided through the paramagnetic support (Fig. 8.17(a)). As the normalized amplitude of the applied magnetic field is increased to $H_a/H_c = 3.0$, magnetic flux starts to enter the superconductor constituent tangentially from both equatorial sides, threading two cylindrical segments of the tube, with refraction of the lines of the magnetic field at the wire's inner surface too, before the interior of the wire accommodates the flux (Fig. 8.17(b)), an effect which, for $H_a/H_c = 4.0$, gets more pronounced still, exhibiting intensified refraction towards the poles (Fig. 8.17(c)). The paramagnetic support here always plays a prominent role in the distribution of the magnetic field. Its shielding capacity defines an effective, reduced field H_m that acts on the wire's superconductor constituent.

The dissimilarity of behaviour carries over to the variation of the half-angle of flux penetration γ with the normalized amplitude of the applied magnetic field H_a/H_c, depicted in Fig. 8.18(a) for a range of values of the relative permeability μ/μ_0: assuming $\mu/\mu_0 = 1$, penetration of magnetic flux starts at $H_a/H_c = \pi/2$, again similar to that for an isolated superconducting tube, according to eqn. (8.26),

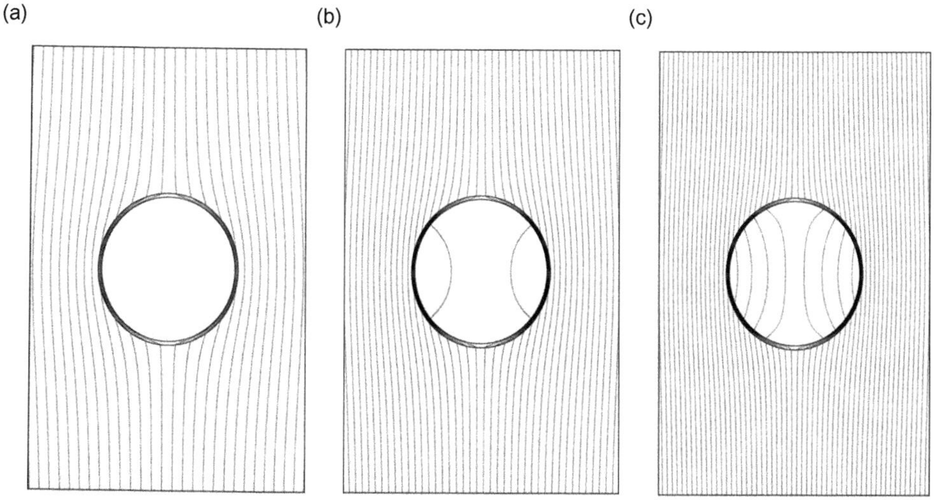

Fig. 8.17 Lines of the magnetic field **H** around and inside a superconducting tubular wire coated with an *outer* paramagnetic support of relative permeability $\mu/\mu_0 = 10$, when the normalized amplitude of the applied magnetic field (a) $H_a/H_c = 2.0$, (b) $H_a/H_c = 3.0$, and (c) $H_a/H_c = 4.0$ (cross-sectional view). The support together with the superconducting tube is indicated by black contour lines. Reproduced with permission from Genenko, Rauh, and Kurdi (2015), *J. Appl. Phys.* 117, pg 243909, AIP Publishing LLC. DOI:10.1063/1.4922982.

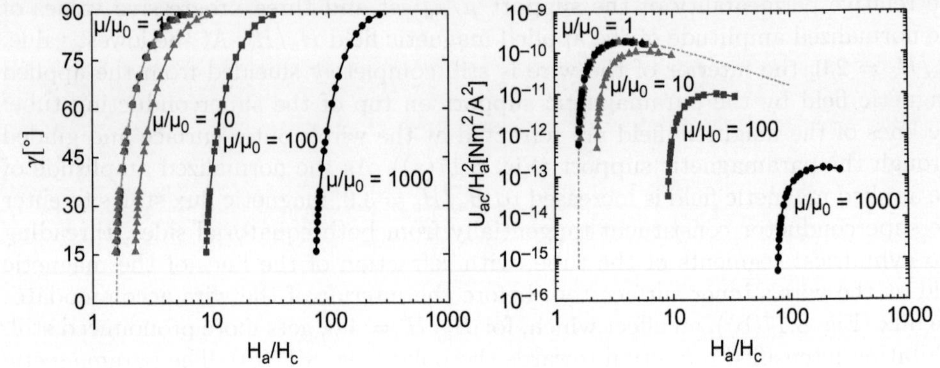

Fig. 8.18 Dependence of (a) the half-angle of flux penetration γ and (b) the normalized hysteretic AC loss $U_{\mathrm{ac}}/H_{\mathrm{a}}^2$ on $H_{\mathrm{a}}/H_{\mathrm{c}}$, the normalized amplitude of the magnetic field applied to the magnetically coated superconducting tubular wire, in the computationally accessible regimes, for four different values of the relative permeability of the *outer* paramagnetic support μ/μ_0 identified on the curves. The full triangles, squares, and circles elucidate the scaling properties of the respective physical observables; the dashed lines represent analytical results for the isolated superconducting tube. Reproduced with permission from Genenko, Rauh, and Kurdi (2015), *J. Appl. Phys.* 117, pg 243909, AIP Publishing LLC. DOI:10.1063/1.4922982.

with a monotonic rise and a tendency towards saturation in a fully flux-filled state, as $H_{\mathrm{a}}/H_{\mathrm{c}}$ augments. Increasing μ/μ_0 to account for a paramagnetic support produces a duplication of this course, yet rescaled to higher values of $H_{\mathrm{a}}/H_{\mathrm{c}}$, a fact that directly reflects the enhanced shielding capacity of the support. Corresponding traits recur in the variation of the normalized hysteretic AC loss $U_{\mathrm{ac}}/H_{\mathrm{a}}^2$ with the normalized amplitude of the applied magnetic field $H_{\mathrm{a}}/H_{\mathrm{c}}$ shown in Fig. 8.18(b). The familiar threshold for the onset of the hysteretic AC loss appears, followed by a sharp monotonic rise with a turn to a shallow maximum, given the relative permeability of a vacuum environment. Increasing μ/μ_0 to model a paramagnetic support thus yields a shift in the hysteretic AC loss to higher values of $H_{\mathrm{a}}/H_{\mathrm{c}}$, but also a considerable reduction in extent. It should be noted that these results can be calibrated to high accuracy against the expressions

$$\gamma = \gamma\,(H_{\mathrm{m}}/H_{\mathrm{c}}) \quad \text{with } H_{\mathrm{a}}/H_{\mathrm{m}} = 1 + 0.048\,(\mu/\mu_0 - 1) \tag{8.28}$$

and

$$U_{\mathrm{ac}}/H_{\mathrm{a}}^2 = 8\pi\mu_0 r_{\mathrm{s}}^2\,\chi_{\mathrm{m}} f\,(H_{\mathrm{m}}/H_{\mathrm{c}}) \quad \text{with } \log\chi_m = -0.325\log^2(\mu/\mu_0)\,, \tag{8.29}$$

the chosen values of the implied geometrical parameters of the wire, turning to the functional dependences of the isolated superconducting tube, eqns. (8.26) and (8.27).

8.4.3 Discussion and conclusion

There are two sources bringing about modifications of the normalized hysteretic AC loss U_{ac}/H_{a}^2 in a superconducting tube for a given paramagnetic support: changes in the amplitude of the applied magnetic field H_{a} and adjustments in the orientation of the local magnetic field $\mathbf{H}$ which acts on the superconducting tube, a steep angle with the surface of the tube, resulting in an enhancement of the loss. The case of an inner paramagnetic support clearly testifies to the latter fact. On the other hand, both sources favour a reduction in U_{ac}/H_{a}^2 in the case of an outer paramagnetic support by depressing and guiding the magnetic flux, consistent with the findings of a previous analysis, which noted the factorization of the shielding effects of the superconducting and paramagnetic constituents (Lousberg et al. 2010). The decrease in the hysteretic AC loss in this configuration certainly involves a more subtle interaction between the two types of constituents. Shielding the transverse applied magnetic field by the paramagnetic support alone would generate a homogeneous magnetic field inside the tube, just as for a massive cylindrical set, the inverse of its reduced strength $H_{\mathrm{a}}/H_{\mathrm{m}}$ revealing a nonlinear dependence on the relative permeability μ/μ_0 (Genenko, Yampolskii, and Pan 2004). The magnetic field in the interior of the tubular wire configuration discussed here, however, proves to be inhomogeneous and strongly redistributed due to the presence of the superconductor constituent, the inverse of the effective, reduced field $H_{\mathrm{a}}/H_{\mathrm{m}}$ extracted from the numerical simulations demonstrating a perfectly linear dependence on μ/μ_0, as eqn. (8.28) states. The variation in the shielding capacity of the outer paramagnetic support with μ/μ_0 in turn admits tuning the normalized hysteretic AC loss $U_{\mathrm{ac}}/H_{\mathrm{a}}^2$ by a judicious choice of μ/μ_0.

Attention should be paid to the circumstance that for a transverse applied magnetic field, the shielding effect of an outer paramagnetic support in the superconducting tubular wire configuration is usually strong and strictly advantageous regarding the hysteretic AC loss, unlike that of a paramagnetic support in a planar bilayer superconductor/paramagnet heterostructure, which is comparatively weak and either beneficial or detrimental, depending on the geometrical and material characteristics of the magnetic environment and the amplitude of the applied magnetic field (cf. section 8.3.1). The reason for this significant difference is of a geometrical sort. The component of the magnetic field penetrating into the flat bilayer heterostructure is oriented normal to the plane of the structure; due to the paramagnetic support guiding the magnetic flux, it may be either enhanced or reduced. In contrast, the component of the magnetic field penetrating into the superconducting tube at its equatorial line is oriented tangential to the surface of the tube; an outer paramagnetic support guiding the magnetic flux can only promote this grazing orientation, thus reducing the field effect on the superconductor constituent.

In conclusion, the finite-element simulations presented here reveal effects of a paramagnetic support on hysteretic AC losses in a tubular superconductor/paramagnet heterostructure subject to an oscillating transverse magnetic field. Accordingly, one-sided magnetic shielding of the superconducting tube by a coaxial paramagnetic support gives rise to a slight increase in hysteretic AC losses compared to those for a vacuum environment, when the support is placed inside; however, a spectacular shielding effect with a possible reduction in hysteretic AC losses by orders of magnitude ensues, depending on the magnetic permeability and the amplitude of the applied magnetic field, when the support is placed outside.

9
Epilogue

The main subject of this book has been current distributions in and magnetic field distributions around thin superconductor strips surrounded by paramagnetic or superconductor environments of various shapes. The Meissner states of these strips were investigated in Chapters 3, 5, 6, and 8 by means of the potential theory. This treatment is based on the fact that the strips are in the flux-free state and involves no specific characteristics of superconductors but the edge barrier against magnetic flux penetration. The partially flux-filled states of superconductor strips investigated in Chapters 4, 5, and 8 were studied on the grounds of the Bean model of the critical state. This is based on the concept of pinning magnetic vortices common, in principal, for all known types of low- and high-temperature superconductors, though the structure of vortices and mechanisms for pinning may differ depending on the symmetry of the order parameter and specific features of the materials. The Bean model is—in its classical formulation—a macroscopic model with the only characteristic parameter being the critical current density j_c, which measures the strength of the pinning. Thus, the description of the main geometrical effects arising in the electrodynamics of macroscopic superconductor/magnet hybrid structures is based on simple phenomenological concepts. These were sufficient to reveal strong enhancements in the dissipation-free current above the critical current of isolated strips and consequent drastic reduction in AC losses possible in hybrid structures of certain geometries.

In Chapter 6, the effects of finite geometrical dimensions and finite magnetic permeability of the shields were systematically investigated. The main conclusion of these studies was that the geometrical factors are more important when utilizing the shielding effect than the permeability. For the bulk shields, the shielding effect saturates at moderately large permeability values, whereas in the case of thin magnetic shields this effect becomes considerable only for large permeability. In both cases, the shape of the shields plays a crucial role in the favourable or deteriorative effect of shielding in the superconducting state.

The question of whether the shielding effects established on planar superconductors may be applied to bulk superconductors was studied in Chapter 7. To find out the lower critical field and the conditions of the initial entry of magnetic flux in superconductor/paramagnet hybrid structures, consideration was extended beyond the above macroscopic approach by using the generalized London theory. In contrast

Electromagnetics of Superconductor/Paramagnet Heterostructures. Yuri Genenko and Hermann Rauh, Oxford University Press.
 DOI: 10.1093/9780191782855.003.0009

to planar superconductors, overcritical transport-current states were not found in this case; nevertheless, a strong favourable effect of the shielding on the behaviour of superconductors in an external field was established.

After all, in Chapter 8 superconductor strips in paramagnetic environments of arbitrary forms and constant permeabilities were studied by means of finite-element calculations using electrostatic–magnetostatic analogues in both cases of a transport current and an applied external magnetic field. These studies confirmed previous analytical and numerical findings and developed an effective tool for optimizing the geometry of the hybrid structures. The latter approach seems to be a promising way to further improve superconductor/paramagnet hybrid structures for their potential applications. For a more realistic description of their electromagnetic behaviour the models should henceforth include a field-dependent critical current as well as nonlinear, hysteretic, and conducting properties of typically used para- and ferromagnetic materials. This task has already been accomplished to a large extent in a number of works cited in this book; see, for example, Shen, Grilli, and Coombs (2020) and Statra, Menana, and Douine (2022) and references therein. Nevertheless, from an engineering standpoint it is already clear that the geometry of hybrid structures is as crucial for the performance of superconducting devices as are the shapes of the wings and the fuselage of an aircraft for its performance. This analogy proves even more meaningful when taking into account the previously discussed close relationship between the flow of magnetic flux in magnetostatics and the flow of a perfect incompressible fluid in hydrodynamics.

Concerning the existence of overcritical states one important note should be made. This phenomenon was predicted based on the classical formulation of Bean's model of the critical state (Bean 1964). Applied to the case of thin strips (Norris 1972; Brand and Indenbom 1993; Zeldov, Clem, McElfresh, and Darwin 1994) this model predicts that, in the partially flux-filled critical state, the central, magnetic flux-free region of a superconductor strip may carry the major part of the transport current. This area turns out to be particularly sensitive to the magnetic surroundings and makes it possible to tailor the distribution of the transport current, as was shown in Chapters 4, 5, and 8. Such a phenomenon, however, can hardly be expected in common numerical simulations for the following reasons. In numerical modelling, the empirical macroscopic relationship between the average current density J and the average electric field E of the form $E = E_c(J/J_c)^n$, with experimentally determined parameters E_c, J_c, and n, is usually applied locally as the material law relating the local current density and the local electric field. This assumption entails the appearance of the electric field caused by the moving magnetic flux everywhere where the current is present, that contradicts the above mentioned picture of the critical state in thin films. Thus, in numerical simulations, a wide dissipation-free region of a thin film carrying a large current can hardly exist. Nevertheless, the magneto-optical observations of highly overcritical currents in the magnetic flux-free regions of the magnetically shielded YBCO film subject to a transverse magnetic field (Jarzina, Jooss, and Freyhardt 2002) give strong evidence supporting the classical concept of the critical state in thin films and, thus, keep the perspective of realization of the overcritical transport currents.

system x, y, z, as Fig. 7.5 depicts. The magnetic dipole moment of the vortex loop is defined by

$$\mathbf{m}_V = \frac{1}{2} \int dv \, (\mathbf{r} \times \mathbf{j}) \tag{A.22}$$

with the density of the induced supercurrent $\mathbf{j}$, the volume integral extending over the superconductor's domain. From eqn. (A.22), the single nontrivial component of the magnetic dipole moment $m_{V,y}$ for the vortex loop situated in the x, y-plane reads

$$m_{V,y} = \frac{1}{2} \int dv \, (z j_x - x j_z) \,. \tag{A.23}$$

Integration by parts with respect to x transforms the first term in brackets of eqn. (A.23) into

$$\int dv \, (z j_x) = - \int\limits_{-\infty}^{\infty} dz \int\limits_{-\infty}^{\infty} dy \int\limits_{0}^{\infty} dx \left(z x \frac{\partial j_x}{\partial x} \right) , \tag{A.24}$$

observing that current flow across the superconductor/paramagnet interface does not exist, $j_x = 0$ at $x = 0$, and also that current flow in the x-direction wanes rapidly away from the core of the vortex loop, $j_x \to 0$ for $x \to \infty$. Likewise, integration by parts with respect to z transforms the second term in brackets of eqn. (A.23) into

$$\int dv \, (x j_z) = - \int\limits_{-\infty}^{\infty} dz \int\limits_{-\infty}^{\infty} dy \int\limits_{0}^{\infty} dx \left(x z \frac{\partial j_z}{\partial z} \right) , \tag{A.25}$$

noting that current flow in the z-direction wanes rapidly away from the core of the vortex loop, $j_z \to 0$ for $z \to \pm\infty$. By referring to the stationary continuity equation $\nabla \cdot \mathbf{j} = 0$, the sum of the left-hand sides of eqns. (A.24) and (A.25) adopts the form

$$\int dv \, (z j_x + x j_z) = \int\limits_{-\infty}^{\infty} dz \int\limits_{-\infty}^{\infty} dy \int\limits_{0}^{\infty} dx \left(z x \frac{\partial j_y}{\partial y} \right) . \tag{A.26}$$

Clearly, as current flow in the y-direction wanes rapidly away from the core of the vortex loop, $j_y \to 0$ for $y \to \pm\infty$, it follows that eqn. (A.26), upon integration with respect to y, makes it possible to conclude

$$\int dv \, (z j_x) = - \int dv \, (x j_z) \,. \tag{A.27}$$

Use of eqn. (A.27) in eqn. (A.23) thus yields the representation

$$m_{V,y} = - \int\limits_{-\infty}^{\infty} dz \int\limits_{-\infty}^{\infty} dy \int\limits_{0}^{\infty} dx \, x j_z(x, y, z) \tag{A.28}$$

in terms of the current density $j_z(x, y, z)$.

A variant of eqn. (A.28) obtains by recourse to Ampère's law, eqn. (2.19), focusing on the magnetic induction of the vortex loop instead,

$$m_{V,y} = -\frac{1}{\mu_0} \int_{-\infty}^{\infty} dz \int_{-\infty}^{\infty} dy \int_{0}^{\infty} dx \, x \left(\frac{\partial B_{V,y}}{\partial x} - \frac{\partial B_{V,x}}{\partial y} \right). \tag{A.29}$$

Integration by parts with respect to x of the first term in brackets of eqn. (A.29) gives

$$\int_{0}^{\infty} dx \, x \left(\frac{\partial B_{V,y}}{\partial x} \right) = - \int_{0}^{\infty} dx \, B_{V,y}, \tag{A.30}$$

observing that the magnetic induction in the y-direction decays rapidly away from the core of the vortex loop, $B_{V,y} \to 0$ for $x \to \infty$. Likewise, integration by parts with respect to y of the second term in brackets of eqn. (A.29) states that

$$\int_{-\infty}^{\infty} dy \, x \left(\frac{\partial B_{V,x}}{\partial y} \right) = 0, \tag{A.31}$$

noting that the magnetic induction in the x-direction decays rapidly away from the core of the vortex loop, $B_{V,x} \to 0$ for $y \to \pm\infty$. Use of eqns. (A.30) and (A.31) in eqn. (A.29) admits the representation

$$m_{V,y} = \frac{1}{\mu_0} \int_{-\infty}^{\infty} dz \int_{-\infty}^{\infty} dy \int_{0}^{\infty} dx \, B_{V,y}(x, y, z) \tag{A.32}$$

in terms of the magnetic induction $B_{V,y}(x, y, z)$. If the latter quantity is expressed through its two-dimensional Fourier transform $\tilde{B}_{V,y}^{k_y,k_z}(x)$ via eqn. (7.44), eqn. (A.32) takes on the succinct shape

$$m_{V,y} = \lim_{k_z \to 0} \lim_{k_y \to 0} \frac{1}{\mu_0} \int_{0}^{\infty} dx \, \tilde{B}_{V,y}^{k_y,k_z}(x), \tag{A.33}$$

which, by recalling eqn. (7.45b), itself yields the representation

$$m_{V,y} = \frac{1}{\mu_0} \int_{0}^{\infty} dx \, \tilde{V}_y^{k_y,k_z}(x) \left[1 - \exp\left(-x/\lambda_L\right)\right] \Big|_{k_y = k_z = 0} \tag{A.34}$$

in terms of the two-dimensional Fourier transform of the vorticity $\tilde{V}_y^{k_y,k_z}(x)$. This, remarkably, proves $m_{V,y}$ to be independent of μ and H_a! Substituting eqn. (7.47b) in eqn. (A.34) and performing the integration finally gives the magnetic dipole moment of the vortex loop,

$$m_{\mathrm{V},y} = (\pi a \Phi_0/\mu_0)\left[I_1(a/\lambda_{\mathrm{L}}) - L_1(a/\lambda_{\mathrm{L}})\right] \tag{A.35}$$

(Gradshteyn and Ryzhik 1994). I_1 and L_1 herein denote the modified Bessel and Struve functions of the first kind of order one, respectively. A useful approximation of eqn. (A.35) deserves mentioning whereby

$$m_{\mathrm{V},y} \simeq \pi a^2 \Phi_0/2\mu_0\lambda_{\mathrm{L}}; \quad a \ll \lambda_{\mathrm{L}}, \tag{A.36}$$

which reveals that, for small values of a, the magnetic dipole moment is controlled by the area of the vortex loop. Conversely, an asymptotic form of eqn. (A.35) is

$$m_{\mathrm{V},y} \simeq 2a\Phi_0/\mu_0; \quad a \gg \lambda_{\mathrm{L}} \tag{A.37}$$

(Abramowitz and Stegun 1972), which links the magnetic dipole moment for large values of a to the diameter of the vortex loop.

A.4 Magnetic vortex loop at a curved superconductor/paramagnet interface

We conclude by studying a magnetic vortex loop of arbitrary shape at the curved interface between an infinitely extended, cylindrical type-II superconductor filament with radius r_{s} in the marginal Meissner state and a coaxial paramagnetic sheath with thickness d_{m} and permeability μ subject to an applied magnetic field of strength H_{a} in the y-direction of a Cartesian coordinate system x, y, z, as Fig. 7.7 depicts. Here, the magnetic dipole moment of the vortex loop $\mathbf{m}_{\mathrm{V}}$ embodies a contribution from the superconducting filament, $\mathbf{m}_{\mathrm{VF}}$, due to the current establishing the vortex itself and a contribution from the paramagnetic sheath, $\mathbf{m}_{\mathrm{VS}}$, due to the vortex-induced magnetic field,

$$\mathbf{m}_{\mathrm{V}} = \mathbf{m}_{\mathrm{VF}} + \mathbf{m}_{\mathrm{VS}}. \tag{A.38}$$

The presentation that follows implies symmetry of the core of the magnetic vortex loop with respect to the x, z-plane.

The *superconductor contribution* to the magnetic dipole moment of the vortex loop is defined by

$$\mathbf{m}_{\mathrm{VF}} = \frac{1}{2}\int dv\,(\mathbf{r} \times \mathbf{j}) \tag{A.39}$$

with the density of the induced supercurrent $\mathbf{j}$, the volume integral extending over the filament's domain. From eqn. (A.39), the single nontrivial Cartesian component of the magnetic dipole moment $m_{\mathrm{VF},y}$ for the vortex loop situated in the x, y-plane reads as

$$m_{\mathrm{VF},y} = \frac{1}{2} \int dv \left((zj_r) \cos\phi - (zj_\phi) \sin\phi - (rj_z) \cos\phi \right), \tag{A.40}$$

introducing cylindrical coordinates r, ϕ, z adapted to the magnetically sheathed filament. Integration by parts with respect to r transforms the first term in brackets of eqn. (A.40) into

$$\int dv\, (zj_r) \cos\phi = -\int_{-\infty}^{\infty} dz \int_{-\pi}^{\pi} d\phi \int_{0}^{r_s} dr\, zr \left(j_r + r\frac{\partial j_r}{\partial r} \right) \cos\phi, \tag{A.41}$$

observing that current flow through the circumference of the filament does not exist, $j_r = 0$ at $r = r_s$. Likewise, integration by parts with respect to ϕ transforms the second term in brackets of eqn. (A.40) into

$$\int dv\, (zj_\phi) \sin\phi = \int_{-\infty}^{\infty} dz \int_{-\pi}^{\pi} d\phi \int_{0}^{r_s} dr \left(rz\frac{\partial j_\phi}{\partial \phi} \right) \cos\phi, \tag{A.42}$$

noting that since current flow along the azimuth of the filament is unique, the values of j_ϕ at $\phi = \pm\pi$ agree. By referring to the stationary continuity equation $\nabla \cdot \mathbf{j} = 0$, the difference of the left-hand sides of eqns. (A.41) and (A.42) adopts the form

$$\int dv \left((zj_r) \cos\phi - (zj_\phi) \sin\phi \right) = \int_{-\infty}^{-\infty} dz \int_{-\pi}^{\pi} d\phi \int_{0}^{r_s} dr\, zr \left(r\frac{\partial j_z}{\partial z} \right) \cos\phi. \tag{A.43}$$

Clearly, as current flow in the z-direction wanes rapidly away from the core of the vortex loop, $j_z \to 0$ for $z \to \pm\infty$, it follows that eqn. (A.43), upon integration with respect to z, makes it possible to conclude that

$$\int dv \left((zj_r) \cos\phi - (zj_\phi) \sin\phi \right) = -\int dv\, (rj_z) \cos\phi. \tag{A.44}$$

Use of eqn. (A.44) in eqn. (A.40) thus yields the representation

$$m_{\mathrm{VF},y} = -\int_{-\infty}^{\infty} dz \int_{-\pi}^{\pi} d\phi \int_{0}^{r_s} dr\, r^2 j_z(r, \phi, z) \cos\phi \tag{A.45}$$

in terms of the current density $j_z(r, \phi, z)$.

A variant of eqn. (A.45) obtains by recourse to Ampère's law, eqn. (2.19), focusing on the magnetic induction of the vortex loop instead,

$$m_{\mathrm{VF},y} = \frac{1}{\mu_0} \int\limits_{-\infty}^{\infty} dz \int\limits_{-\pi}^{\pi} d\phi \int\limits_{0}^{r_{\mathrm{s}}} dr\, r \left(B_{\mathrm{V},\phi}(r,\phi,z) + \frac{\partial B_{\mathrm{V},r}(r,\phi,z)}{\partial \phi} \right) \cos\phi$$

$$- \frac{1}{\mu_0} \int\limits_{-\infty}^{\infty} dz \int\limits_{-\pi}^{\pi} d\phi\, r^2 B_{\mathrm{V},\phi}(r,\phi,z) \cos\phi \Big|_{r=r_{\mathrm{s}}}. \qquad \text{(A.46)}$$

If the cylindrical components of the magnetic induction $B_{\mathrm{V},r}(r,\phi,z)$ and $B_{\mathrm{V},\phi}(r,\phi,z)$ are expressed through their two-dimensional Fourier transforms $\tilde{B}^{n,k}_{\mathrm{V},r}(r)$ and $\tilde{B}^{n,k}_{\mathrm{V},\phi}(r)$ via eqn. (7.69), with inherent symmetry properties due to the assumed geometry of the loop taken into account, eqn. (A.46) adopts the succinct shape

$$m_{\mathrm{VF},y} = \frac{2\pi i}{\mu_0} \lim_{k \to 0} \left\{ \int\limits_{0}^{r_{\mathrm{s}}} dr\, r \left(\tilde{B}^{1,k}_{\mathrm{V},r}(r) - i \tilde{B}^{1,k}_{\mathrm{V},\phi}(r) \right) + i r^2 \tilde{B}^{1,k}_{\mathrm{V},\phi}(r) \Big|_{r=r_{\mathrm{s}}} \right\}. \qquad \text{(A.47)}$$

The *paramagnet contribution* to the magnetic dipole moment of the vortex loop is defined by

$$\mathbf{m}_{\mathrm{VS}} = \int dv\, \mathbf{M} \qquad \text{(A.48)}$$

with the local magnetization $\mathbf{M}$, the volume integral extending over the paramagnet's domain. Making recourse to the constitutive relation $\mathbf{B}_{\mathrm{V}} = \mu_0(\mathbf{H}_{\mathrm{V}} + \mathbf{M})$ as well as to the assumption $\mathbf{B}_{\mathrm{V}} = \mu \mathbf{H}_{\mathrm{V}}$ and implementing the representation of the magnetic field $\mathbf{H}_{\mathrm{V}}$, eqn. (7.65), thus yields the single nontrivial Cartesian component of the magnetic dipole moment $m_{\mathrm{VS},y}$ for the vortex loop situated in the x,y-plane

$$m_{\mathrm{VS},y} = -(\mu/\mu_0 - 1) \int dv\, \left((\nabla \psi_{\mathrm{V}})_r \sin\phi + (\nabla \psi_{\mathrm{V}})_\phi \cos\phi \right). \qquad \text{(A.49)}$$

The two integral terms on the right-hand side of eqn. (A.49) read as

$$\int dv\, (\nabla \psi_{\mathrm{V}})_r \sin\phi = \int\limits_{-\infty}^{\infty} dz \int\limits_{-\pi}^{\pi} d\phi \int\limits_{r_{\mathrm{s}}}^{r_{\mathrm{s}}+d_{\mathrm{m}}} dr \left(r \frac{\partial \psi_{\mathrm{V}}}{\partial r} \right) \sin\phi \qquad \text{(A.50)}$$

and

$$\int dv\, (\nabla \psi_{\mathrm{V}})_\phi \cos\phi = \int\limits_{-\infty}^{\infty} dz \int\limits_{-\pi}^{\pi} d\phi \int\limits_{r_{\mathrm{s}}}^{r_{\mathrm{s}}+d_{\mathrm{m}}} dr \left(\frac{\partial \psi_{\mathrm{V}}}{\partial \phi} \right) \cos\phi. \qquad \text{(A.51)}$$

If the scalar magnetic potential $\psi_{\mathrm{V}}(r,\phi,z)$ is expressed through its two-dimensional Fourier transform $\tilde{\psi}^{n,k}_{\mathrm{V}}(r)$ via eqn. (7.83), with the inherent symmetry property due to the assumed geometry of the loop taken into account, eqn. (A.50) admits that

$$\int_{-\infty}^{\infty} dz \int_{-\pi}^{\pi} d\phi \int_{r_{\rm s}}^{r_{\rm s}+d_{\rm m}} dr \left(r \frac{\partial \psi_{\rm V}}{\partial r} \right) \sin\phi = \lim_{k \to 0} 2\pi i \int_{r_{\rm s}}^{r_{\rm s}+d_{\rm m}} dr \left(r \frac{\partial \tilde{\psi}_{\rm V}^{1,k}}{\partial r} \right) \tag{A.52}$$

and eqn. (A.51) yields

$$\int_{-\infty}^{\infty} dz \int_{-\pi}^{\pi} d\phi \int_{r_{\rm s}}^{r_{\rm s}+d_{\rm m}} dr \left(\frac{\partial \psi_{\rm V}}{\partial \phi} \right) \cos\phi = \lim_{k \to 0} 2\pi i \int_{r_{\rm s}}^{r_{\rm s}+d_{\rm m}} dr\, \tilde{\psi}_{\rm V}^{1,k}(r). \tag{A.53}$$

Thus, eqns. (A.50) to (A.53) enable the succinct form

$$\int dv\, \left((\nabla \psi_{\rm V})_r \sin\phi + (\nabla \psi_{\rm V})_\phi \cos\phi \right) = \lim_{k \to 0} 2\pi i \int_{r_{\rm s}}^{r_{\rm s}+d_{\rm m}} dr \frac{\partial}{\partial r} \left(r \tilde{\psi}_{\rm V}^{1,k}(r) \right), \tag{A.54}$$

which readily integrates to give

$$m_{{\rm VS},y} = -(\mu/\mu_0 - 1) \lim_{k \to 0} 2\pi i \left(r \tilde{\psi}_{\rm V}^{1,k}(r) \right) \Big|_{r=r_{\rm s}}^{r=r_{\rm s}+d_{\rm m}} \tag{A.55}$$

in conjunction with eqn. (A.49).

Altogether, eqns. (A.38), (A.47), and (A.55) afford the representation

$$\begin{aligned} m_{{\rm V},y} =& \frac{2\pi i}{\mu_0} \lim_{k \to 0} \Bigg\{ \int_0^{r_{\rm s}} dr\, r \left(\tilde{B}_{{\rm V},r}^{1,k}(r) - i \tilde{B}_{{\rm V},\phi}^{1,k}(r) \right) + i r^2 \tilde{B}_{{\rm V},\phi}^{1,k}(r) \Big|_{r=r_{\rm s}} \\ & -(\mu - \mu_0) \left(r \tilde{\psi}_{\rm V}^{1,k}(r) \right) \Big|_{r=r_{\rm s}}^{r=r_{\rm s}+d_{\rm m}} \Bigg\}, \end{aligned} \tag{A.56}$$

linking the magnetic dipole moment to an integral and select values of the Fourier transforms of the vortex magnetic induction and potential field.

Appendix B

Electromagnetic energy associated with magnetic vortices in superconductor/paramagnet heterostructures

To calculate the electromagnetic energy due to the presence of magnetic vortices of arbitrary length and shape in the superconducting constituent of a superconductor/ paramagnet heterostructure, W_V, we proceed from the generalized London approach, section 2.1.3, together with the classical theory of magnetostatic fields. Let the domains occupied by the superconducting and paramagnetic constituents be denoted by V_s and V_m, respectively, and let their bounding surfaces S_s and S_m have outwardly directed normal vector area elements $\mathbf{ds}_{\mathrm{s}}$ and $\mathbf{ds}_{\mathrm{m}}$. Moreover, let the residual vacuum domain be denoted by V_{v}, and let its bounding surface S_{v} have outwardly directed normal vector area element $\mathbf{ds}_{\mathrm{v}}$. The magnetic induction in the various areas shall be named as $\mathbf{B}_{\mathrm{s}}$ and $\mathbf{B}_{\mathrm{m}}$ or $\mathbf{B}_{\mathrm{v}}$, and the magnetic field as $\mathbf{H}_{\mathrm{s}}$ and $\mathbf{H}_{\mathrm{m}}$ or $\mathbf{H}_{\mathrm{v}}$. Accordingly, there are three contributions that together make up W_V:

$$W_V = W_s + W_{\mathrm{m}} + W_{\mathrm{v}}. \tag{B.1}$$

The term W_{s} arises from the field energy due to the magnetic induction of the vortices and the kinetic energy of the supercurrents linked to the field in the domain occupied by the superconducting constituent, neglecting the contribution of the infinitesimally thin normal vortex cores. Calling upon eqn. (2.24) with eqn. (2.19) thus admits the volume integral representation

$$W_{\mathrm{s}} = \frac{1}{2\mu_0} \int\limits_{V_{\mathrm{s}}} dv \left(\mathbf{B}_{\mathrm{s}}^2 + \lambda_{\mathrm{L}}^2 \left(\nabla \times \mathbf{B}_{\mathrm{s}} \right)^2 \right). \tag{B.2}$$

The term W_{m} accounts for the magnetic energy in the domain occupied by the paramagnetic constituent; thanks to the prevailing linear relation between the magnetic induction and the magnetic field, it takes the form

$$W_{\mathrm{m}} = \frac{1}{2} \int\limits_{V_{\mathrm{m}}} dv \left(\mathbf{B}_{\mathrm{m}} \cdot \mathbf{H}_{\mathrm{m}} \right) \tag{B.3}$$

(Jackson 1998). The term W_v accomodates the magnetic energy in the vacuum domain; it correspondingly reads as

$$W_\text{v} = \frac{1}{2} \int\limits_{V_\text{v}} dv \, (\mathbf{B}_\text{v} \cdot \mathbf{H}_\text{v}) \,. \tag{B.4}$$

When using the identity for the magnetic induction in the superconductor domain,

$$(\nabla \times \mathbf{B}_\text{s})^2 = \mathbf{B}_\text{s} \cdot (\nabla \times (\nabla \times \mathbf{B}_\text{s})) + \nabla \cdot (\mathbf{B}_\text{s} \times (\nabla \times \mathbf{B}_\text{s})) \,, \tag{B.5}$$

together with the generalized London equation (2.20) and exploiting the divergence theorem, eqn. (B.2) can be restated to include a surface integral term as well

$$W_\text{s} = \frac{1}{2\mu_0} \int\limits_{V_\text{s}} dv \, (\mathbf{V} \cdot \mathbf{B}_\text{s}) + \frac{\lambda_\text{L}^2}{2\mu_0} \int\limits_{S_\text{s}} \mathbf{ds}_\text{s} \cdot (\mathbf{B}_\text{s} \times (\nabla \times \mathbf{B}_\text{s})) \,, \tag{B.6}$$

displaying the vorticity $\mathbf{V}$ which, in general, is given by eqn. (2.23). Furthermore, when introducing the scalar magnetic potential ψ_m as a solution of the Laplace equation $\Delta\psi_\text{m} = 0$ to represent the magnetic field in the paramagnet domain $\mathbf{H}_\text{m} = -\nabla\psi_\text{m}$ and applying the identity

$$\mathbf{B}_\text{m} \cdot \nabla\psi_\text{m} = \nabla \cdot (\mathbf{B}_\text{m} \, \psi_\text{m}) \tag{B.7}$$

along with the divergence theorem, eqn. (B.3) can be put as

$$W_\text{m} = -\frac{1}{2} \int\limits_{S_\text{m}} \mathbf{ds}_\text{m} \cdot (\mathbf{B}_\text{m} \, \psi_\text{m}) \,. \tag{B.8}$$

Likewise, when introducing the scalar magnetic potential ψ_v as a solution of the Laplace equation $\Delta\psi_\text{v} = 0$ to represent the magnetic field in the vacuum domain $\mathbf{H}_\text{v} = -\nabla\psi_\text{v}$ and consulting the identity

$$\mathbf{B}_\text{v} \cdot \nabla\psi_\text{v} = \nabla \cdot (\mathbf{B}_\text{v} \, \psi_\text{v}) \tag{B.9}$$

along with the divergence theorem, eqn. (B.4) can be reformulated as

$$W_\text{v} = -\frac{1}{2} \int\limits_{S_v} \mathbf{ds}_\text{v} \cdot (\mathbf{B}_\text{v} \, \psi_\text{v}) \,. \tag{B.10}$$

Since the connectivity of the superconductor/paramagnet heterostructure is not yet defined, eqn. (B.1) with eqns. (B.6), (B.8), and (B.10) reflects the most general form of the electromagnetic energy associated with the magnetic vortices. It can, however, be further simplified by taking into account that all, or part, of the surface of the superconducting constituent S_s may coincide either with the surface of the paramagnetic constituent S_m—giving rise to the superconductor/paramagnet interface $S_\text{s,m}$—or with the surface of the vacuum domain S_v—forming the superconductor/vacuum interface $S_\text{s,v}$—or with both. If moreover all, or part, of the surface of the paramagnetic constituent $S_{\rm m}$ coincides with the surface of the vacuum domain $S_{\rm v}$, the paramagnet/vacuum interface $S_{\rm m,v}$ as well comes into play. This also is the interface where the scalar magnetic potentials $\psi_{\rm m}$ and $\psi_{\rm v}$ should match to render their introduction in eqns. (B.8) and (B.10) unique. In any case, $\psi_{\rm m}$ and/or $\psi_{\rm v}$ must decay to 0 far from the vortex cores if their domains of definition $V_{\rm m}$ and/or $V_{\rm v}$ extend unlimitedly.

Let us revert to eqn. (B.6) and start with inspecting its second term. Clearly, the requirement of continuity of the tangential components of the magnetic field across the superconductor/paramagnet interface $S_{\rm s,m}$ and/or the superconductor/vacuum interface $S_{\rm s,v}$ exists. Representing the magnetic field through the scalar magnetic potential

$$\psi = \begin{cases} \psi_{\rm m} & \text{on } S_{\rm s,m} \\ \psi_{\rm v} & \text{on } S_{\rm s,v} \end{cases} \tag{B.11}$$

and using some vector analysis, we therefore obtain

$$\int\limits_{S_{\rm s}} d\mathbf{s}_{\rm s} \cdot (\mathbf{B}_{\rm s} \times (\nabla \times \mathbf{B}_{\rm s})) = \mu_0 \int\limits_{S_{\rm s}} d\mathbf{s}_{\rm s} \cdot (\psi\, \nabla \times (\nabla \times \mathbf{B}_{\rm s}) - \nabla \times (\psi\, \nabla \times \mathbf{B}_{\rm s}))\,. \tag{B.12}$$

Fortunately, the second term of the integral on the right-hand side of eqn. (B.12) does not contribute to the integral on the left-hand side (Kogan 1994). This is easily seen when splitting the surface of the superconducting constituent $S_{\rm s}$ into two arbitrary parts, $S_{\rm s\pm}$, divided by the closed contour $\partial S_{\rm s}$. Exploiting Stokes' theorem and applying Ampère's law, eqn. (2.19), at first yields

$$\int\limits_{S_{\rm s\pm}} d\mathbf{s}_{\rm s} \cdot (\nabla \times (\psi\, \nabla \times \mathbf{B}_{\rm s})) = \pm\mu_0 \int\limits_{\partial S_{\rm s}} d\mathbf{l}_{\rm s} \cdot (\mathbf{j}_{\rm s}\, \psi)\,. \tag{B.13}$$

Integration over the total surface $S_{\rm s}$ hence leads to mutual cancellation of the two results of eqn. (B.13). Utilizing this finding for eqn. (B.12) and applying the generalized London equation (2.20) thus enables both terms of eqn. (B.6) to be cast together into the form

$$W_{\rm s} = \frac{1}{2\mu_0} \int\limits_{V_{\rm s}} dv\, (\mathbf{V} \cdot \mathbf{B}_{\rm s}) + \frac{1}{2} \int\limits_{S_{\rm s,m}} d\mathbf{s}_{\rm s} \cdot ((\mathbf{V} - \mathbf{B}_{\rm s})\, \psi_{\rm m}) + \frac{1}{2} \int\limits_{S_{\rm s,v}} d\mathbf{s}_{\rm s} \cdot ((\mathbf{V} - \mathbf{B}_{\rm s})\, \psi_{\rm v})\,. \tag{B.14}$$

In a similar vein, when remembering that all, or part, of the surface of the paramagnetic constituent $S_{\rm m}$ may represent the superconductor/paramagnet interface $S_{\rm s,m}$ (where $d\mathbf{s}_{\rm s} = -d\mathbf{s}_{\rm m}$) or the paramagnet/vacuum interface $S_{\rm m,v}$, or both, and ensuring continuity of the normal component of the magnetic induction across $S_{\rm s,m}$, eqn. (B.8) can be rewritten as

$$W_{\rm m} = \frac{1}{2} \int\limits_{S_{\rm s,m}} d\mathbf{s}_{\rm s} \cdot (\mathbf{B}_{\rm s}\, \psi_{\rm m}) - \frac{1}{2} \int\limits_{S_{\rm m,v}} d\mathbf{s}_{\rm m} \cdot (\mathbf{B}_{\rm m}\, \psi_{\rm m})\,. \tag{B.15}$$

Likewise, when recalling that all, or part, of the surface of the vacuum domain S_v may form the paramagnet/vacuum interface $S_\mathrm{m,v}$ (where $\mathrm{d}\mathbf{s}_\mathrm{v} = -\mathrm{d}\mathbf{s}_\mathrm{m}$) or the superconductor/vacuum interface $S_\mathrm{s,v}$, or both, and ensuring continuity of the normal component of the magnetic induction across $S_\mathrm{m,v}$, eqn. (B.10) can be transcribed as

$$W_\mathrm{v} = \frac{1}{2}\int\limits_{S_\mathrm{s,v}} \mathrm{d}\mathbf{s}_\mathrm{s} \cdot (\mathbf{B}_\mathrm{s}\, \psi_\mathrm{v}) + \frac{1}{2}\int\limits_{S_\mathrm{m,v}} \mathrm{d}\mathbf{s}_\mathrm{m} \cdot (\mathbf{B}_\mathrm{m}\, \psi_\mathrm{v})\,. \tag{B.16}$$

If, finally, matching of ψ_m and ψ_v at the paramagnet/vacuum interface $S_\mathrm{m,v}$ is ensured, eqn. (B.1) with eqns. (B.14), (B.15), and (B.16) adopts the succinct form

$$W_\mathrm{V} = \frac{1}{2\mu_0}\int\limits_{V_\mathrm{s}} dv\, (\mathbf{V} \cdot \mathbf{B}_\mathrm{s}) + \frac{1}{2}\int\limits_{S_\mathrm{s,m}} \mathrm{d}\mathbf{s}_\mathrm{s} \cdot (\mathbf{V}\, \psi_\mathrm{m}) + \frac{1}{2}\int\limits_{S_\mathrm{s,v}} \mathrm{d}\mathbf{s}_\mathrm{s} \cdot (\mathbf{V}\, \psi_\mathrm{v})\,. \tag{B.17}$$

Since, apart from the vorticity $\mathbf{V}$ itself, only the magnetic induction $\mathbf{B}_\mathrm{s}$ and the scalar magnetic potentials ψ_m and ψ_v delineating the constituents of the superconductor/paramagnet heterostructure enter herein, eqn. (B.17) can be evaluated most conveniently.

Three special cases deserve mentioning. If the superconductor/paramagnet interface $S_\mathrm{s,m}$ is the only interface with the superconductor present, eqn. (B.17) reduces to

$$W_\mathrm{V} = \frac{1}{2\mu_0}\int\limits_{V_\mathrm{s}} dv\, (\mathbf{V} \cdot \mathbf{B}_\mathrm{s}) + \frac{1}{2}\int\limits_{S_\mathrm{s,m}} \mathrm{d}\mathbf{s}_\mathrm{s} \cdot (\mathbf{V}\, \psi_\mathrm{m})\,. \tag{B.18}$$

Conversely, if the superconductor/vacuum interface $S_\mathrm{s,v}$ is the only interface with the superconductor present, eqn. (B.17) reduces to

$$W_\mathrm{V} = \frac{1}{2\mu_0}\int\limits_{V_\mathrm{s}} dv\, (\mathbf{V} \cdot \mathbf{B}_\mathrm{s}) + \frac{1}{2}\int\limits_{S_\mathrm{s,v}} \mathrm{d}\mathbf{s}_\mathrm{s} \cdot (\mathbf{V}\, \psi_\mathrm{v})\,. \tag{B.19}$$

If there is no interface at all, as for a superconductor domain of unlimited extent, eqn. (B.17) simplifies to

$$W_\mathrm{V} = \frac{1}{2\mu_0}\int\limits_{V_\mathrm{s}} dv\, (\mathbf{V} \cdot \mathbf{B}_\mathrm{s})\,, \tag{B.20}$$

covering the result for a straight magnetic vortex of infinite length, eqn. (2.26).

References

Abramowitz, M. and Stegun, I. A. (1972). *Handbook of Mathematical Functions.* Dover, New York.
Abrikosov, A. A. (1957). *Sov. Phys.—JETP* **5**, 1174.
Abrikosov, A. A. (1988). *Fundamentals of the Theory of Metals.* North-Holland, Amsterdam.
Ainslie, M. D., Yuan, W. and Flack, T. J. (2013). *IEEE Trans. Appl. Supercond.* **23**, 4700104.
Aladyshkin, A. Y., Mel'nikov, A. S., Shereshevsky, I. A. and Tokman, I. D. (2001). *Physica* C **361**, 67.
Aladyshkin, A. Y., Silhanek, A. V., Gillijns, W. and Moshchalkov, V. V. (2009). *Supercond. Sci. Technol.* **22**, 053001.
Aladyshkin, A. Y., Volodin, A. P. and Moshchalkov, V. V. (2010). *J. Appl. Phys.* **108**, 033911.
Alamgir, A. K. M., Gu, C. and Han, Z. (2005). *Physica* C **432**, 153.
Aliev, Y. M. and Silin, V. P. (1993). *Sov. Phys.—JETP* **77**, 142.
Aliev, Y. M., Silin, V. P. and Uryupin, S. A. (1992). *Superconductivity* **5**, 230.
Amemiya, N. and Nakahata, M. (2007). *Physica* C **463-465**, 775.
Amemiya, N., Nakahata, M., Fujiwara, N. and Shiohara, Y. (2010). *Supercond. Sci. Technol.* **23**, 014022.
Anlage, S. M. (2011). *J. Opt.* **13**, 024001.
Arfken, G. B., Weber, H. J. and Harris, F. E. (2012). *Mathematical Methods for Physicists: A Comprehensive Guide.* Academic Press, New York.
Ashkin, M. (1979). *J. Appl. Phys.* **50** 7060.
Babaei Brojeny, A. A. and Clem, J. R. (2003). *Phys. Rev.* B **68**, 174514.
Badía-Majós, A. and López, C. (2001). *Phys. Rev. Lett.* **87**, 127004.
Badía-Majós, A. and López, C. (2014). *Supercond. Sci. Technol.* **25**, 104004.
Badía-Majós, A., López, C. and Ruiz, H. S. (2009). *Phys. Rev.* B **80**, 144509.
Bardeen, J. (1962). *Rev. Mod. Phys.* **34**, 667.
Bardeen, J., Cooper, L. N. and Schrieffer, J. R. (1957). *Phys. Rev.* **108**, 1175.
Bardeen, J. and Stephen, M. J. (1965). *Phys. Rev.* **140**, 1197.
Barone, A. and Paternò, G. (1982). *Physics and Applications of the Josephson Effect.* Wiley, New York.
Batygin, V. V. and Toptygin, I. N. (1978). *Problems in Electrodynamics.* Academic Press, London.
Bean, C. P. (1962). *Phys. Rev. Lett.* **8**, 250.
Bean, C. P. (1964). *Rev. Mod. Phys.* **36**, 31.
Bean, C. P. and Livingston, J. D. (1964). *Phys. Rev. Lett.* **12**, 14.
Bednorz, J. G. and Müller, K. A. (1986). *Z. Phys.* B **64**, 189.

Belevtsov, L. V., Genenko, Y. A., Ivanchenko, Y. M. and Medvedev, Y. V. (1991). *Sov. Low Temp. Phys.* **17**, 655.

Benkraouda, M. and Clem, J. R. (1996). *Phys. Rev.* B **53**, 5716.

Berrospe-Juarez, E., Trillaud, F., Zermeño, V. M. R. and Grilli, F. (2021). *Supercond. Sci. Technol.* **34**, 044002.

Blatter, G., Geshkenbein, V. B. and Larkin, A. I. (1992). *Phys. Rev. Lett.* **68**, 875.

Bondarenko, S. I., Shablo, A. A., Pavlov, P. P. and Perepelkin, S. S. (2002). *Physica* C **372–6**, 158.

Brandt, E. H. (1992). *Phys. Rev.* B **46**, 8628.

Brandt, E. H. (1995a). *Phys. Rev. Lett.* **74**, 3025.

Brandt, E. H. (1995b). *Rep. Prog. Phys.* **58**, 1465.

Brandt, E. H. and Indenbom, M. (1993). *Phys. Rev.* B **48**, 12893.

Brandt, E. H. and Mikitik, G. P. (2000). *Phys. Rev. Lett.* **85**, 4164.

Brandt, E. H. and Mikitik, G. P. (2007). *Phys. Rev.* B **76**, 064526.

Buckel, W. and Kleiner, R. (2012). *Supraleitung.* Wiley-VCH, Weinheim.

Campbell, A. M. (2007). *Supercond. Sci. Technol.* **20**, 292.

Campbell, A. M. and Evetts, J. E. (1972). *Critical Currents in Superconductors.* Taylor and Francis, London.

Canfield, P. C., Bud'ko, S. L. and Finnemore, D. K. (2003). *Physica* C **385**, 1.

Cap, F. F. (2003). *Mathematical Methods in Physics and Engineering with MATHEMATICA*. Chapman and Hall, New York.

Carr, W. J. (2001). *AC Loss and Macroscopic Theory of Superconductors.* Taylor and Francis, London.

Chang, W. H. (1979). *J. Appl. Phys.* **50**, 8129.

Claassen, J. H. (2007). *Supercond. Sci. Technol.* **20**, 35.

Claassen, J. H. and Thieme, C. L. H. (2008). *Supercond. Sci. Technol.* **21**, 105003.

Clem, J. R. (1991). *Phys. Rev.* B **43**, 7837.

Clem, J. R. (2008). *Phys. Rev.* B **77**, 134506.

Clem, J. R. (2009). *Phys. Rev.* B **80**, 184517.

Clem, J. R. and Malozemoff, A. P. (2010). *Supercond. Sci. Technol.* **23**, 034014.

Clem, J. R., Pe, T. and Benkraouda, M. (1997). *Physica* C **282–7**, 311.

Clem, J. R. and Pérez-González, A. (1984). *Phys. Rev.* B **30**, 5041.

Cook, R. D., Malkus, D. S., Plesha, M. E. and Witt, R. J. (2002). *Concepts and Applications of Finite-Element Analysis.* Wiley, New York.

Dasgupta, T., Derov, J. S. and Sridhar, S. (2001). *J. Appl. Phys.* **90**, 2915.

Deaver, B. S. and Fairbank, W. M. (1961). *Phys. Rev. Lett.* **7**, 43.

Del-Valle, N., Navau, C., Sanchez, A. and Chen, D. X. (2011). *Appl. Phys. Lett.* **98**, 202506.

Del-Valle, N., Navau, C., Sanchez, A. and Dinner, R. B. (2012). *AIP Adv.* **2**, 022166.

Delves, L. M. and Mohamed, J. L. (1985). *Computational Methods for Integral Equations.* Cambridge University Press, Cambridge, UK.

Díaz, A., Méchin, L., Berghuis, P. and Evetts, J. E. (1998a). *Phys. Rev.* B **58**, R2960.

Díaz, A., Méchin, L., Berghuis, P. and Evetts, J. E. (1998b). *Phys. Rev. Lett.* **80**, 3855.

Doll, R. and Näbauer, M. (1961). *Phys. Rev. Lett.* **7**, 51.

Doss, J. D. (1989). *Engineer's Guide to High-Temperature Superconductivity.* Wiley, New York.

Drozdov, A. P., Eremets, M. I., Troyan, I. A., Ksenofontov, V. and Shylin, S. I. (2015). *Nature* **525**, 73.

Duckworth, R. C., Thompson, J. R., Gouge, M. J., Lue, J. W., Ijaduola, A. O., Yu, D. and Verebelyi, D. T. (2003). *Supercond. Sci. Technol.* **16**, 1294.

Efetov, K. B. (1979). *Sov. Phys.—JETP* **49**, 905.

Erdélyi, A., Magnus, W., Oberhettinger, F. and Tricomi, F. G. (1953). *Higher Transcendental Functions.* McGraw-Hill, New York.

Essmann, U. and Träuble, H. (1967). *Phys. Lett.* A **24**, 526.

Fareed, M. U. and Ruiz, H. S. (2021). *Materials* **14**, 6204.

Fareed, M., Robert, B. and Ruiz, H. S. (2019). *IEEE Trans. Appl. Supercond.* **29**, 5900705.

Farinon, S., Fabbricatore, P., Gömöry, F., Greco, M. and Seiler, E. (2005). *IEEE Trans. Appl. Supercond.* **15**, 2867.

Feigel'man, M. V., Geshkenbein V. B. and Larkin, A. I. (1990). *Physica* C **167**, 177.

Ferrell, R. A. and Prange, R. E. (1963). *Phys. Rev. Lett.* **10**, 479.

Fiory, A. T., Hebard, A. F., Mankievich, P. M. and Howard, R. E. (1988). *Appl. Phys. Lett.* **52**, 2165.

Friedman, A., Shaked, N., Perel, E., Sinvani, M., Wolfus, Y. and Yeshurun, Y. (1999). *Cryogenics* **39**, 53.

Gadala, M. S. (2020). *Finite Elements for Engineers with ANSYS Applications.* Cambridge University Press, Cambridge, UK.

Gaevski, M. E., Bobyl, A. V., Shantsev, D. V., Galperin, Y. M., Johansen, T. H., Baziljevich, M., Bratsberg, H. and Karmanenko, S. F. (1999). *Phys. Rev.* B **59**, 9655.

Galaiko, V. P. (1966). *Sov. Phys.—JETP* **23**, 878.

Gallop, J. C. (1991). *SQUIDs, the Josephson Effects and Superconducting Electronics.* Hilger, Bristol.

Gammel, P. L., Bishop, D. J., Dolan, G. J., Kwo, J. R., Murray, C. A., Schneemeyer, L. F. and Waszcak, J. V. (1987). *Phys. Rev. Lett.* **59**, 2592.

Gardner, M. W., Govorkov, S. A., Liang, R., Bonn, D. A., Carolan, J. F. and Hardy, W. N. (1998). *J. Appl. Phys.* **83**, 3714.

Genenko, Y. A. (1992). *Sverkhprovodimost (KIAE)* **5**, 1402 [*Superconductivity (Moscow)* **5**, 1376].

Genenko, Y. A. (1994). *Phys. Rev.* B **49**, 6950.

Genenko, Y. A. (1995). *Phys. Rev.* B **51**, 3686.

Genenko, Y. A., Snezhko, A. V., Troche, P., Hoffmann, J. and Freyhardt, H. C. (1998). *Phys. Rev.* B **57**, 1164.

Genenko, Y. A. (2002a). *Physica* C **372-376**, 1847.

Genenko, Y. A. (2002b). *Phys. Rev.* B **66**, 184520.

Genenko, Y. A. (2002c). *Phys. Stat. Sol* (a) **189**, 469.

Genenko, Y. A. (2004). *Physica* C **401**, 210.

Genenko, Y. A. and Rauh, H. (2003). *Appl. Phys. Lett.* **82**, 2115.

Genenko, Y. A. and Rauh, H. (2006). *J. Phys.: Conf. Ser.* **43**, 568.

Genenko, Y. A. and Rauh, H. (2007). *Physica* C **460–2**, 1264.
Genenko, Y. A. and Rauh, H. (2010). *Supercond. Sci. Technol.* **23**, 075007.
Genenko, Y. A., Rauh, H. and Krüger, P. (2011). *Appl. Phys. Lett.* **98**, 152508.
Genenko, Y. A., Rauh, H., Krüger, P. and Narayanan, N. (2009). *Supercond. Sci. Technol.* **22**, 055001.
Genenko, Y. A., Rauh, H. and Kurdi, S. (2015). *J. Appl. Phys.* **117**, 243909.
Genenko, Y. A., Rauh, H. and Snezhko, A. (2001). *Supercond. Sci. Technol.* **14**, 699.
Genenko, Y. A., Rauh, H. and Snezhko, A. (2002). *Physica* C **372–6**, 1389.
Genenko, Y. A., Rauh, H. and Snezhko, A. (2003). In *Supraleitung und Tieftemperaturtechnik* (eds. J. Fröhlingsdorf, H. Hoffschulz and F. Sicking) m. 7. VDI Technologiezentrum, Düsseldorf.
Genenko, Y. A., Rauh, H. and Yampolskii, S. V. (2005). *J. Phys.: Condens. Matter* **17**, L93.
Genenko, Y. A. and Snezhko, A. (2002). *J. Appl. Phys.* **92**, 357.
Genenko, Y. A., Snezhko, A. and Freyhardt, H. C. (2000). *Phys. Rev.* B **62**, 3453.
Genenko, Y. A., Snezhko, A., Troche, P., Hoffmann, J. and Freyhardt, H. C. (1998). *Phys. Rev.* B **57**, 1164.
Genenko, Y. A., Snezhko, A. and Usoskin, A. (2004a). *IOP Conf. Ser.* **181**, 2599.
Genenko, Y. A., Snezhko, A. and Usoskin, A. (2004b). *Physica* C **401**, 236.
Genenko, Y. A., Usoskin, A. and Freyhardt, H. C. (1999). *Phys. Rev. Lett.* **83**, 3045.
Genenko, Y. A., Usoskin, A., Snezhko, A. and Freyhardt, H. C. (2000a). *Physica* C 341–8, 1063.
Genenko, Y. A., Usoskin, A., Snezhko, A. and Freyhardt, H. C. (2000b). *IOP Conf. Ser.* **167**, 483.
Genenko, Y. A., Yampolskii, S. V. and Pan, A. V. (2004). *Appl. Phys. Lett.* **84**, 3921.
Gennes, P. G. de (1994). *Superconductivity of Metals and Alloys.* Addison-Wesley, New York.
Gheorghe, D. G., Wijngaarden, R. J., Bell, C. and Aarts, J. (2009). *Phys. Rev.* B **80**, 144518.
Gillijns, W., Milošević, M. V., Silhanek, A. V., Moshchalkov, V. V. and Peeters, F. M. (2007). *Phys. Rev.* B **76**, 184516.
Ginzburg, V. L. and Landau, L. D. (1950). *Zh. Eksp. Teor. Fiz.* **20**, 1064.
Glowacki, B. A. and Majoros, M. (2000). *Supercond. Sci. Technol.* **13**, 483.
Glowacki, B. A. and Majoros, M. (2009). *J. Phys.: Condens. Matter* **21**, 254206.
Gömöry, F. (2006). *Appl. Phys. Lett.* **89**, 072506.
Gömöry, F., Solovyov, M. and Šouc, J. (2015). *Supercond. Sci. Technol.* **28**, 044001.
Gömöry, F., Solovyov, M., Šouc, J., Navau, C., Prat-Camps, J. and Sanchez, A. (2012). *Science* **335**, 1466.
Gömöry, F., Šouc, J., Vojenčiak, M., Alamgir, A. K. M., Han, Z. and Gu, C. (2007). *Appl. Phys. Lett.* **90**, 092506.
Gömöry, F., Šouc, J., Vojenčiak, M., Seiler, E., Klinčok, B., Ceballos, J. M., Pardo, E., Sanchez, A., Navau, C., Farinon, S. and Fabbricatore, P. (2006). *Supercond. Sci. Technol.* **19**, S60.

Gömöry, F., Vojenčiak, M., Pardo, E., Solovyov, M. and Šouc, J. (2010). *Supercond. Sci. Technol.* **23**, 034012.

Gömöry, F., Vojenčiak, M., Pardo, E. and Šouc, J. (2009). *Supercond. Sci. Technol.* **22**, 034017.

Gradshteyn, I. S. and Ryzhik, I. M. (1994). *Table of Integrals, Series and Products.* Academic Press, New York.

Grilli, F., Ashworth, S. P. and Civale, L. (2007). *J. Appl. Phys.* **102**, 073909.

Grilli, F. and Kario, A. (2016). *Supercond. Sci. Technol.* **29**, 083002.

Grilli, F., Pardo, E., Stenvall, A., Nguyen, D. N., Yuan, W. and Gömöry, F. (2014). *IEEE Trans. Appl. Supercond.* **24**, 8200433.

Gu, C., Alamgir, A. K. M., Qu, T. and Han, Z. (2007). *Supercond. Sci. Technol.* **20**, 133.

Gudoshnikov, S. A., Deryuzhkina, Y. V., Rudenchik, P. E., Sitnov, Y. S., Bondarenko, S. I., Shablo, A. A., Pavlov, P. P., Kalabukhov, A. S., Snigirev, O. V. and Seidel, P. (2001). *IEEE Trans. Appl. Supercond.* **11**, 219.

Gurevich, A. (1992). *Phys. Rev.* B **46**, 3187.

Gurevich, A. (1993). *Phys. Rev.* B **48**, 12857.

Gurevich, A., Mints, R. G. and Rakhmanov, A. L. (1997). *The Physics of Composite Superconductors.* Begell House, New York.

He, A., Xue, C., Yong, H. D. and Zhou, Y. H. (2014). *Supercond. Sci. Technol.* **27**, 025004.

Heese, H. (2004). *Math. Meth. Appl. Sci.* **27**, 1725.

Hong, Z., Jiang, Q., Pei, R., Campbell, A. M. and Coombs, T. A. (2007). *Supercond. Sci. Technol.* **20**, 331.

Horvat, J., Soltanian, S. and Yeoh, W. K. (2005). *Supercond. Sci. Technol.* **18**, 682.

Horvat, J., Wang, X. L., Soltanian, S. and Dou, S. X. (2002). *Appl. Phys. Lett.* **80**, 829.

Horvat, J., Yeoh, W. K. and Miller, L. M. (2005). *Appl. Phys. Lett.* **87**, 102503.

Huebener, R. P. (2010). *Magnetic Flux Structures in Superconductors.* Springer, Berlin.

Indenbom, M., Kronmüller, H., Li, T. W., Kes, P. H. and Menovsky, A. A. (1994). *Physica* C **222**, 203.

Ivanchenko, Y. M. and Soboleva, T. K. (1990a). *Phys. Lett.* A **147**, 65.

Ivanchenko, Y. M. and Soboleva, T. K. (1990b). *Sov. Phys.—Solid St.* **32**, 1181.

Ivanchenko, Y. M., Belevtsov, L. V., Genenko, Y. A. and Medvedev, Y. V. (1992). *Physica* C **193**, 291.

Jackson, J. D. (1998). *Classical Electrodynamics.* Wiley, New York.

Jarzina, H., Jooss, C. and Freyhardt, H. C. (2002). *J. Appl. Phys.* **91**, 3775.

Johansen, T. H., Baziljevich, M., Bratsberg, H., Galperin, Y. M., Lindelof, P. E., Shen, Y. and Vase, P. (1996). *Phys. Rev.* B **54**, 16264.

Jooss, C., Brinkmeier, E. and Heese, H. (2005). *Phys. Rev.* B **72**, 144516.

Jooss, C., Forkl, A., Warthmann, R., Habermeier, H. U., Leibold, B. and Kronmüller, H. (1996). *Physica* C **266**, 235.

Jooss, C., Warthmann, R., Forkl, A. and Kronmüller, H. (1998). *Physica* C **299**, 215.

Josephson, B. D. (1962). *Phys. Lett.* **1**, 251.

Kajikawa, K., Hayashi, T., Yoshida, R., Iwakuma, M. and Funaki, K. (2003). *IEEE Trans. Appl. Supercond.* **13**, 3630.

Kajikawa, K., Kawano, T., Osaka, R., Nakamura, T., Sugano, M., Takahashi, M. and Wakuda, T. (2010). *Supercond. Sci. Technol.* **23**, 045026.

Kamerlingh Onnes, H. (1911). *Comm. Phys. Lab. Univ. Leiden* **120**, 124.

Kapolka, M. and Ruiz, H. S. (2022). *Sci. Rep.* **12**, 7030.

Ketterson, J. B. and Song, S. N. (1999). *Superconductivity.* Cambridge University Press, Cambridge, UK.

Kim, Y. B., Hempstead, C. F. and Strnad, A. R. (1964). *Rev. Mod. Phys.* **36**, 43.

Kogan, V. G. (1994). *Phys. Rev.* B **49**, 15874.

Kogan, V. G., Dobrovitski, V. V., Clem, J. R., Mawatari, Y. and Mints, R. G. (2001). *Phys. Rev.* B **63**, 144501.

Kondoh, J. and Umeda, M. (2002). *Physica* C **382**, 16.

Koshelets, V. P. and Shitov, S. V. (2000). *Supercond. Sci. Technol.* **13**, R53.

Koshelev, A. E. (1992). *Physica* C **191**, 219.

Kováč, P., Hušek, I., Melišek, T., Ahoranta, M., Šouc, J., Lehtonen, J. and Gömöry, F. (2003). *Supercond. Sci. Technol.* **16**, 1195.

Krüger, P. A. C., Grilli, F. and Farinon, S. (2011). *Physica* C **471**, 1083.

Krüger, P., Grilli, F., Vojenčiak, M., Zermeño, V. M. R., Demencik, E. and Farinon, S. (2013). *Appl. Phys. Lett.* **102**, 202601.

Kunchur, M. N., Christen, D. K. and Phillips, J. M. (1993). *Phys. Rev. Lett.* **70**, 998.

Kunzler, J. E., Buehler, E., Hsu, F. S. L. and Wernick, J. H. (1961). *Phys. Rev. Lett.* **6**, 89.

Kupriyanov, M. Y. and Likharev, K. K. (1975). *Sov. Phys.—Solid St.* **16**, 1835.

Kuzovlev, Y. E. (1997). *Physica* C **292**, 117.

Kuzovlev, Y. E. and Lomtev, A. I. (1997). *Sov. Phys.—JETP* **84**, 986.

Laiho, R., Lähderanta, E., Sonin, E. B. and Traito K. B. (2003). *Phys. Rev.* B **67**, 144522.

Lancaster, M. J. (2006). *Passive Microwave Device Applications of High-Temperature Superconductors.* Cambridge University Press, Cambridge, UK.

Landau, L. D. (1930). *Z. Phys.* **64**, 629.

Landau, L. D., Lifshitz, E. M. and Pitaevskii, L. P. (1995). *Electrodynamics of Continuous Media.* Elsevier, Oxford.

Lange, M., Van Bael, M. J., Bruynseraede, Y. and Moshchalkov, V. V. (2003). *Phys. Rev. Lett.* **90**, 197006.

Larbalestier, D., Gurevich, A., Feldman, D. M. and Polyanskii, A. (2001). *Nature* **414**, 368.

Laurinčikas, A. and Garunkštis, R. (2002). *The Lerch Zeta Function.* Kluwer Academic, Dordrecht.

Lehner, G. (2010). *Elektromagnetische Feldtheorie.* Springer, Berlin.

Leonhardt, U. (2006). *Science* **312**, 1777.

Likharev, K. K. (1971). *Radiophys. Quantum Electron.* **14**, 722.

London, F. (1948). *Phys. Rev.* **74**, 562.

London, F. (1950). *Superfluids: Macroscopic Theory of Superconductivity.* Wiley, New York.

London, F. and London, H. (1935). *Proc. R. Soc. London* A **149**, 71.
London, H. (1963). *Phys. Lett.* A **6**, 162.
Loudon, J. C. and Midgley, P. A. (2009). *Ultramicroscopy* **109**, 700.
Lousberg, G. P., Fagnard, J. F., Ausloos, M., Vanderbemden, P. and Vanderheyden, B. (2010). *IEEE Trans. Appl. Supercond.* **20**, 33.
Lyuksyutov, I. F. and Pokrovsky, V. L. (2005). *Adv. Phys.* **54**, 67.
Ma, G. T. and Rauh, H. (2013). *Supercond. Sci. Technol.* **26**, 105001.
Maggio-Aprile, I., Renner, C., Erb, A., Walker, E. and Fischer, ø. (1995). *Phys. Rev. Lett.* **75**, 2754.
Majoros, M. (1996). *Physica* C **272**, 62.
Majoros, M., Glowacki, B. A. and Campbell, A. M. (2000a). *Physica* C **334**, 129.
Majoros, M., Glowacki, B. A. and Campbell, A. M. (2000b). *Physica* C **338**, 251.
Majoros, M., Sumption, M. D., Collings, E. W. and van der Laan, D. C. (2014). *Supercond. Sci. Technol.* **27**, 125008.
Malozemoff, A. P., Fleshler, S., Rupich, M., Thieme, C. L. H., Li, X., Zhang, W., Otto, A., Maguire, J., Folts, D., Yuan, J., Kraemer, H. P., Schmidt, W., Wohlfart, M. and Neumüller, H. W. (2008). *Supercond. Sci. Technol.* **21**, 034005.
Mawatari, Y. (1996). *Phys. Rev.* B **54**, 13215.
Mawatari, Y. (2008). *Phys. Rev.* B **77**, 104505.
Mawatari, Y. (2009). *Phys. Rev.* B **80**, 184508.
Mawatari, Y. (2011). *Phys. Rev.* B **83**, 134512.
Mawatari, Y. and Clem, J. R. (2001). *Phys. Rev. Lett.* **86**, 2870.
Meissner, W. and Ochsenfeld, R. (1933). *Naturwiss.* **21**, 787.
Meister, E. (1983). *Randwertaufgaben der Funktionentheorie.* Teubner, Stuttgart.
Mikheenko, P. N. and Kuzovlev, Y. E. (1993). *Physica* C **204**, 229.
Mikhlin, S. G. (1964). *Integral Equations.* Pergamon Press, Oxford.
Mikitik, G. P., Mawatari, Y., Wan, A. T. S. and Sirois, F. (2013). *IEEE Trans. Appl. Supercond.* **23**, 8001920.
Miyagi, D., Amadutsumi, Y. and Takahashi, N. (2007). *IEEE Trans. Appl. Supercond.* **17**, 3167.
Miyagi, D., Umabuchi, M., Takahashi, N. and Tsukamoto, O. (2007). *Physica* C **463–5**, 785.
Miyagi, D., Umabuchi, M., Takahashi, N. and Tsukamoto, O. (2008a). *IEEE Trans. Appl. Supercond.* **18**, 1297.
Miyagi, D., Umabuchi, M., Takahashi, N. and Tsukamoto, O. (2008b). *IEEE Trans. Appl. Supercond.* **18**, 1374.
Miyagi, D., Yunoki, Y., Umabuchi, M., Takahashi, N. and Tsukamoto, O. (2008). *Physica* C **468**, 1743.
Monaco, R., Boffa, M., Beneduce, C., Cucolo, A. M. and Cucolo, M. C. (2000). *Physica* C **341–8**, 2685.
Moon, P. and Spencer, D. E. (1988). *Field Theory Handbook.* Springer, Berlin.
Morelle, M. and Moshchalkov, V. V. (2006). *Appl. Phys. Lett.* **88**, 172507.
Morse, P. M. and Feshbach, H. (1953). *Methods of Theoretical Physics.* McGraw-Hill, New York.
Muchowski, E. and Schmid, A. (1972). *Z. Phys.* **255**, 187.

Müller, K. H. (1997). *Physica* C **289**, 123.
Muskhelishvili, N. I. (2008). *Singular Integral Equations.* Dover Publications, New York.
Narayana, S. and Sato, Y. (2012). *Adv. Mater.* **24**, 71.
Neumüller, H. W., Schmidt, W., Kinder, H., Freyhardt, H. C., Stritzker, B., Wördenweber, R. and Kirchhoff, V. (1997). *J. Alloys Compd.* **251**, 366.
Newhouse, V. L. (1969). In *Superconductivity*, Vol. 2 (ed. R. D. Parks), p. 1308. Dekker, New York.
Nguyen, D. N., Ashworth, S. P. and Willis, J. O. (2009). *J. Appl. Phys.* **106**, 093913.
Nguyen, D. N., Ashworth, S. P., Willis, J. O., Sirois, F. and Grilli, F. (2010). *Supercond. Sci. Technol.* **23**, 025001.
Nguyen, D. N., Sastry, P. V. P. S. S. and Schwartz, J. (2007). *J. Appl. Phys.* **101**, 053905.
Nicolaides, R. and Walkington, N. (1996). *MAPLE: A Comprehensive Introduction.* Cambridge University Press, Cambridge, UK.
Niu, M., Yong, H. D., Xia, J. and Zhou, Y. H. (2019). *J. Supercond.* Nov. *Magn.* **32**, 499.
Norris, W. T. (1970). *J. Phys.* D **3**, 489.
Ogawa, J., Fukui, S., Oka T., Yamaguchi, M., Sato, T., Kume, S., Shinkai, K. and Sato, S. (2008). *IEEE Trans. Appl. Supercond.* **18**, 1353.
Orlando, T. P. and Delin, K. A. (1991). *Foundations of Applied Superconductivity.* Addison-Wesley, New York.
Ott, H. W. (2009). *Electromagnetic Compatibility Engineering.* Wiley, New York.
Pan, A. V. and Dou, S. X. (2004). *J. Appl. Phys.* **96**, 1146.
Pan, A. V., Zhou, S. and Dou, S. X. (2004). *Supercond. Sci. Technol.* **17**, S410.
Pan, A. V., Zhou, S., Liu, H. K. and Dou, S. X. (2003). *Supercond. Sci. Technol.* **16**, L33.
Pearl, J. (1964). *Appl. Phys. Lett.* **5**, 65.
Peña-Roche, J., Genenko, Y. A. and Badía-Majós, A. (2016). *Appl. Phys. Lett.* **109**, 092601.
Pence, T. J. and Wichman, I. S. (2020). *Essential Mathematics for Engineers and Scientists.* Cambridge University Press, Cambridge, UK.
Pendry, J. B., Schurig, D. and Smith, D. R. (2006). *Science* **312**, 1780.
Petukhov, B. V. and Chechetkin, V. R. (1973). *Zh. Exp. Teor. Fiz.* **65**, 1653.
Plourde, B. L. T., Van Harlingen, D. J., Besseling, R., Hesselberth, M. B. S. and Kes, P. H. (2000). *Physica* C 341–8, 1023.
Poole, C. P., Farach, H. A., Creswick, R. J. and Prozorov, R. (2007). *Superconductivity.* Elsevier, Oxford.
Portis, A. M. (1993). *Electrodynamics of High-Temperature Superconductors.* World Scientific, Singapore.
Press, W. H., Teukolsky, S. A., Vetterling, W. T. and Flannery, B. P. (2007). *Numerical Recipes in FORTRAN: The Art of Scientific Computing.* Cambridge University Press, Cambridge, UK.
Prigozhin, L. (1996). *J. Comput. Phys.* **129**, 190.
Prigozhin, L. (1997). *IEEE-TAS* **7**, 3866.

Prigozhin, L. and Sokolovsky, V. (2023). *IEEE-TAS* **33**, 6601310.

Prudnikov, A. P., Brychkov, Y. A. and Marychev, O. I. (1986). *Integrals and Series.* Gordon and Breach, New York.

Pryor, R. W. (2011). *Multiphysics Modelling using COMSOL: A First Principles Approach.* Jones and Bartlett, London.

Ranecky, S. T., Watanabe, H., Ogawa, J., Oka, T., Gölden, D., Alff, L. and Genenko, Y. A. (2017). *Supercond. Sci. Technol.* **30**, 085005.

Rauh, H. and Genenko, Y. A. (2004). *Physica* C **401**, 286.

Rauh, H., Genenko, Y. A. and Rieck, C. T. (2004). *J. Phys.: Condens. Matter* **16**, S2715.

Rauh, H. and Ma, G. T. (2013). *J. Appl. Phys.* **114**, 193902.

Rauh, H., Stoneham, A. M. and Choy, T. C. (1992). In *Materials Modelling: From Theory to Technology* (eds. C. A. English, J. R. Matthews, H. Rauh, A. M. Stoneham and R. Thetford), p. 201. IOP, Bristol.

Rhoderick, E. H. and Wilson, E. M. (1962). *Nature* **194**, 1167.

Richards, D. (2002). *Advanced Mathematical Methods with MAPLE.* Cambridge University Press, Cambridge, UK.

Riley, K. F., Hobson, M. P. and Bence, S. J. (2010). *Mathematical Methods for Physics and Engineering.* Cambridge University Press, Cambridge, UK.

Rose-Innes, A. C. and Rhoderick, E. H. (1994). *Introduction to Superconductivity.* Pergamon Press, Oxford.

Ruiz, H. S. and Badía-Majós, A. (2013). *J. Appl. Phys.* **113**, 193906.

Ruiz, H. S., Badía-Majós, A., Genenko, Y. A., Rauh, H. and Yampolskii, S. V. (2012). *Appl. Phys. Lett.* **100**, 112602.

Samokhvalov, A. V. (1995). *Zh. Exp. Teor. Fiz.* **108**, 1091.

Sanchez, A., Del-Valle, N., Navau, C. and Chen, D. X. (2010). *Appl. Phys. Lett.* **97**, 072504.

Sanchez, A., Navau, C., Prat-Camps, J. and Chen, D. X. (2011). *New J. Phys.* **13**, 093034.

Schafroth, M. R. (1960). In *Solid State Physics*, Vol. 10 (eds. F. Seitz and D. Turnbull), p. 293. Academic Press, New York.

Schmidt, V. V. (1997). *The Physics of Superconductors.* Springer, Berlin.

Schuster, T., Indenbom, M., Kuhn, H., Brandt, E. H. and Konczykowski, M. (1994). *Phys. Rev. Lett.* **73**, 1424.

Seidel, P. (2015). *Applied Superconductivity: Handbook on Devices and Applications.* Wiley–VCH, Weinheim.

Sheahen, T. P. (1994). *Introduction to High-Temperature Superconductivity.* Plenum Press, New York.

Shen, F., Grilli, F. and Coombs, T. (2020). *Supercond. Sci. Technol.* **33**, 033002.

Shubnikov, L. V., Hotkevich, V. I., Shepelev, Y. D. and Ryabinin, Y. N. (1937). *Zh. Exp. Teor. Fiz.* **7**, 221.

Silhanek, A. V., Gillijns, W., Moshchalkov, V. V., Metlushko, V. and Ilic, B. (2006). *Appl. Phys. Lett.* **89**, 182505.

Silhanek, A. V., Verellen, N., Metlushko, V., Gillijns, W., Gozzini, F., Ilic, B. and Moshchalkov, V. V. (2008). *Physica* C **468**, 563.

Snider, E., Dasenbrock-Gammon, N., McBride, R., Wang, X. J., Meyers, N., Lawler, K. V., Zurek, E., Salamat, A. and Dias, R. P. (2021). *Phys. Rev. Lett.* **126**, 117003.

Soltanian, S., Wang, X. L., Kusevic, I., Babic, E., Li, A. H., Qin, M. J., Horvat, J., Liu, H. K., Collings, E. W., Lee, E., Sumption, M. D. and Dou, S. X. (2001). *Physica* C **361**, 84.

Song, M., Xia, Y. and Ma, T. (2019). *J. Supercond.* Nov. *Magn.* **32**, 3143.

Souza Silva, C. C., de Silhanek, A. V., Van de Vondel, J., Gillijns, W., Metlushko, V., Ilic, B. and Moshchalkov, V. V. (2007). *Phys. Rev. Lett.* **98**, 117005.

Statra, Y., Menana, H. and Douine, B. (2022). *IEEE Trans. Appl. Supercond.* **24**, 8200433.

Suenaga, M., Iwakuma, M., Sueyoshi, T., Izumi, T., Mimura, M., Takahashi, Y. and Aoki, Y. (2008). *Physica* C **468**, 1714.

Suenaga, M. and Li, Q. (2006). *Appl. Phys. Lett.* **88**, 262501.

Sumption, M. D., Collings, E. W., Lee, E., Wang, X. L., Soltanian, S. and Dou, S. X. (2002a). *Physica* C **378–81**, 894.

Sumption, M. D., Collings, E. W., Lee, E., Wang, X. L., Soltanian, S., Dou, S. X. and Tomsic, M. (2002b). *Physica* C **382**, 98.

Sumption, M. D., Lee, E., Collings, E. W., Wang, X. L. and Dou, S. X. (2001). *Adv. Cryog. Eng. (Mats.)* **48**, 824.

Tamm, I. E. (1979). *Fundamentals of the Theory of Electricity.* Mir Publishers, Moscow.

Tikhonov, A. N. and Samarskii, A. A. (1967). *Partial Differential Equations of Mathematical Physics.* Holden-Day, San Francisco.

Tilley, D. R. and Tilley, J. (2003). *Superfluidity and Superconductivity.* IOP, Bristol.

Tinkham, M. (1964). *Phys. Rev. Lett.* **13**, 804.

Tinkham, M. (2004). *Introduction to Superconductivity.* Dover, New York.

Tinkham, M. and Lobb, C. B. (1989). *Physical Properties of the New Superconductors.* Academic Press, New York.

Touitou, N., Bernstein, P., Hamet, J. F., Simon, C., Méchin, L., Contour, J. P. and Jacquet, E. (2004). *Appl. Phys. Lett.* **85**, 1742.

Ullmaier, H. (1975). *Irreversible Properties of Type-II Superconductors.* Springer, Berlin.

Umabuchi, M., Miyagi, D., Takahashi, N. and Tsukamoto, O. (2008). *Physica* C **468**, 1739.

Usoskin, A., Rutt, A., Knoke, J., Krauth, H. and Arndt, T. (2005). *IEEE Trans. Appl. Supercond.* **15**, 2604.

Vlasko-Vlasov, V. K., Colauto, F., Benseman, T., Rosenmann, D. and Kwok, W. K. (2017a). *Phys. Rev.* B **96**, 214510.

Vlasko-Vlasov, V. K., Colauto, F., Buzdin, A. I., Rosenmann, D., Benseman, T. and Kwok, W. K. (2017b). *Phys. Rev.* B **95**, 144504.

Vlasko-Vlasov, V. K., Divan, R., Rosenmann, D., Leishman, A., Welp, U., te Velthuis, S. G. E. and Kwok, W. K. (2022a). *J. Supercond.* Nov. *Magn.* **35**, 1555.

Vlasko-Vlasov, V. K., Palacious, E., Rosenmann, D., Pearson, J., Jia, Y., Wang, Y. L., Welp, U. and Kwok, W. K. (2015). *Supercond. Sci. Technol.* **28**, 035006.
Vlasko-Vlasov, V. K., Rydh, A., Divan, R., Rosenmann, D., Glatz, A. and Kwok, W. K. (2022b). *J. Magn. Magn. Mater.* **557**, 169476.
Vodolazov, D. Y., Gribkov, B. A., Gusev, S. A., Klimov, A. Y., Nozdrin, Y. N., Rogov, V. V. and Vdovichev, S. N. (2005). *Phys. Rev.* B **72**, 064509.
Waldram, J. R. (1996). *Superconductivity of Metals and Cuprates.* IOP, Bristol.
Wan, X. X., Huang, C. G., Yong, H. D. and Zhou, Y. H. (2015). *AIP Adv.* **5**, 117139.
Weinstock, H. (2000). *Applications of Superconductivity.* Springer, Berlin.
Wijngaarden, R. J., Spoelder, H. J. W., Surdeanu, R. and Griessen, R. (1996). *Phys. Rev.* B **54**, 6742.
Witt, T. J. (1988). *Phys. Rev. Lett.* **61**, 1423.
Wolsky, A. M. and Campbell, A. M. (2008). *Supercond. Sci. Technol.* **21**, 075021.
Yampolskii, S. V. and Genenko, Y. A. (2005). *Phys. Rev.* B **71**, 134519.
Yampolskii, S. V. and Genenko, Y. A. (2014a). *Appl. Phys. Lett.* **104**, 033501.
Yampolskii, S. V. and Genenko, Y. A. (2014b). *Appl. Phys. Lett.* **104**, 143504.
Yampolskii, S. V., Genenko, Y. A. and Rauh, H. (2004a). *IOP Conf. Ser.* **181**, 2335.
Yampolskii, S. V., Genenko, Y. A. and Rauh, H. (2004b). *Physica* C **415**, 151.
Yampolskii, S. V., Genenko, Y. A. and Rauh, H. (2007). *Physica* C **460–2**, 1262.
Yampolskii, S. V., Genenko, Y. A., Rauh, H. and Snezhko, A. (2006). *J. Phys.: Conf. Ser.* **43**, 576.
Yampolskii, S. V., Yampolskaya, G. I. and Rauh, H. (2006). *Europhys. Lett.* **74**, 334.
Yampolskii, S. V., Yampolskaya, G. I. and Rauh, H. (2007). *Physica* C **460-462**, 1200.
Yoshida, T., Endo, J., Kasai, H., Harada, K., Osakabe, N., Tonomura, A. and Pozzi, G. (1999). *J. Appl. Phys.* **85**, 1228.
Yurchenko, V. V., Shantsev, D. V., Galperin, Y. M., Alamgir, A. K. M., Han, Z. and Johansen, T. H. (2007). *Physica* C **460–2**, 803.
Zeldov, E., Clem, J. R., McElfresh, M. and Darwin, M. (1994). *Phys. Rev.* B **49**, 9802.
Zeldov, E., Larkin, A. I., Geshkenbein, V. B., Konczykowski, M., Majer, D., Khaykovich, B., Vinokur, V. M. and Shtrikman, H. (1994). *Phys. Rev. Lett.* **73**, 1428.
Zhang, M., Kvitkovic, J., Pamidi, S. V. and Coombs, T. A. (2012). *Supercond. Sci. Technol.* **25**, 125020.
Zhitlukhina, E., Poláčková, M., Volkov, S., Grancic, B., Gregor, M., Plecenik, T., Belogolovskii, M., Seidel, P. and Plecenik, A. (2023) *Appl. Nanosci.* **13**, 4771.

Author index

Subject index